Spatial Analysis: Modeling and Trends

Spatial Analysis: Modeling and Trends

Editor: Harry Garrison

MURPHY & MOORE
www.murphy-moorepublishing.com

Murphy & Moore Publishing,
1 Rockefeller Plaza,
New York City, NY 10020, USA

Visit us on the World Wide Web at:
www.murphy-moorepublishing.com

ISBN: 978-1-63987-502-3 (Hardback)

Cataloging-in-Publication Data

Spatial analysis : modeling and trends / edited by Harry Garrison.
p. cm.
Includes bibliographical references and index.
ISBN 978-1-63987-502-3
1. Spatial analysis (Statistics). 2. Geography--Statistical methods. I. Garrison, Harry.
G70.3 .S63 2022
910.21--dc23

Table of Contents

Preface

Spatial analysis, also called spatial statistics, consists of the formal techniques which study objects using their topological, geometric or geographic properties. It is a type of geographical analysis that attempts to explain human behavioral patterns and its spatial expression in terms of mathematics and geometry. Spatial analysis involves a range of techniques using various analytical methods. These techniques are applied in diverse fields such as astronomy and chip fabrication engineering. Many of the models deal with micro-economics and predict the spatial patterns which ought to occur. Spatial analysis is applied to structures at the human scale. This book is a compilation of chapters that discuss the most vital concepts and emerging trends in this field. It elucidates the concepts and innovative models around prospective developments with respect to spatial analysis. This book will serve as a reference to a broad spectrum of readers.

This book has been the outcome of endless efforts put in by authors and researchers on various issues and topics within the field. The book is a comprehensive collection of significant researches that are addressed in a variety of chapters. It will surely enhance the knowledge of the field among readers across the globe.

It gives us an immense pleasure to thank our researchers and authors for their efforts to submit their piece of writing before the deadlines. Finally in the end, I would like to thank my family and colleagues who have been a great source of inspiration and support.

Editor

1

Modelling Driving Forces of Urban Growth with Fuzzy Sets and GIS

Khalid Al-Ahmadi

Abstract

Urban growth occurs in conjunction with a series of decision-making processes and is, on the whole, not deterministic but rather is the outcome of competing local demands and uncontrolled, chaotic processes. Fuzzy sets theory is ideally suited to treat the complexity and uncertainties in the decision-making process. This chapter presented an example of how fuzzy sets can be applied to model urban growth driving forces within geographical information system environment. The mathematical models for measuring, computing and defining 10 fuzzy urban growth factors to form fuzzy driving forces of urban growth in Riyadh City, Saudi Arabia, were discussed. Four factors were considered as the driving forces for urban growth in Riyadh City: the transport support factor, urban agglomeration and attractiveness factor, topographical constraints factor, and planning policies and regulations factor. The urban growth factors were established using fuzzy set theory, which quantified the effect of distance decay using fuzziness. This approach is a transparent method of interpreting the curve of distance decay using linguistic variables. This feature does not exist in the linear, negative exponential or inverse power functions. The results indicate that fuzzy accessibility and fuzzy urban density factors are capable of mimicking and representing the uncertainty in the behaviour of the human decision-making process in land development in a very efficient manner.

Keywords: urban growth, fuzzy sets, GIS, driving forces, Riyadh, Saudi Arabia

1. Introduction

Urban growth takes place in conjunction with a series of decision-making processes. This is a feature of the complexity and uncertainties of reality and decision-making in planning. While matters of high policy regarding urban growth are determined by senior urban planners and elected personnel, site selection is usually a subjective process which follows individuals' preferences. This process is, on the whole, not deterministic, but is the outcome of competing local demands and uncontrolled and chaotic decision-making processes. This approach is founded on the supposition that people in general use linguistic constructs for the evaluation of environmental or social situations [1]. It is, therefore, valid to specify rules on the basis of studied judgment, as opposed to definitive attributes. Heuristic decision-making processes would thus guide the rules of urban growth [2]. As the complexity of an urban system increases, the utility of traditional mathematical methods decreases. Where urban growth decisions are noted for a high level of complexity and uncertainty, this is more relevant [1]. Proponents of this approach contend that for handling uncertainties and complexity in decision-making, fuzzy sets and fuzzy logic are ideally suited [3, 4].

Uncertainty, under the fuzzy approach, is assumed to have an inherently vague nature such that it becomes identified through 'possibility' and not 'probability', which is the norm in statistical calculations [2]. Modelling urban development, then, is made possible by integrating heuristic decision-making processes with simulation such as cellular automata (CA). 'Heuristic' is taken to mean that the transition rule is not decided on the basis of numbers but rather on linguistic variables using fuzzy sets that describe certain preconditions of a decision. Development propensities are, then, assessed not using explicit mathematical formula but by adopting indistinct and subjective natural language statements using linguistic variables [1, 2]. For example, where a house is in close proximity to a road, the accessibility can be rated as being 'Very good access'. If a system is represented by a series of 'linguistic variables' such as 'good', 'very good' or 'moderate' [2, 5–8], a concept of degree of membership is defined in a fuzzy set. This is employed in illustrating the extent of uncertainties within the fuzzy set. The processes of 'fuzzification' and 'defuzzification' make it possible to model the behaviour of human thought in decision-making in an appropriately multifaceted manner [3]. Fuzzy set theory is used in many different fields of study to address a range of imprecise problems and issues [9]. Since mathematics deals with ideal figures, it does not always provide accurate representations for modelling the behaviour of complex systems. Zadeh [10] proposed a solution through a different set of computations. The cloudy quantities or mathematics of fuzzy are not realized with respect to statistical probability distributions. While the fuzzy set theory is conceptualised to deal with mathematical problems which have only approximate values and in which events are vague, fuzzy logic is applied to show how fuzzy variables interact in a complex system [3].

Several applications of fuzzy set theory for spatial planning have been attempted in recent years. Notable amongst these have been the multi-dimensional evaluation of areas on the land market [1]. Leung [11] applied fuzzy linear programming for urban land-use planning. Fuzzy clustering methods have been employed in transport modelling [12] and in geo-demographic [13]. Davidson *et al.* [14] and Hall *et al.* [15] have made use of applications of fuzzy set theory and fuzzy logic in land suitability analysis. In Expert and Spatial Decision Support Systems, Leung and Leung [16] have proposed an expert system shell for an intelligent knowledge-based geographic information system (GIS) that is governed by a fuzzy logic inference engine. For long-term agricultural planning, Davidson *et al.* [14] developed a fuzzy logic-based land evaluation. Indeed, academic literature investigating the possible integration of fuzzy reasoning and fuzzy query into GIS has flourished since the 1990s [6, 7, 17–19].

According to Tobler's so-called first law of geography, 'everything is related to everything else but near things are more related than distant things' ([20], pp. 236). It is implicit that there is a spatial interaction between any two localities, and this interaction can be measured by a distance decay function which is based on the premise that the interaction between two localities declines as the distance increases. Most urban models quantify the distance decay effect of accessibility using either a negative exponential function or an inverse power function [21–24]. These two methods describe, however, distance decay as a rapid decline in accessibility with distance from a centre. There are some difficulties when one tries to interpret these calibrated parameters [25, 26]. However, a quantified distance decay effect of accessibility using fuzzy sets provides a more realistic set of results and an efficient manner in terms of interpretation. The calibrated parameters of fuzzy distance decay can directly be read and understood from the shape of the membership function [27]. Therefore, the meaning of the figure can be read linguistically and describes an individual's preference for selecting an urban area.

A fuzzy cellular urban growth model (FCUGM), which is capable of simulating and predicting the complexities of urban growth for the city of Riyadh, Saudi Arabia, was presented by Al-Ahmadi et al. [27–31]. The work within this chapter has reviewed the driving forces of urban growth in both the scientific literature and in the context of Riyadh, and also reviewed the principles of fuzzy set theory and presented an example of how these principles are applied for modelling urban growth driving forces. Additionally, the mathematical models for measuring, computing and defining fuzzy input variables (accessibility and urban density (UD)) and fuzzy driving forces of urban growth of Riyadh were also discussed.

2. Driving forces of urban growth

Urban growth indicates a transformation of a vacant land or a natural environment to the urban fabrics including residential, industrial and infrastructure developments. It mostly occurred in the firings of urban areas. Urban expansion is driven by many factors such as population growth and urbanization, the development of urban economy, the increase of investment on fixed assets such as infrastructure (utility networks and roads) and the increase of income and living standard in urban areas (better housing, health, social and recreational services) [32]. According to the scientific literatures, these factors are considered to some extent as driving forces in both western and eastern countries. However, the driving factors of urban growth might differ from one place to another, according to political, economic, demographic, social and physical factors.

In terms of the driving forces that have been used in most urban models to model urban growth of cities, the main urban growth forces mentioned already can be dealt with as spatial factors such as (1) travel time or spatial distance to road networks, railways or subways, (2) travel time or spatial distance to a town centre or a metropolitan sub-centres, (3) development density, (4) proximity to protected sources (such as drinking water, forest or wetland), (5) physical properties such as altitude, slope or soil and geology type, (6) proximity to socioeconomic services, commercial or industrial districts, (7) environmental suitability or (8) Government's zoning regulations. Although most of the studies in the literature relate to different contexts such as countries in North America, Europe and Asia, there are very limited studies being applied to investigate the relationship between spatial patterns of urban growth and its divining forces for Middle East cities such as Riyadh.

Given that, in this research, the urban growth driving forces were selected based on three criteria: (1) characteristics of urban growth in Riyadh based on the understanding of the historical documents, master plans, and its attached studies and opinion of expert planners, (2) scientific literature and (3) some empirical structural analysis. According to these three criteria, in this research, four factors are considered as the driving forces for urban growth: (1) transport support factor (TSF), (2) urban agglomeration and attractiveness factor (UAAF), (3) topographical constraints factor (TCF) and (4) planning policies and regulations factor (PPRF). These four driving forces are formed by integrating 10 input variables (in this research). **Table 1** illustrates the input variables and its corresponding driving forces. However, the authors argue that these driving forces are potentially comprehensive and can be applied in most contexts, since they are capable of covering most of the urban growth factors which have previously been used in the study as the driving forces of urban growth. These four driving forces can be seen as general containers, where one can remove or add different input variables for a driving force according to the context (different study areas) and data availability.

This section gives an examination of the four 'driving forces' (TSF, UAAF, TCF and PPRF) and their corresponding 10 'input variables' listed in **Table 1**. Firstly, each driving force will be examined in the context of international scientific literature and, secondly, in the context of Riyadh city.

Raw data	Input files	Input variables	Driving forces
Road network maps	Road network	Accessibility to local road (ALR)	TSF
Road network maps	Road network	Accessibility to main road (AMR)	TSF
Road network maps	Road network	Accessibility to major road (AMJR)	TSF
Satellite images	Classified urban areas	Urban density (UD)	UAAF
Land-use maps	Town centre	Accessibility to town centre (ATC)	UAAF
Land-use maps	Employment centres and Socioeconomic services	Accessibility to employment centres and socioeconomic services (AECSES)	UAAF
DEM	Topographical characteristics	Slope gradient (SG)	TCF
DEM	Topographical characteristics	Altitude (A)	TCF
Master plans	Planning and zoning regulation areas	Planned areas (PA)	PPRF
Master plans	Planning and zoning regulation areas	Excluded areas (EA)	PPRF

TSF: transport support factor; UAAF: urban agglomeration & attractiveness factor; TCF: topographical constraints factor; PPRF: planning policies and regulations factor.

Table 1. The raw data, input files, input variables and driving forces of urban growth in Riyadh.

2.1. Transport support factor

The transport network has proved to be the key factor in the urban evolution process because of its influence over urban form and land price. Although the relationships between the road network and the urban form are rarely apparent, it is considered an intersecting process. Transport is considered as a significant factor in accelerating urban growth and attracting new development. It offers easy and efficient access from a certain location to other supporting activities such as employment centres, socioeconomic services and amenities. An efficient and affordable transport network increases the accessibility to land, and land with good accessibility is usually more attractive for urban development. The use of a transport network as a factor influencing urban growth can be traced back to Alonso's [33] theory of urban land markets. Before that, of course, Van Thunen's [34] theory of agricultural rent, firms and households established the basic economic relationships. These theories can be used to demonstrate that land or locations with good accessibility are more attractive and have a higher market value than less accessible locations.

One of the early attempts in United States to examine the inter-relationships between transport and urban development was by Hansen [35] in which he argued that locations with good accessibility had a higher chance of being developed and at a higher density, than less accessible or remote locations. Torrens [36] argued that developing transport systems and improving accessibility will alter the decisions of landlords, developers, firms and households about where to locate a new business or a home. For example, the private automobile along with good transport networks has made most areas across a city almost evenly appropriate as a place to live or work [37]. But according to travel-cost minimization theories and discrete choice theories, each household pursues a location that yields the greatest utility or satisfaction by minimizing the travel time or travel cost from a place to live to human activities such as working, shopping, education or leisure [36, 38].

Transport is a crucial factor influencing land use and is usually one of the chief factors in contemporary urban land-use models and urban CA models. It can be noted apparently that the accessibility of urban land to the transport network factor has been widely employed in most urban CA models [22, 24, 39–45] With respect to Riyadh, the early development of the city was tightly connected to the development of its road networks. Since 1970, the urban area of Riyadh expanded outwards along the radial roads. Most urban construction was carried out alongside roads for easy access to reduce development costs. Therefore, transport is a vital factor that should be introduced into the model in this research. The concept of accessibility is widely used in urban and regional planning, land use and transport modelling. Accessibility has been defined 'as the ease with which activities at a given destination may be reached from an origin location using a particular mode of transport' ([36], pp. 49). Urban accessibility can be defined 'as the capability to get from place to place and the connectivity of a place with other places' ([46], pp. 1). Accessibility is the spatial constraint on urbanization and can be measured by the distance, travel cost, travel time or several other measures based on these variables. From a spatial perspective, accessibility to different human activities is a major factor in urban development [36, 46].

Thus, the transport support factor (TSF) is chosen in this research as a driving force. In the Riyadh context, it is formed by integrating three types of roads—accessibility to local road

(ALR), accessibility to main road (AMR) and accessibility to major road (AMJR). As such, integration is more appropriate in modelling practices because it reduces the number of variables that lead to a more understandable model.

2.1.1. Accessibility to local road

The road network of Saudi Arabia is classified based on the width. Roads with a width between 3 and 20 m are called local roads. Most residential uses are located along the sides of this class of roads, and the neighbourhood usually has a relatively low urban density (<10 dwellings/hectare). In addition, light commercial uses (groceries) and educational services (primary or intermediate schools) that serve one quarter of the neighbourhood, which extends over an area typically of about 4 km^2, are located on local roads [47]. The availability of a paved local road is a further indicator that land adjacent to a local road is ready for development, that is, mostly covered with key infrastructure such as electricity supply, water and drainage, sewage and telephone lines.

2.1.2. Accessibility to main road

Roads with a width between 30 and 40 m are called main roads. Most mixed uses such as residential and commercial land use are located beside main roads, and the density of the population is commonly average or medium (20–40 dwellings/hectare) [47]. Commercial use includes retail, education (high schools), light industrial uses and services (administrative offices, branch of companies) that serve one neighbourhood (approximately 2 × 2 km) and are usually located on main roads.

2.1.3. Accessibility to major road

Roads with a width more than 40 m are called major roads. They are typically found in commercial areas (firms and establishments), governmental uses (ministries and main administrative offices) and some mixed uses, that is, residential/commercials are located along such roads. The urban density is generally high (>40 dwellings/hectare). Additionally, the high levels of educational services (universities and colleges) and heavy industrial uses (factories) that serve the whole city are usually located on main roads.

2.2. Urban agglomeration and attractiveness factor

UAAF driving force comprises three input variables including urban density (UD), accessibility to employment centres and socioeconomic services (AECSES) and accessibility to town centre (ATC). Of course, other factors can also be included such as accessibility to sub-centres.

2.2.1. Urban density

Proximity to an existing urban area is a significant factor in land development. Non-developed land adjacent to an existing development usually has a higher possibility of imminent development than other land. Land near to urban agglomerations is more attractive for households and developers, because they can take advantage of the existing facilities, infrastructures and commercial and social activities such as electricity, water supply, drainage and sewage, telephone lines and business opportunities. Many urban CA models include urban density or development density, which is the proportion of developed land within a specific area (neighbourhood), as a factor for urban development [22, 24, 39–45]. In this research, the number of neighbouring developed cells that surround a non-developed cell influences the transformation of the state of that non-developed cell. For instance, a central cell with a non-urban state located within a neighbourhood with a high urban density (the majority of cells are urban cells) will be forced to transform into an urban state by its urban neighbours.

2.2.2. Accessibility to employment centres and socioeconomic services

The AECSES is an important factor for residential development because most people wish to minimise commuting distance, time or cost from home to work. This argument can be supported by the notion of Lowry's framework that examines the relationship between residential development and employment centres. Lowry's [48] Model of Metropolis, which is the first operational attempt to implement an urban land-use transport model, proceeds on the assumption that the place of residence is governed by the place of employment, that is, jobs decide where people live, and this idea continues to be at the heart of several of the contemporary urban models [36, 37]. With respect to Riyadh city, in this research, the AECSES factor includes most employment centres (such as administrative offices, firms and establishments) and services (such as schools, hospitals, mosques and community centres). AECSES has been considered as an urban growth factor in much of the urban CA literature [2, 40, 41, 44, 45].

2.2.3. Accessibility to town centre

The ATC factor is important for urban development because the town centre of the capital of Saudi Arabia (SA), Riyadh city, functions as the key administrative and commercial centre not only for Riyadh but for all surrounding regions. According to the bid-rent theory [34], the new development will happen as near as possible to the city centre while at the same time trying to minimize the land cost which controversially may lead to a trend of finding new sites far away from the urban centre. Land development usually takes place around the city centres first, then further away. The town centre provides better facility services, more working and business opportunities, etc. In Riyadh, the town centre includes Government offices, ministries, retail centres and headquarters of major companies. Hence, the city provides opportunities that are unlikely to be found anywhere else in Riyadh city. This does not necessarily mean that living in the town centre is the most appropriate choice but the greater the distance to the town centre, the less attractive is a location for development. This factor is usually included in urban CA models [24, 42, 49].

2.3. Topographical constraints factor

The physical characteristics of a site or a location can be an important factor in deciding whether or not to proceed to develop the site. The physical characteristics typically include the subsurface and surface geology, soil, elevation, slope, vegetation and other factors. Although all of these physical characteristics could be considered for use in modelling urban growth in this research, there is no reason to consider soil, geology or vegetation in the case of Riyadh because most of the soils are sandy, the geology is stable and there is very little vegetation for much of the year. Hence, slope and altitude or elevation are the only physical factors included in the model. But other factors could, of course, be considered and used in the model if they were felt to be important at a different geographical location.

2.3.1. Slope gradient

The slope of the land surface is important because it influences the use that can be made of the land, costs of building houses, access roads and other infrastructure and safety. Urban growth can be constrained by slopes, typically the steeper a slope the greater the constraint. For example, land with steep slopes requires substantially more investment to develop. Land-owners or developers would develop a site if they judge that they can make a profit on a development site with very steep slopes. The profit margins can be calculated as a function of the compromise between input costs (the costs of developing a site) and the anticipated selling price of the development [36]. The steepness of slopes has been considered to be a significant factor influencing the attractiveness of a site and its potential for development in many scientific research papers [22, 39, 43, 49].

2.3.2. Altitude

Altitude can similarly affect the attractiveness of a site for development. If a site is low-lying, it might be subject to flooding; if it is at a relatively high altitude, it might be more or less attractive depending on the local topography, scenery and views. The topography of Riyadh area includes mountains and relatively large variations in altitude from 200 to 1200 m above the sea level. Urban development can occur in areas of steep terrain and high altitude under certain circumstances, for instance, where there is great demand for building land and there is a short supply of relatively flat land, or when people with high income prefer to live in elevated areas in order to secure attractive scenic views from their windows [43]. But the cost of building in general is greater with an increase in elevation. Altitude is therefore one of the prime factors in the model of this research.

2.4. Planning policies and regulations factor

Just as urban growth is generally constrained by physical or site-specific factors, it is also constrained by institutional controls and planning policies. The urban growth process is subject to local planning regulations and policies imposed by the national, regional or local authorities. Planning regulations and policies serve diverse purposes but one of the most important is to control the urban expansion and prevent unplanned urban sprawl. Wegener [50] argued that uncontrolled urban sprawl in the USA was generally regarded to be undesirable, often results in unidentified and undistinguished neighbourhoods, reduces or damages open spaces, annihilates natural habitats and leads to a significant increase in traffic volume. The PPRF has therefore been used in several urban CA models [22, 24, 43]. The PPRF is introduced to the model in this research as two variables: (1) planned area (PA) and (2) excluded area (EA).

2.4.1. Planned area and excluded area

The planned area refers to areas outlined in the Master Plan of Riyadh and depicts the spatial boundary of the area planned for urban growth. The excluded area denotes the area where urban growth cannot be considered either owing to its physical characterises such as river courses, low-lying valleys subject to seasonal flooding, or water bodies, or to cultural, environmental or governmental policies to conserve the land for cultural pursuits such as heritage places, wildlife-protected areas, groundwater areas or pollution mitigation areas. The EA cannot be developed because it is forbidden by law or by urban-zoning regulations and policies.

3. Fuzzy sets

In set theory, a classical set is usually defined as a collection of abstract objects or elements, which are characterized as a finite, distinct, crisp, sharp, countable object that has an unambiguous boundary. The universe of discourse or the set domain is all possible elements which a set can contain [51]. The classical set is defined by distinct or crisp boundaries, that is there is no uncertainty about the objects or location of the boundaries of the set, as shown in **Figure 1a**, where the boundary of the set *A* is an unambiguous line [9].

In classical set theory, a single element is assessed using Boolean logic; the element either belongs or does not belong to the set [52]. This can be expressed with the *characteristic function* for the elements of a given universe to belong to a certain subset of this universe. Mathematically, a set can be expressed as follows:

Let X be a universe of discourse (collection of objects), whose generic element is denoted as x. Thus, X = {x}. A crisp set *A* in X is described by the *characteristic function*, whereby X*A* of *A* is defined as X*A*: X → {0, 1} and expressed mathematically as follows:

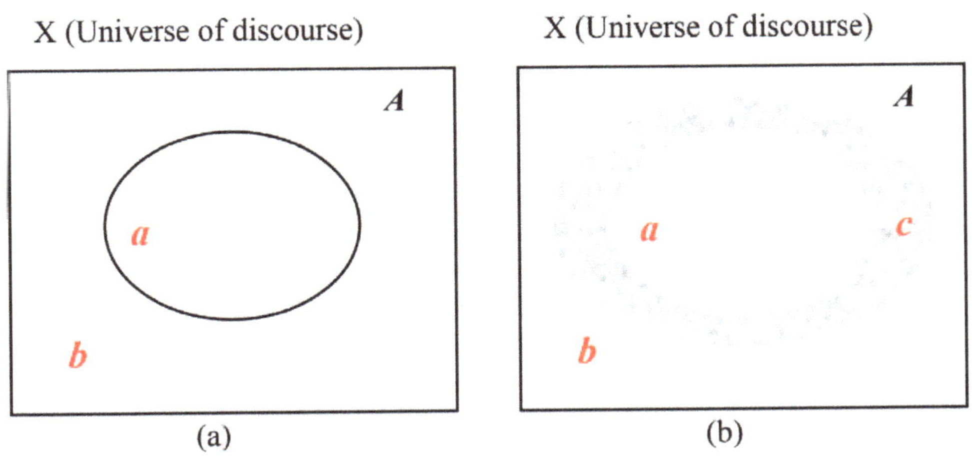

Figure 1. Diagram for (a) crisp set boundary and (b) fuzzy set boundary (after Ross, [9]).

$$XA(x) = \begin{cases} 1 & \text{if } x \in A \\ 0 & \text{if } x \notin A \end{cases} \quad (1)$$

where the symbol, ∈, denotes membership and ∉, non-membership. It maps the universe to the set of only two elements, 0 and 1. One can indicate whether an element belongs or does not belong to a distinct or a crisp set. Some degree of uncertainty can, however, be allowed as to whether or not an element belongs to a set. This can be expressed by the membership of an element to a set by its membership function using fuzzy sets. 'A fuzzy set is described by vague or ambiguous properties; hence its boundaries are ambiguous, as illustrated by the fuzzy boundary for set *A* in **Figure 1b**. This shows the vague, ambiguous boundary of a fuzzy set *A* on the discourse universe X, the shaded boundary representing the boundary region of *A*. In the central (unshaded) region of the fuzzy set, a point *a* is clearly a member of the set, while point b is clearly not a member. However, the membership of point *c*, which is on the boundary region, is ambiguous. If point *a* is represented by 1 (complete membership) and point *b* with 0 (no membership), then point *c* must have some intermediate value of membership (partial membership) in fuzzy set *A* on the interval [0, 1]' ([9], pp. 24).

Fuzzy set theory was inspired and developed by [8, 10, 53] in order to extend classical set theory for handling soft (continuous) rather than crisp classifications. A fuzzy set can be defined as a collection of objects in which the transition from a full membership to a non-membership is gradual rather than abrupt [10]. A set can be regarded as fuzzy if an element can partly belong to it, rather than having to belong completely or not at all. Consequently, fuzzy set theory assigns membership grades to objects which are not restricted to 0 (non-membership) or 1 (full-membership), but which may locate somewhere between 0 and 1 [43]. A fuzzy set can be expressed mathematically as follows:

A fuzzy set A of a universe X is defined by a *membership function* μA (vs. *characteristic function* XA in a crisp set) such that μA: X → [1,0]. A fuzzy set *A* in X is a set of ordered pairs,

$$A = \{(x, \mu A(x)) \mid x \in X\} \quad (2)$$

where μA (x) represent the grade of membership of x in *A*, which associate with each x a real number in [0,1], and the universe X is always a crisp set.

Degree of vagueness or uncertainty is usually exhibited by many phenomena. The boundaries of spatial features, for example, often are not clearly defined. Ideas and impressions like 'steep', 'close' or 'suitable' can be better articulated with degrees of membership to a fuzzy set, rather than with a binary yes/no classification. In human thinking and language, we often use uncertain or vague concepts. Our thinking and language is not binary, that is, black and white, zero or one, yes or no. In real life, we add much more variation to our judgements and classifications. These vague or uncertain concepts are understood and assumed to be fuzzy. We encounter fuzziness almost everywhere in our everyday lives [54]. With respect to this study of urban growth, we might look for a suitable site to develop land for housing. The criteria describing the area that we are looking for could be formulated as shown in **Table 2**.

The site should:

Fuzzy	Crisp
have a *moderate* altitude	have altitude between 800 and 1000 m
have a *favourable* slope	have slope less than 8°
have a *favourable* aspect	have aspect between 175 and 250°
be *nearby* schools	be within 40 m to schools
be *not close* to an airport	not be within 1 km of an airport

Table 2. Example of differences between fuzzy and crisp in terms of site suitability.

All of the conditions mentioned in the fuzzy column are vague and uncertain but correspond to the way we express these conditions in our languages and thinking. Using the classical approach, the fuzzy conditions would be translated into distinct, sharp or crisp classes, such as those in the crisp column. Taking the school criteria as an example, if a location falls within the given criteria (be within 40 m of a school), we would accept it for developing, otherwise it would be excluded from analysis (**Figure 2a**). If, however, we allow degrees of membership to our classes, we can accommodate those locations that just miss a criterion by a few metres. They may get a low degree of membership, but will be included in the analysis as shown in **Figure 2b**, where locations that have distance to school within 40 m assigned full membership (1.0), and those locations above 40 and below 70 m still have membership but decrease gradually as the distance increases, and those locations with distance over 70 m assigned no-membership (0.0), that is, they are definitely not nearby schools.

The fuzzy set membership function is a crucial component of fuzzy set theory and is one of the building blocks of a fuzzy model because it is used to define the model variables, which are based upon imprecise and abstract concepts, for example, a 'young' or 'old age', or a 'short' or 'long' distance [55]. Although the shape and form of the membership function should accurately reflect the semantics of the underlying concept in order to represent the real world as closely as possible, fuzzy systems tend to be extremely tolerant of variations in fuzzy set representations, which makes the resulting models highly robust [55, 56]. Different membership functions

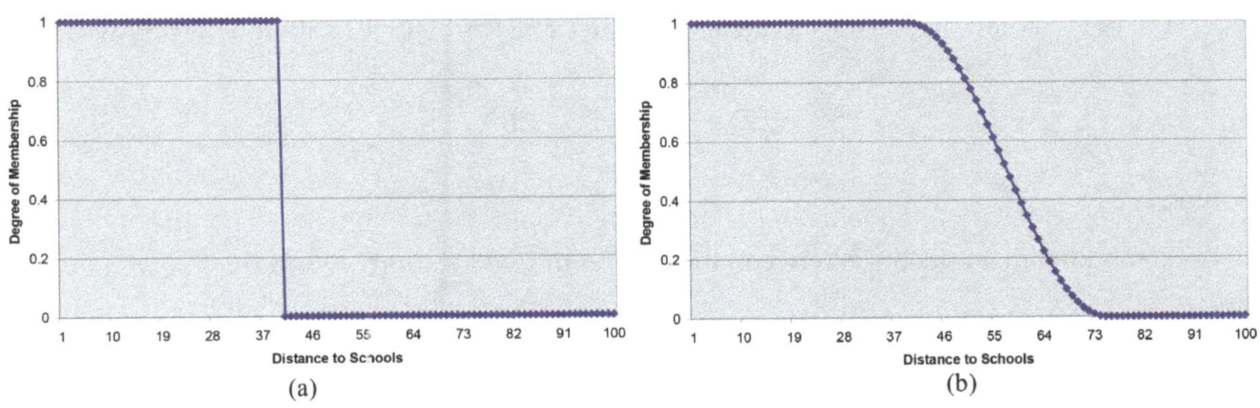

Figure 2. (a) Crisp set of school criteria and (b) fuzzy set of school criteria.

represent different fuzzy sets, even though they may have a similar context. A membership function can generally assume any shape although the surface tends to be a continuous line from the left to the right edge of the set. Engineering applications tend to use one of the following parameterized membership functions: Gaussian, triangular, trapezoidal or sinusoidal [51]. The trapezoidal and sinusoidal functions can present different shapes: symmetric, monotonically decreasing or monotonically increasing. The different shapes of membership functions are controlled by different parameters in terms of the number and place. There are several different ways of deciding where and how many membership functions to place on a variable domain. The most important requirement is that they overlap, but the optimal number will be determined by experiment and the complexity of the problem. Ross [9] outlines a series of methods which are used to decide on the placement of membership functions, including the use of intuition or expert knowledge, genetic algorithms and neural networks. For example, triangular is widely used in engineering applications, whereas symmetric sinusoidal is commonly used for spatial or geographical applications.

4. Results and discussion

4.1. Fuzzy accessibility

Accessibility to local road (ALR), accessibility to main road (AMR), accessibility to major road (AMJR), accessibility to town centre (ATC) and accessibility to employment centres and socioeconomic services (AECSES) were created by computing accessibility to these variables using a Euclidean distance algorithm [27]. The FCUGM tends, however, not to use the direct result of applying this technique. This is because the Euclidean distance is based on a spatial linear function as opposed to the nature of urban development. This results in the FCUGM transforming the linear results of applying the Euclidean method into a non-linear fuzzy measure of accessibility using fuzzy set theory.

The first step for measuring accessibility is to calculate the Euclidian distance. This can be defined as the straight-line distance between two points on a plane [57]. Euclidean distance, or distance 'as the crow flies', can be calculated using Pythagoras's theorem as expressed in Eq. (3). The Euclidian distance was calculated using ESRI ArcGIS 9.2 software as follows [57]:

$$ED_{AB} = \sqrt{AC^2 + CB^2} \tag{3}$$

where

ED_{AB} is the Euclidean distance between the centres of two locations A and B;

C is the hypotenuse of the right triangle;

After the Euclidean distance has been calculated for a particular input variable, the resultant layer is weighted and standardized (1–100) using the FCUGM according to Eq. (4).

$$WSED_{ijk} = \frac{ED_{ijk} - \min[ED_{ijk}]}{\max[ED_{ijk}] - \min[ED_{ijk}]}^{*} W_k{}^{*}100 \tag{4}$$

where

$WSED_{ijk}$ is the weighted standardized score of Euclidian distance of a cell *ij* for input variable *k* (falling within the range 1–100) ED_{ijK} is the original value of Euclidian distance of a cell *ij* for input variable *k*

Wk is the parameter weight for input variable *k*

After the Euclidean distance has been calculated for a particular input variable, the resultant layer is weighted and standardized. This was followed by transforming the weighted standardized accessibility layer into a fuzzy accessibility measure using the FCUGM. Although there are several types of membership functions (fuzzy sets) such as triangular, trapezoidal, Gaussian, bell-shaped or sinusoidal, the author adopted the sinusoidal function for the FCUGM. The rationale for this choice was based on consideration of three criteria: (1) non-linear parameterization, (2) shape of curve or slope that was relatively easy to interpret and (3) the function that is smoothly and gradually descending. There are different shapes of the sinusoidal function; however, the decrease monotonically has been employed in this research to calculate the accessibility of input variables. This is because according to the results of the structural analysis conducted on the input variables [31], a relatively similar shape was found as monotonically decreasing. The sinusoidal function is defined by Eq. (5) and is controlled by the two parameters *P1* and *P2* which generally control the slope of the curve as shown in **Figure 3**. *P1* is equivalent to the maximal elemental value of the domain that acquired a membership degree of 1.0 (definite membership to a particular set), while *P2* is equivalent to the minimal elemental value of the domain which acquired a membership degree of 0 (no membership). For a given elemental value, *sedijk*, in the domain, the membership value can be computed as follows:

$$\mu Fuzzy\ A\ (wsed_{ij}) = \begin{cases} 1 & 0 < wsed_{ijk}, < P1 \\ \tfrac{1}{2}\,(1 + \cos(\pi\,(wsed_{ijk} - P1)\,/\,(P2 - P1)) & P1 \leq wsed_{ijk} \leq P2 \\ 0 & wsed_{ijk} > P2 \end{cases} \tag{5}$$

where

wsedijk is the weighted standardized score of Euclidean distance of a cell *ij* for input variable *k*, and (falling within the range 1–100)

P1 and *P2* are the parameters that quantify the distance decay effects of accessibility from a cell *ij* to input variable *k*.

μFuzzy A (*wsedij*) is the fuzzy accessibility (e.g. to local roads) of a cell *ij* for input variable *k* (falling within the range 1–100);

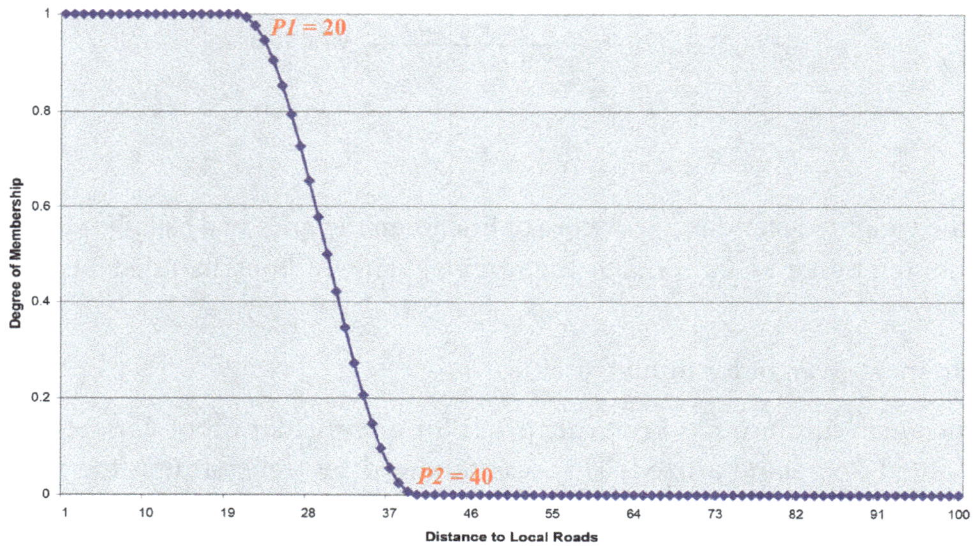

Figure 3. Accessibility to main road using fuzzy set.

The two parameters *P1* and *P2* are critically important for urban growth because they reflect people's behavioural inclination over decisions about living *near* or *far* from a certain variable. In other words, the two parameters *P1* and *P2* tend to quantify the distance decay effect of a particular variable, that is, to what extent people prefer to live near or far from local, main or major roads, employment centres or the town centre. In the FCUGM, as the distance between cell *xij* and an input variable *k* (such as local roads) increases, the accessibility to that variable decreases, and the quantity of that decreases is determined by the two parameters *P1* and *P2* which are calibrated [27].

Using the approach described earlier, five input variables of the FCUGM are generated through computing fuzzy accessibility to their features including local road, main road, major road, employment centres and socioeconomic services and town centre. Firstly, the linear Euclidean accessibility was generated for the five input variables; **Figure 4** shows the resultant linear accessibility of the five input variables.

The process of measuring fuzzy accessibility (Eq. 3) is performed using the 'Fuzzifier' model in the SIM-FCUGM [27–29]. As explained previously, the fuzzy accessibility is controlled by two fuzzy distance decay parameters denoted as *P1* and *P2*. As has been argued previously, *P1* and *P2* are of great importance as they reflect people's behavioural inclination over decisions about living *near* or *far* from certain variables (e.g. local roads). Thus, in order to investigate the effects that *P1* and *P2* might play in terms of spatial accessibility, different values of these two parameters were tested on fuzzy accessibility to major roads. **Figure 5** shows an example of the original linear Euclidean accessibility to major roads along with nine fuzzy accessibility maps in accordance with three different values of *P1* and three different values of *P2*. It can be seen from the original linear Euclidean (**Figure 5a**) that the accessibility has a linear effect at a distance. This means that as the distance to the major roads gets larger, the accessibility gets lower. This has the knock-on effect of meaning that land with low accessibility will have a low potential of being developed, which is not reflecting the behaviour of decision-makers in terms

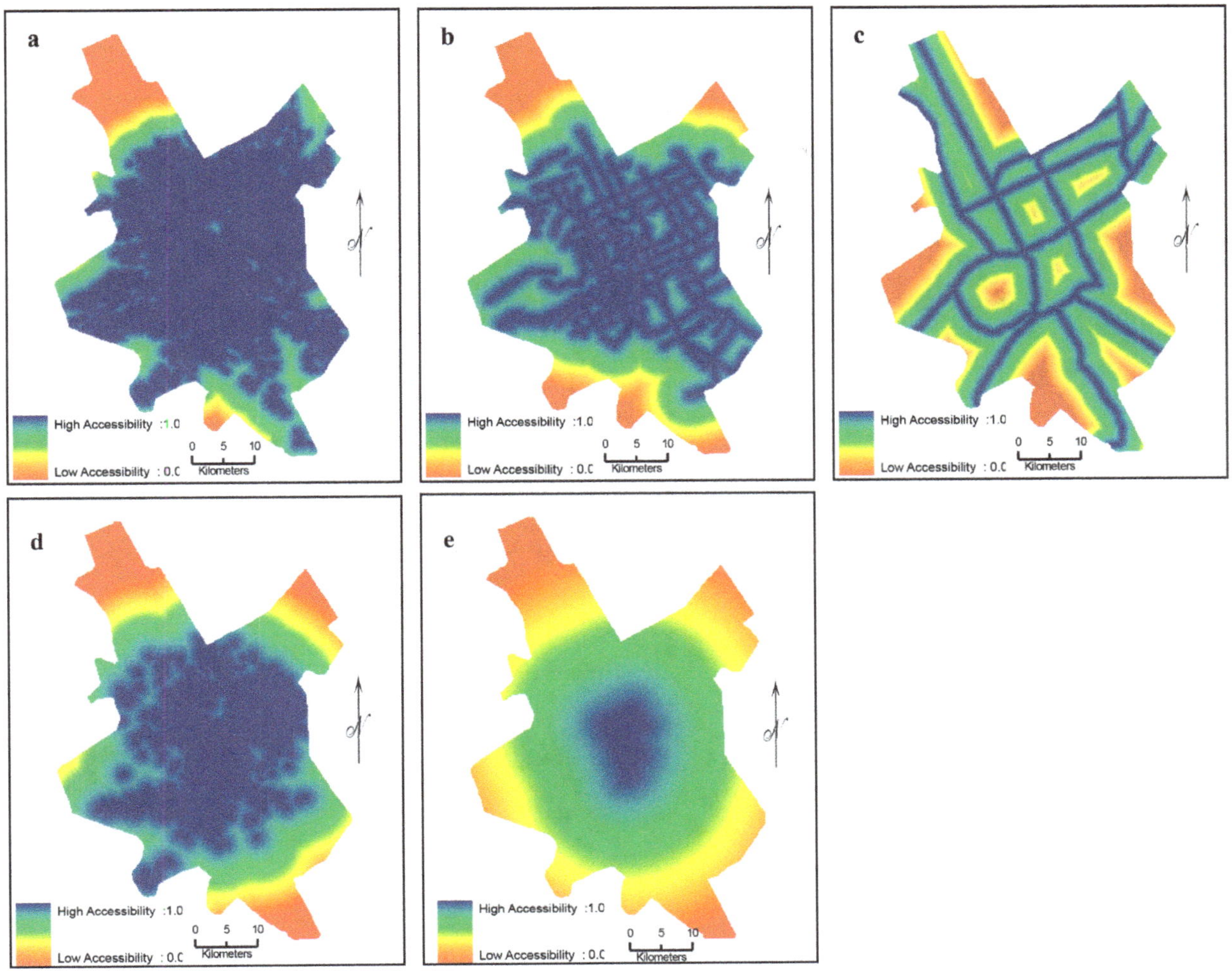

Figure 4. Euclidean linear accessibility to input variables: local roads (ALR), main roads (AMR), major roads (AMJR), employment centres and socioeconomic services (AECSES) and town centre (TC) of Riyadh city in 2005 (a–e, respectively).

of developing lands. In contrast, it can be noted from the nine fuzzy maps that both high and low accessibility areas might have a non-linear development potentiality by trading off values of *P1* and *P2* parameters.

By examining both maps and figures in **Figures 5** and **6**, it can be inferred that the fuzzy accessibility to several features can be characterized through three parts of membership including 'full', 'part' and 'no' memberships as illustrated in **Figure 5a** and **6a**. Areas with 'full' (dark blue) and 'no' (brown) membership denote the areas with the highest and lowest accessibilities, respectively, so they are crisp (0 or 1) and *compact* to either class. In contrast, 'part' membership areas are those in which the accessibility decreases gradually from 'full' to 'no' membership, and these areas are *smooth*. The value of P1 controls the amount of areas that have 'full' membership, while P2 controls the areas of 'no' membership. As the value of P1 increases, the areas of high accessibility increase; in contrast, as the value of P2 decreases, the areas of low accessibility increase. The three maps b, c and d of **Figure 5** and the three figures b, c and d of **Figure 6**, for example, show that the size of areas with high accessibility (blue)

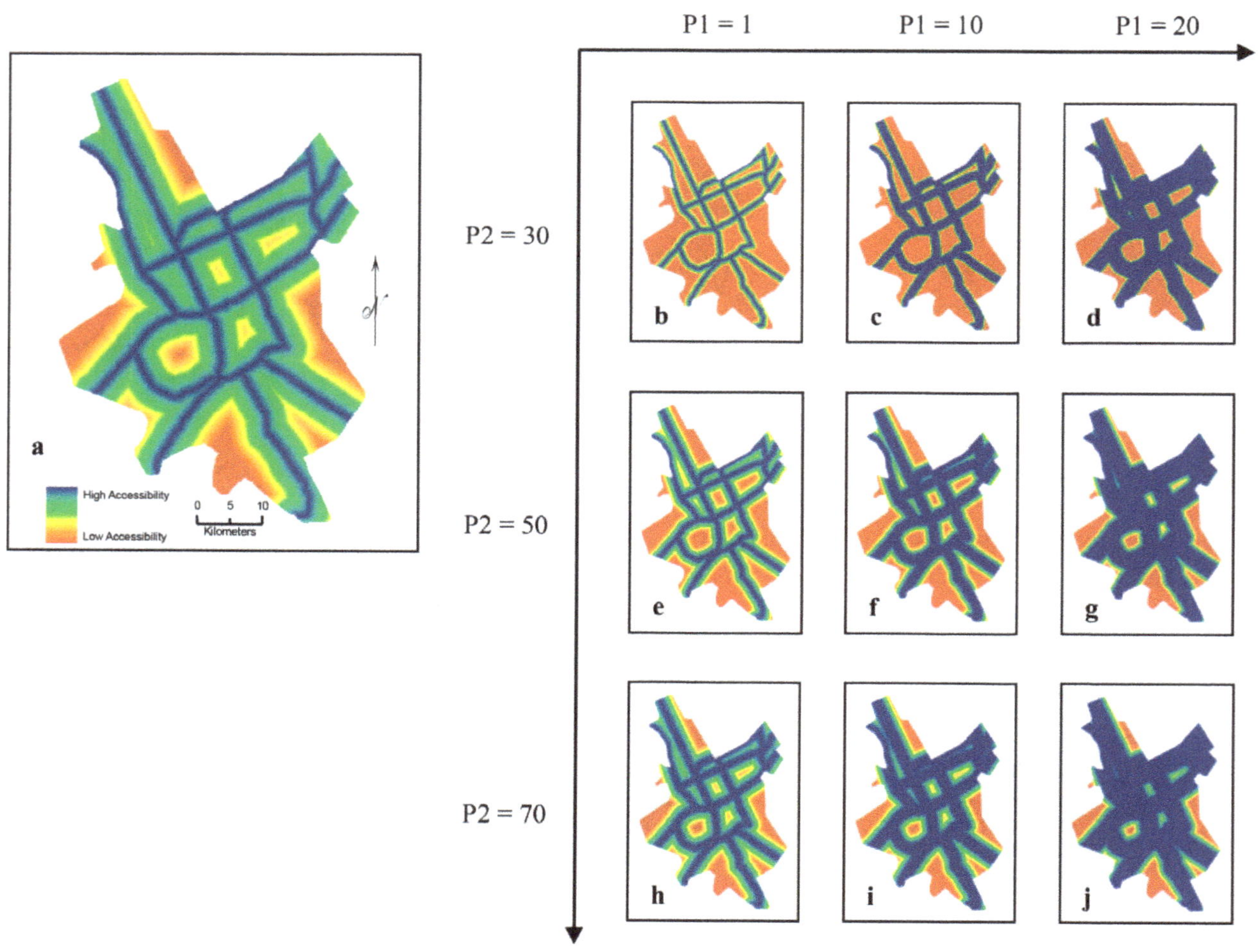

Figure 5. Original linear Euclidean accessibility (a) and fuzzy accessibility to major roads (b–j).

getting larger as the values of P1 increase. The three maps and figures b, e and h, on the contrary, show that the size of areas with low accessibility (brown) gets larger as the values of P2 decrease. Interestingly, the diagonal from the top-left to bottom-right exhibits the levels to which both parameters are traded off against each other. Additionally, the difference between P1 and P2 values controls the size of areas in which the accessibility surface decreases smoothly, which represents the area of 'part' membership.

It can be seen from both **Figures 5** and 6 that fuzzy accessibility has a high capability to mimic and represent the uncertainty in the behaviour of the human decision-making process in land development in a very efficient way. For example, one can represent the behaviour of people in three aspects 'full membership, 'part membership' and 'no membership' for those people who prefer to live 'near', 'fairly near' and 'far' from a certain feature, respectively. Thus, the numerical distances using fuzzy sets are converted to linguistic words and can be modified easily by parameters P1 and P2. This feature does not exist in the linear, negative exponential or inverse power functions. While the latter two functions can generate non-linear accessibility, the meanings of their parameters are somewhat difficult to interpret. It can be seen that as the difference between the two parameters gets larger, the size of areas with 'part' membership

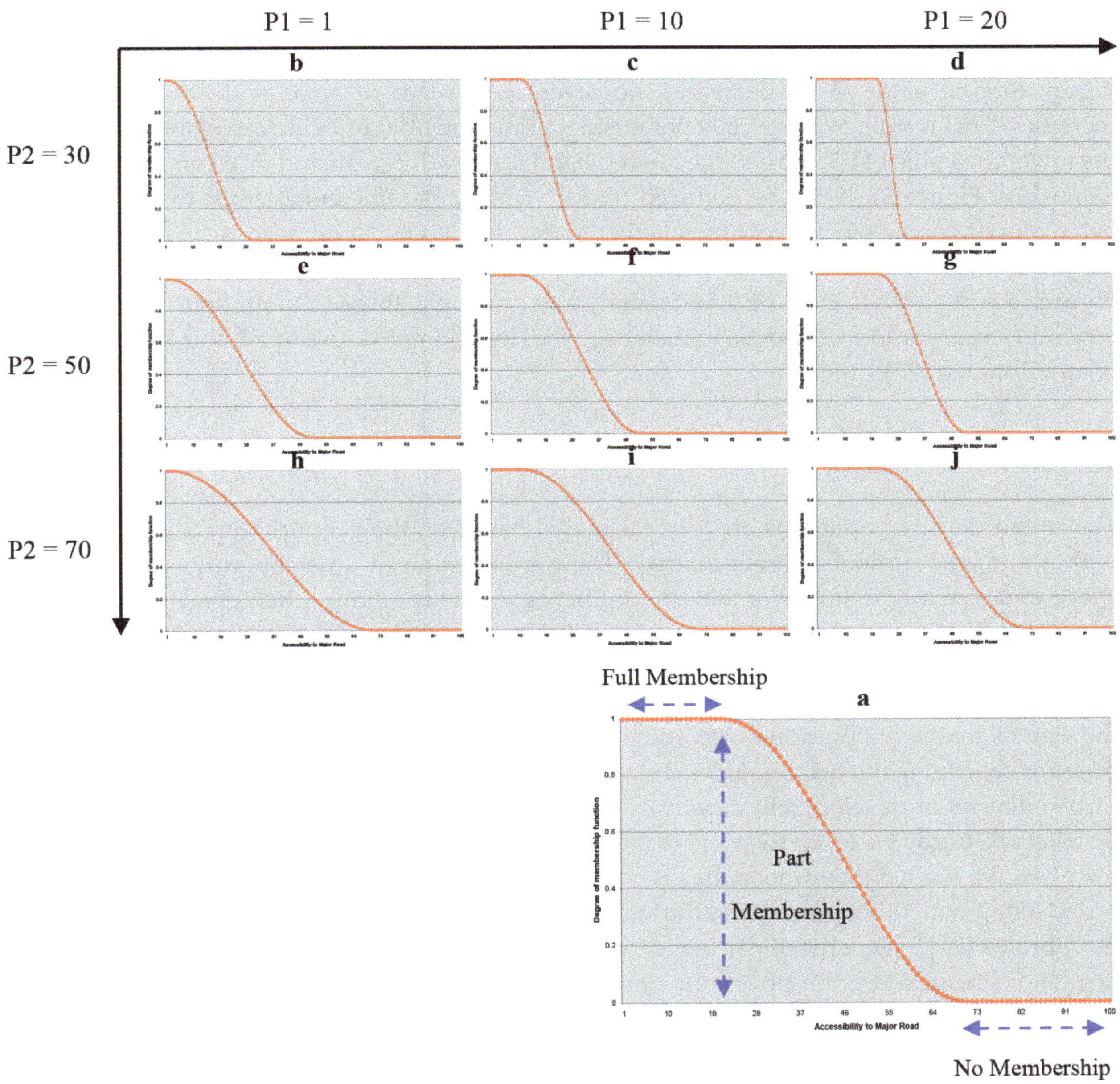

Figure 6. Fuzzy membership functions of the nine maps shown in **Figure 5**.

gets greater. The three maps and figures b, e and h, for example, show the gradual increase of the smooth areas, represented by green to yellow areas in maps and areas under the curve in figures. In particular, the b and h maps represent the lowest (30–1 = 29) and highest (70–1 = 69) differences between P1 and P2, respectively, which as a result decrease and increase the smooth accessibility areas.

The lower value of P1 results in compact areas with high accessibility but only for those cells situated close to the feature of interest. This implies that only areas located very near to major roads have a high potential for development compared to other areas (see maps b, e and h in **Figure 5** and figures b, e and h in **Figure 6**).

In contrast, the higher value of P1 expands the size of areas with high accessibility further outward from the feature of interest (see maps d, g and i in **Figure 5** and figures d, g and i in **Figure 6**). The value of P2 decreases, however, as the size of areas with low accessibility increases. This results in more cells possessing a low potential of being developed. Therefore, the lower the value of P2, the less the areas of land for development and vice versa (see maps b, e and h in **Figure 5**). It can be deduced that P1 and P2 parameters attempt to represent the behaviour of people when deciding whether to develop land near or far from an input variable (in the above example of major roads). The diagonal from the top-left to bottom-right of **Figures 5** and **6** shows three different behaviours of people, those who prefer to develop land *very close*, *moderately close* and *far* to major roads. The optimal values of P1 and P2 parameters were calibrated [27].

4.2. Fuzzy urban density

The urban density variable is broadly calculated based on the proportion of developed cells within a neighbourhood (kernel matrix). The configuration of a neighbourhood, such as its shape and size might, however, have an influence on the resulting urban density layer. It has been argued that the circular shape of a neighbourhood has no bias in any direction as opposed to rectangular or square neighbourhoods which generate significant distortions [58]. As a result, the FCUGM adopts a circular shape of neighbourhood (kernel matrix). In terms of the size of the neighbourhood, according to [59], there are no rules of thumb or theoretical basis for specifying the neighbourhood size when computing the urban agglomeration variable (urban density or development density). In most urban CA models, the neighbourhood size is decided arbitrarily based on the view of the modeller or the objective of the study. In contrast, in the FCUGM, the neighbourhood size is considered as a parameter that is calibrated and optimized along with other parameters during the calibration process. The size of a neighbourhood is calibrated [27] to generate the urban density variable, and its optimized size value reflects the degree of pressure resulting from the developed urban density within a neighbourhood of a non-developed cell. The urban density input variable is computed using Eq. (6).

$$UD_{ij} = \frac{\sum D_{ijk}}{\pi\zeta^2} \tag{6}$$

where

UD_{ij} is the development urban density of a cell ij;

D_{ijk} is the number of urban developed cells in the neighbourhood k of the cell ij and ζ is the radius parameter of the circular neighbourhood k.

Once the urban density has been calculated, the resultant layer is weighted and standardized (1–100) by the FCUGM using Eq. (7).

$$WSUD_{ij} = \frac{UD_{ij} - \min[UD_{ij}]}{\max[UD_{ij}] - \min[UD_{ij}]}^{*} W_k{}^{*}100 \tag{7}$$

where

$WSUD_{ij}$ is the weighted standardized score of the urban density of a cell *ij*;

UD_{ij} is the original value of urban density of a cell *ij* and

Wk is the parameter weight for input variable *k*.

This is followed by transforming the weighed standardized urban density layer into a fuzzy urban density using the FCUGM. The membership function type, shape and parameters that are used for measuring the fuzzy accessibility are applied for the fuzzy urban density as well and calculated as shown in Eq. (8).

$$\mu Fuzzy\ UD\ (wsud_{ij}) = \begin{cases} 0 & wsudi_j < P1 \\ \frac{1}{2}\ (1 - \cos(\pi\ (wsud_{ij} - P1) / (P2 - P1)) & P1 \leq wsud_{ij} \leq P2 \\ 1 & wsud_{ij} > P2 \end{cases} \quad (8)$$

where

Fuzzy UD is the standardized score of urban density of cell *ij* for input variable *k* (falling within the range 1–100);

wsudij is the weighted standardized score of the urban density of a cell *ij* and

*P*1 and *P*2 are the parameters that quantify the distance decay effects of urban density for a cell *ij*.

The only difference is that urban density is calculated using monotonically increasing sinusoidal function, rather than decreasing one as used for calculating the accessibility, because increasing shape could better fit the effect of urban density on urban expansion as found in the structural analysis [31]. By the monotonically increasing function, those high values of the domain (in this case urban density values) acquired a membership degree of 1.0, and because urban expansion took place in relatively high density urban areas, such shape is employed. As previously discussed, developers prefer to build near to existing developed land to take advantage of available services, facilities and infrastructure. The urban density input variable is computed by calculating the proportion of developed lands within a neighbourhood. This process is performed using the 'Fuzzifier' in the SIM-FCUGM [27–29]. Urban density is a smooth surface controlled by the size of a neighbourhood along with two distance decay parameters P1 and P2. In the FCUGM, the size of a neighbourhood ranges between 2 and 40 cells, with a cell size of 20 m.

Figure 7 shows the urban density of Riyadh city in 2005, with five different sizes of neighbourhood including 2.5, 5, 10, 20 and 40. It can be seen from **Figure 7** that as the neighbourhood increases in size, the urban density surface becomes smoother and the size of areas with high urban density also increases. In contrast, as the size of the neighbourhood decreases, the surface becomes more compact and areas with low urban density increase. This is mainly because a large neighbourhood size has a higher potential for urban cells to be located within it. One might assume that the small size of the neighbourhood restricts high urban density

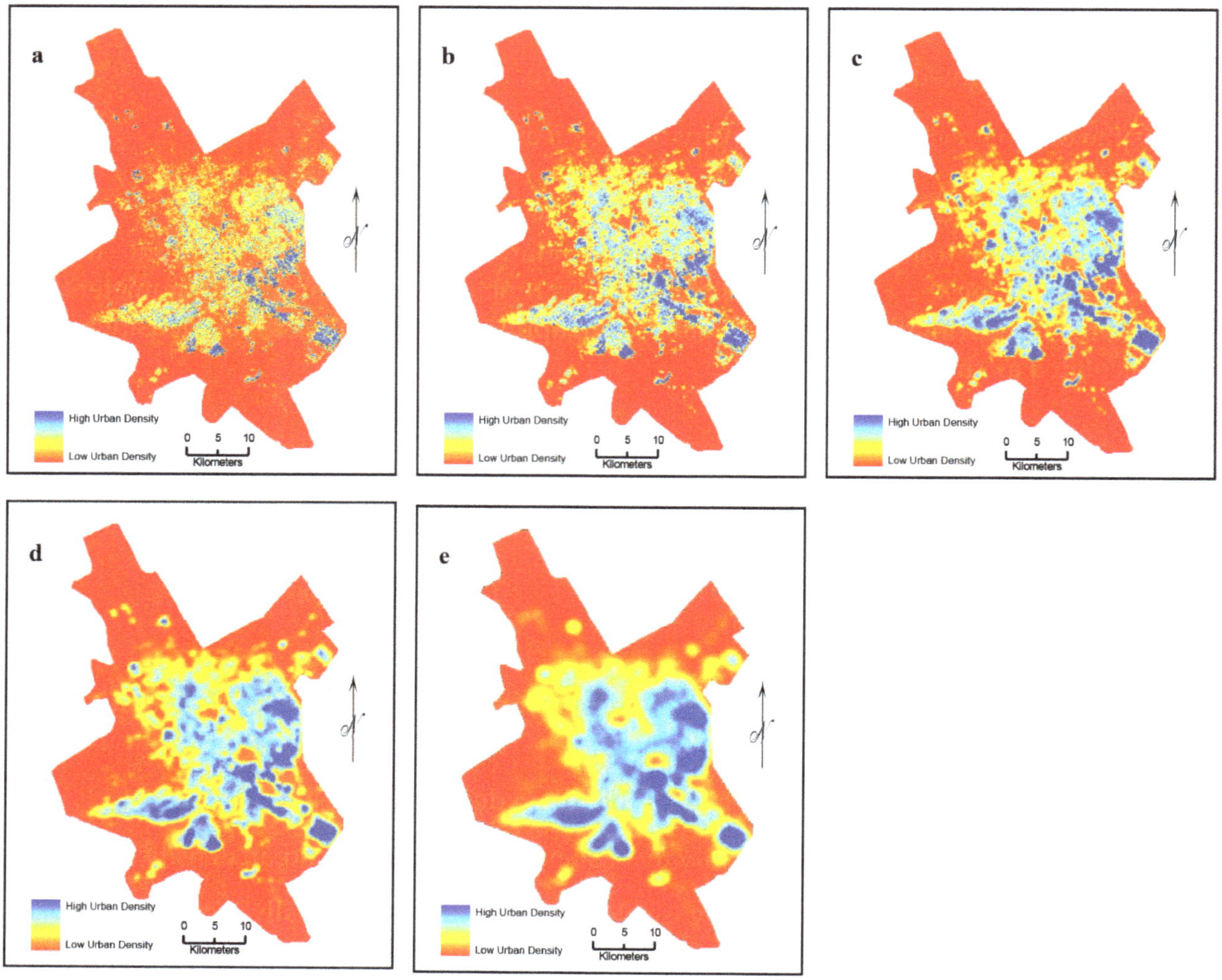

Figure 7. Urban density of Riyadh city in 2005 with five different neighbourhood sizes 2.5, 5, 10, 20 and 40 (a–e), respectively.

areas to being close to existing built-up areas. On the contrary, a large neighbourhood results in high urban density areas expanding outwards from the existing built-up areas.

In terms of the distance decay parameters P1 and P2, the two parameters for *urban density* have the same general effect as accessibility with little difference in interpretation of the parameters' effects. According to **Figure 8**, as the value of P2 decreases, only areas in close proximity to the very high urban density increase (dark blue areas in maps b, c and d decrease with a decrease of P2 value and become more compact). As the value of P1 decreases, areas with a very high density increase all over the city (dark blue areas in maps b, e and h increase with a decrease of P1 value and become more scattered). The difference between P1 and P2 values controls the size of areas in which the urban density surface decreases smoothly (sky blue to yellow).

4.3. Fuzzy constraints factors

The slope and altitude input variables are derived directly from digital elevation model (DEM) using ESRI ArcGIS 9.2 software [60]. This step is followed by three procedures for weighting,

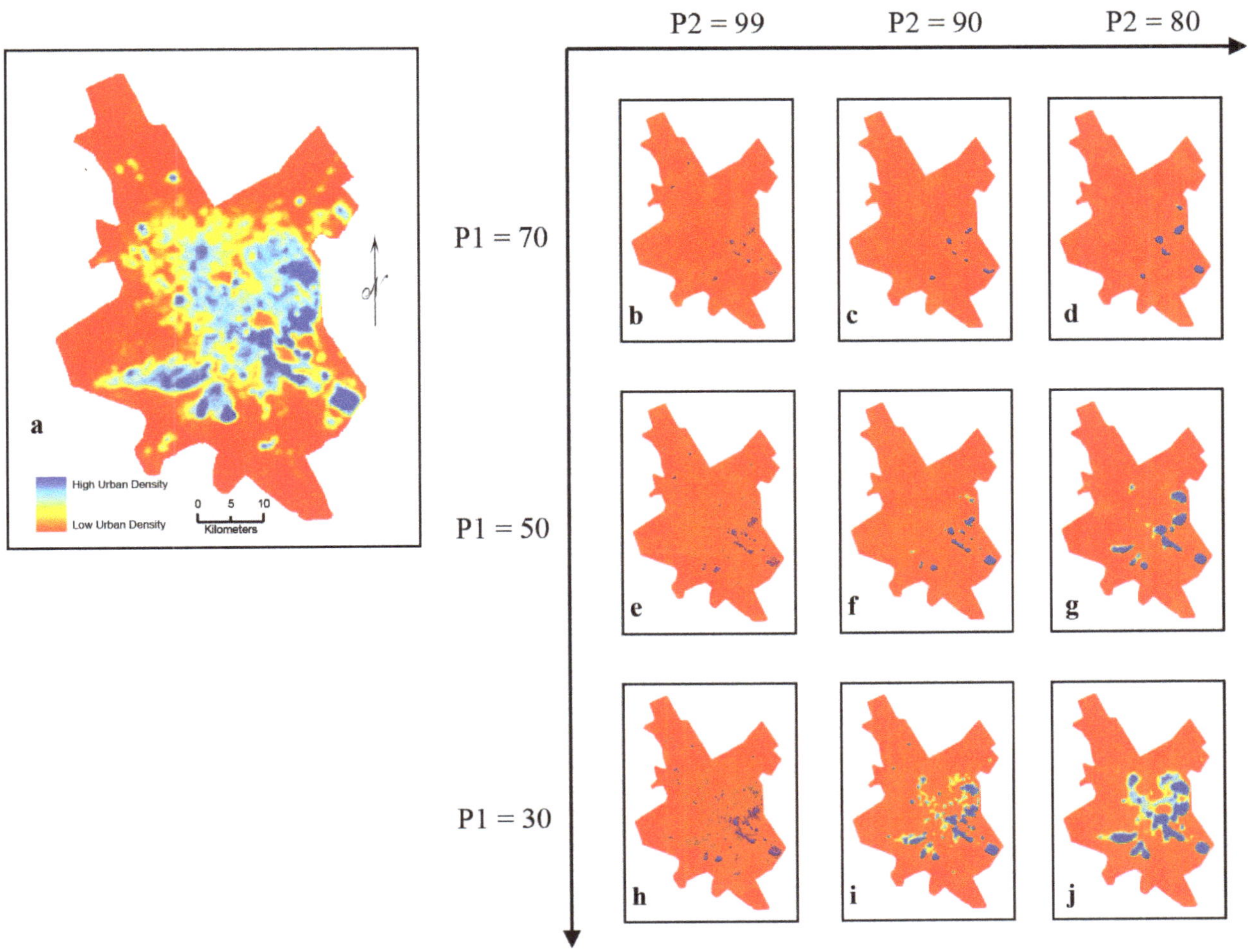

Figure 8. (a) Original and (b–j) fuzzy urban density.

standardization and fuzziness using Eq. (5). The physical topography of Riyadh confines to some extent the direction of urban growth. This is because the majority of urban development takes place in areas with a relatively moderate altitude and flat slope, while those areas with a steep slope and a high altitude are left without development. However, the range of slope and altitude variables varies according to the values of distance decay parameters P1 and P2, which together identify three areas including the preferable range of altitude and slope (highly suitable areas), the non-preferable (non-suitable areas) and the partly preferable (moderately suitable areas). These classifications reflect the behaviour of people developing land. As can be seen from **Figures 9** and **10**, as the value of P2 decreases, those non-suitable areas with a high altitude or a steep slope (red) increase. The highly suitable areas with a low altitude or a flat slope (blue), on the contrary, increase as the values of P1 increase. The size of partly suitable areas increases as the difference between P1 and P2 values increases. The d, i and b maps of **Figures 9** and **10** show the largest areas of highly non-suitable areas, highly suitable areas and the partly suitable areas, respectively. The optimal values of P1 and P2 parameters were calibrated [27].

The planning area (PA) layer was generated by on-screen digitizing of boundaries of the new areas where planning applications were approved and urban development was allowed. The sources of planning areas are the Master Plans 1, 2, 3 of Riyadh city which were established by

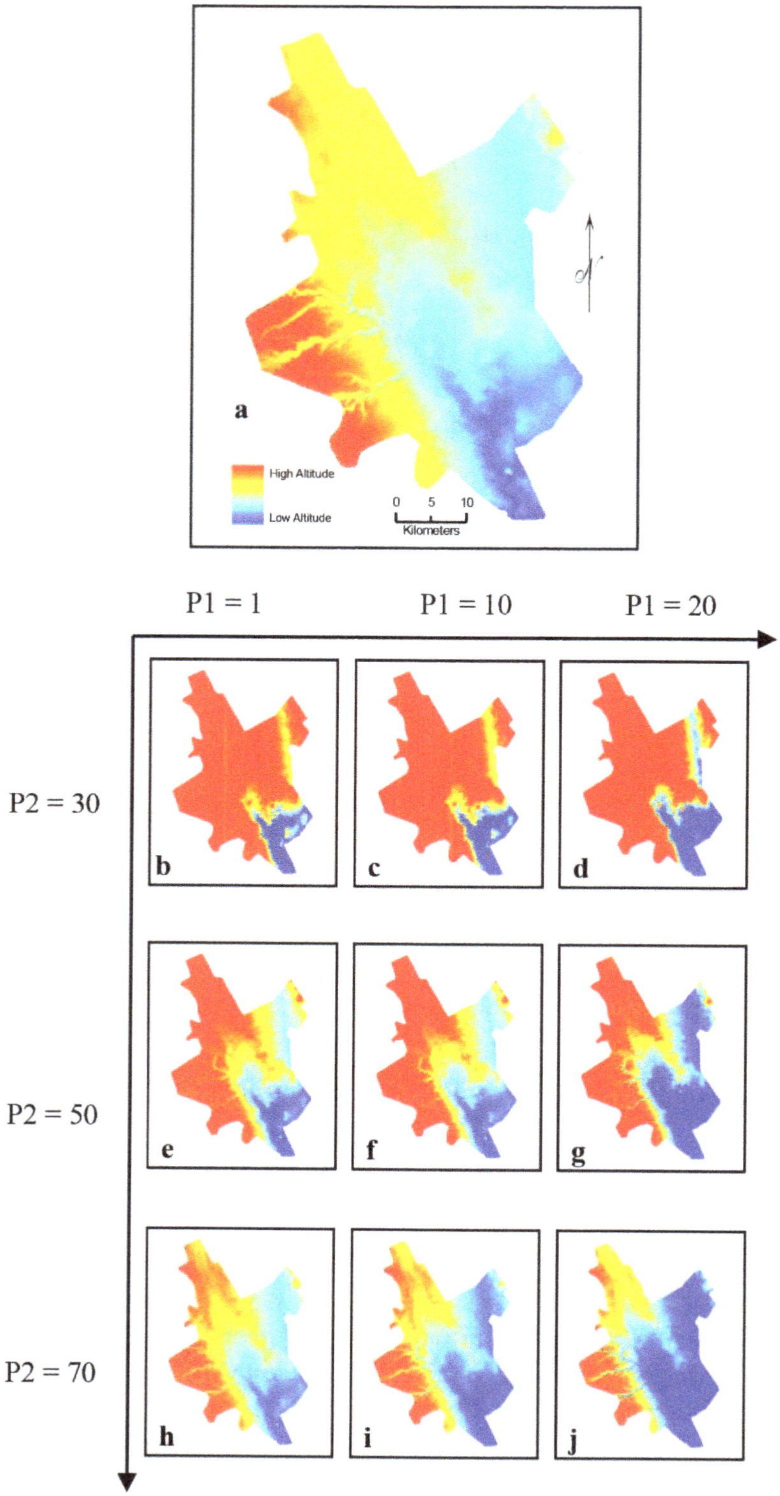

Figure 9. Original altitude (a) and fuzzy altitude (b–j).

the Arriyadh Development Authority (ADA). The PA layer has a binary value of 1 for 'Planned Area' and 0 for 'non-planned area'. The excluded area (EA) layer was created from several sources such as slope images, altitude images, classified remote sensing images to derive

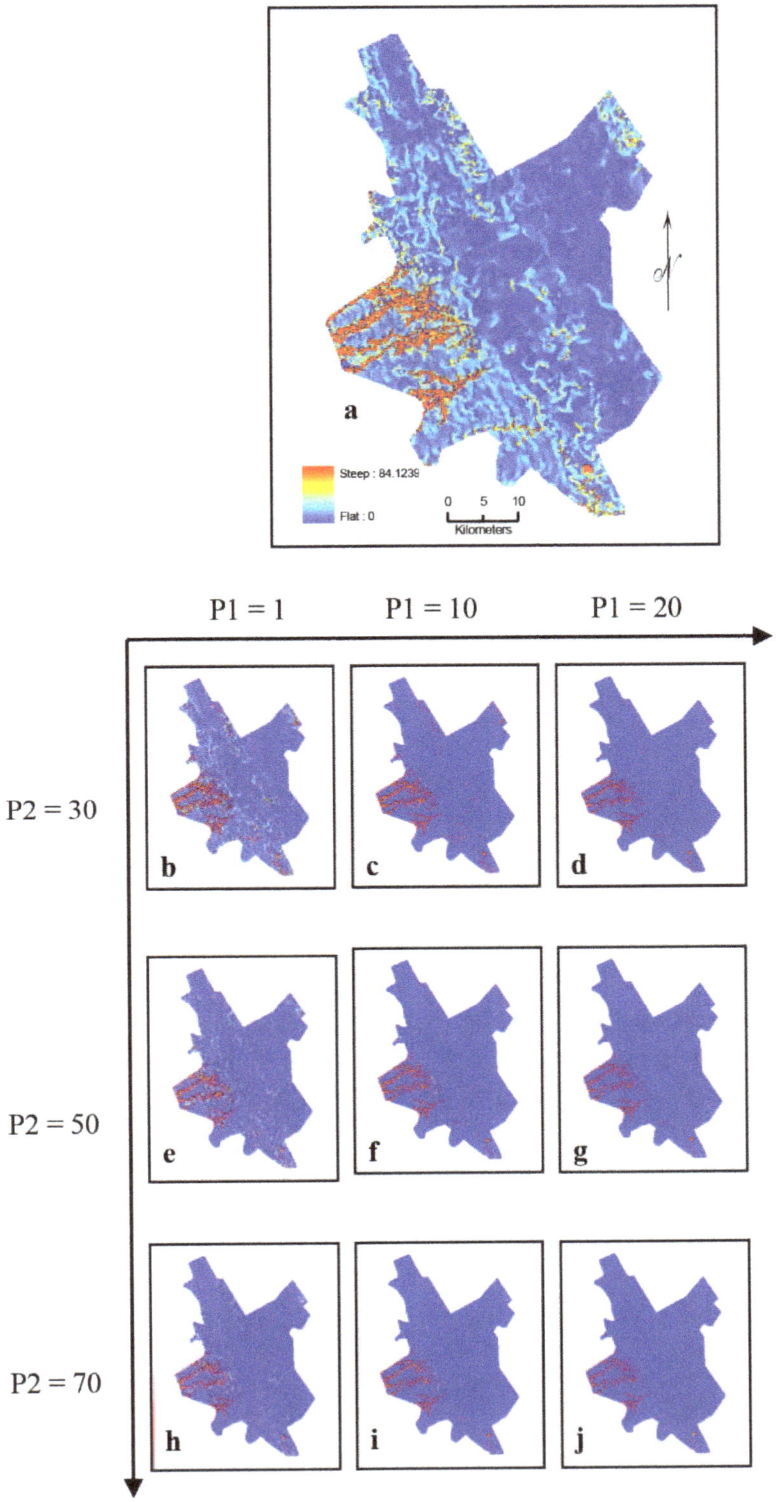

Figure 10. Original slope (a) and fuzzy slope (b–j).

agricultural areas and zoning regulation areas to reflect those in the master plan. Then, all of these layers were integrated and overlaid into one layer called EA, which has a binary value 1 for 'non-excluded area' and 0 for 'excluded area'.

4.4. Urban growth driving forces as fuzzy variables

This section demonstrates and discusses the results of computing the input variables that form the four driving forces of urban growth. The four driving forces, (1) transport support factor (TSF), (2) urban agglomeration and attractiveness factor (UAAF), (3) topographical constraints factor (TCF) and (4) planning policies and regulations factor (PPRF), will be converted into linguistic variables (**Figure 11**) and then entered into the FCUGM [28, 29] for modelling the spatial pattern of urban growth for Riyadh.

4.4.1. Transport support factor (TSF)

The TSF is based on three *input variables:* (1) weighted fuzzy local road accessibility (WFLRA), (2) weighted fuzzy main road accessibility (WFMRA) and (3) weighted fuzzy major road accessibility (WFMJRA). The TSF is generated in two stages: (1) integration of input variables and (2) definition of fuzzy sets. Under (1), the three *input variables* are integrated into one layer, called Transport *Tij*, using a linear combination approach as shown in Eq. (9).

$$T_{ij} = WFLRA_{ij} + WFMRA_{ij} + WFMJRA_{ij} \tag{9}$$

Under (2), the Transport layer is used to create the TSF by transforming this layer into linguistic (fuzzy) variable by defining a series of three overlapping fuzzy sets. Although there is no rule for specifying the number of fuzzy sets for a linguistic (fuzzy) variable, most of the fuzzy modelling studies used three or five sets depending on the complexity of the problem. Since the TSF includes three input variables, which have the same spatial characteristics (all of them related to road networks), three fuzzy sets were selected. The TSF is composed of three fuzzy sets: Weak, Moderate and Strong. Each set describes a different condition of TSF. **Figure 11** depicts the placement and shape of these three fuzzy sets (membership functions) and Eqs. (10)–(12) provide the mathematical models used to calculate them. The shape of the

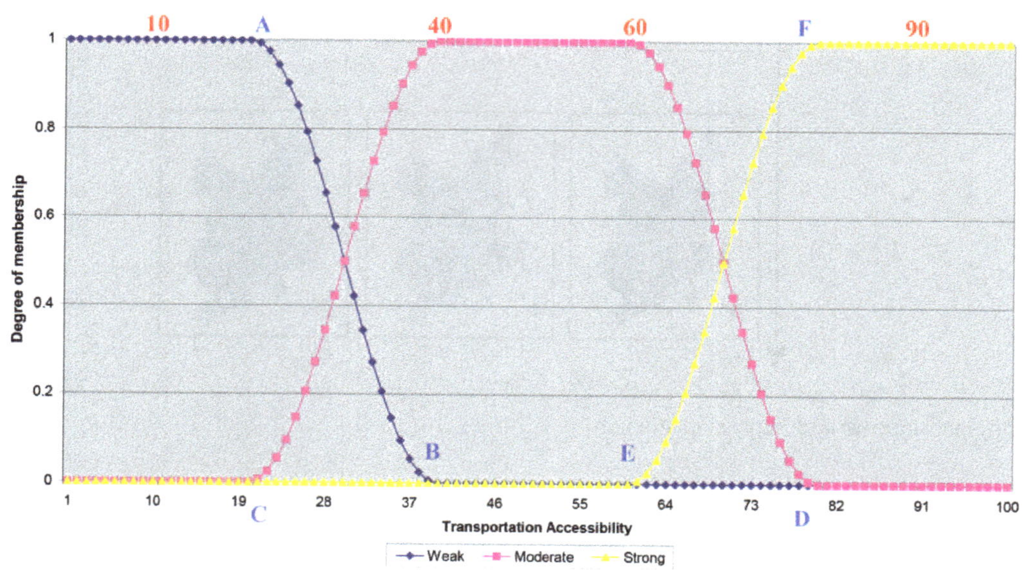

Figure 11. Membership functions of TSF driving force.

fuzzy sets follows a sinusoidal function. The fuzzy sets Weak, Moderate and Strong show a monotonically decreasing, symmetric and monotonically increasing functions, respectively. Although the shape of these fuzzy sets is usually decided by intuition or expert knowledge, in the FCUGM, they are optimized [27]. The authors believe that deciding the shape of the fuzzy sets in a subjective manner would result in restricting interpretations of the results.

The FCUGM optimizes the positions of a fuzzy set by fixing the central value of that set and calibrating the width or the range. This approach can be regarded as a hybrid and has two advantages: (1) it integrates the automatic self-tuned and expert knowledge techniques which could minimize the complexity of the problem (calibration urban CA model) and (2) decreases the number of parameters to be calibrated which reduces the complexity of the calibration process itself and the computation time required. According to **Figure 11**, the letters a–f are parameters that control the width of the fuzzy sets to be calibrated, while the numbers 10, 40, 60 and 90 are the fixed central values.

Linguistic Expression *Membership Function*

$$\mu Weak\ (Tij) = \begin{cases} 1 & T_{ij} < A \\ \frac{1}{2}(1 - \cos(\pi\,(T_{ij} - A) / (B - A) & A \leq T_{ij} \leq B \\ 0 & T_{ij} > B \end{cases} \quad (10)$$

$$\mu Moderate\ (Tij) = \begin{cases} 0 & Ti_j < C \\ \frac{1}{2}(1 + \cos(\pi\,(T_{ij} - C) / (40 - C) & C \leq T_{ij} \leq 40 \\ 1 & 40 < T_{ij} < 60 \\ \frac{1}{2}(1 - \cos(\pi\,(T_{ij} - 60) / (D - 60) & 60 \leq T_{ij} \leq D \\ 0 & T_{ij} > D \end{cases} \quad (11)$$

$$\mu Strong\ (Tij) = \begin{cases} 0 & T_{ij} < E \\ \frac{1}{2}(1 - \cos(\pi\,(T_{ij} - E) / (F - E) & E \leq T_{ij} \leq F \\ 1 & T_{ij} > F \end{cases} \quad (12)$$

4.4.2. Urban agglomeration and attractiveness factor (UAAF)

The urban agglomeration and attractiveness factor (UAAF) is formed using three *input variables* including weighted fuzzy urban density (WFUD), weighted fuzzy town centre accessibility (WFTCA) and weighted fuzzy employment centres and socioeconomic services accessibility (WFECSESA). The UAAF is generated in two stages: (1) integration of input variables and (2) definition of fuzzy sets. Under (1), the three *input variables* are integrated into one layer, called Urban Agglomeration and Attractiveness *UAAij*, using a linear combination approach as shown in Eq. (13).

$$UAA_{ij} = WFUD_{ij} + WFTCA_{ij} + WFECSESA_{ij} \tag{13}$$

Under (2), the urban agglomeration and attractiveness layer is used to create the UAAF by transforming this layer into linguistic (fuzzy) variable, UAAF, by defining a series of five overlapping fuzzy sets. The motivation of selecting five sets for UAAF, rather than three as TSF, is that UAAF is formed by integrating three input variables that have different characteristics (urban density, employment centres and socioeconomic services and town centre). Thus, the degree of vagueness or uncertainty of this driving force is higher than TSF, meaning that it could be more appropriate for the UAAF driving force to be expressed with a higher number of fuzzy sets. Thus, the UAAF is composed of five fuzzy sets namely Very Low, Low, Medium, High and Very High. Each set describes a different condition of UAAF. **Figure 12** depicts the placement and shape of these five fuzzy sets (membership functions) and Eqs. (14)–(18) provide the mathematical models used to calculate those sets. In **Figure 12**, the letters A–J are parameters that control the width of the fuzzy sets which are to be calibrated [27], while the numbers 25.5, 27.5, 47.5, 52.5, 72.5 and 77.5 are the fixed central values.

Linguistic Expression *Membership Function*

$$\mu_{Very\ Low}(UAA_{ij}) = \begin{cases} 0 & UAA_{ij} \leq 0 \\ 1 & 0 < UAA_{ij} < A \\ \tfrac{1}{2}(1 + \cos(\pi\,(UAA_{ij} - A) / (B - A) & A \leq UAA_{ij} \leq B \\ 0 & UAA_{ij} > B \end{cases} \tag{14}$$

$$\mu_{Low}(UAA_{ij}) = \begin{cases} 0 & UAA_{ij} < C \\ \tfrac{1}{2}(1 + \cos(\pi\,(UAA_{ij} - C) / (25.5 - C) & C \leq UAA_{ij} \leq 25.5 \\ 1 & 25.5 < UAA_{ij} < 27.5 \\ \tfrac{1}{2}(1 - \cos(\pi\,(UAA_{ij} - 27.5) / (D - 27.5) & 27.5 \leq UAA_{ij} \leq D \\ 0 & UAA_{ij} > D \end{cases} \tag{15}$$

$$\mu_{Medium}(UAA_{ij}) = \begin{cases} 0 & UAA_{ij} < E \\ \tfrac{1}{2}(1 + \cos(\pi\,(UAA_{ij} - E) / (47.5 - E) & E \leq UAA_{ij} \leq 47.5 \\ 1 & 47.5 < UAA_{ij} < 52.5 \\ \tfrac{1}{2}(1 - \cos(\pi\,(UAA_{ij} - 52.5) / (F - 52.5) & 52.5 \leq UAA_{ij} \leq F \\ 0 & UAA_{ij} > F \end{cases} \tag{16}$$

$$\mu_{High}(UAA_{ij}) = \begin{cases} 0 & UAA_{ij} < G \\ \tfrac{1}{2}(1 + \cos(\pi\,(UAA_{ij} - G) / (72.5 - G) & G \leq UAA_{ij} \leq 72.5 \\ 1 & 72.5 < UAA_{ij} < 77.5 \\ \tfrac{1}{2}(1 - \cos(\pi\,(UAA_{ij} - 77.5) / (H - 77.5) & 77.5 \leq UAA_{ij} \leq H \\ 0 & UAA_{ij} > H \end{cases} \tag{17}$$

$$\mu_{Very\ High}(UAA_{ij}) = \begin{cases} 0 & UAA_{ij} \leq I \\ \tfrac{1}{2}(1 + \cos(\pi\,(UAA_{ij} - I) / (J - 1) & I \leq UAA_{ij} \leq J \\ 1 & UAA_{ij} > J \end{cases} \tag{18}$$

4.4.3. Topographical constraints factor (TCF)

The TCF was based on two *input variables:* (1) weighted slope (WSG) and (2) weighted altitude (WA). The TCF is generated in two stages: (1) input variables integration and (2) definition of fuzzy sets. Under (1), the two *input variables* are integrated into one layer, called Topographical Constraints *TCij*, using a linear combination approach as expressed in Eq. (19).

$$TC_{ij} = WSG_{ij} + WA_{ij} \tag{19}$$

The topographical constraints layer was used to create the TCF by transforming this layer into a linguistic (fuzzy) variable by defining a series of three overlapping fuzzy sets. TCF is composed of three fuzzy sets namely slightly favourable (SM), moderately favourable (MF) and considerably

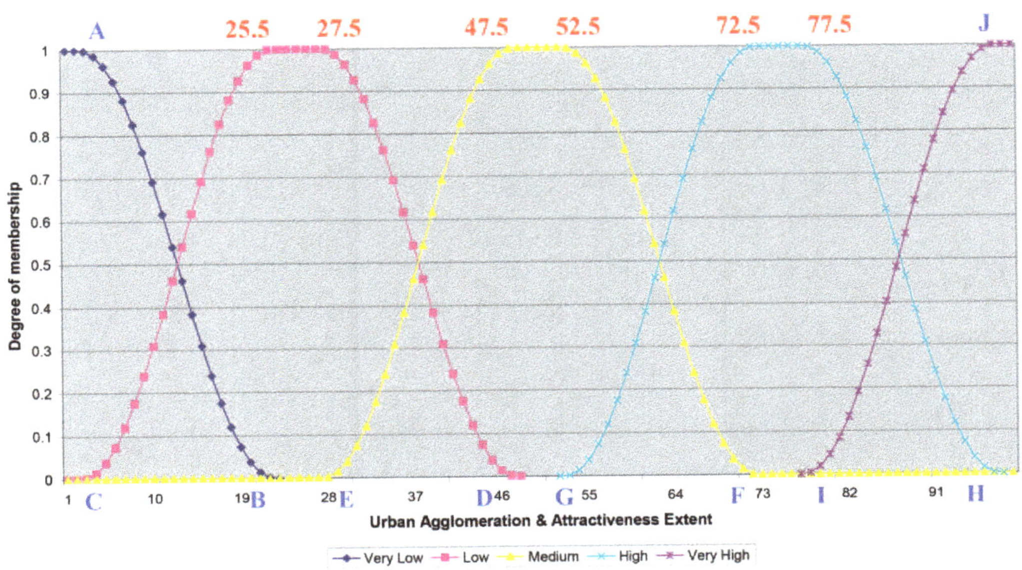

Figure 12. Membership functions of UAAF driving force.

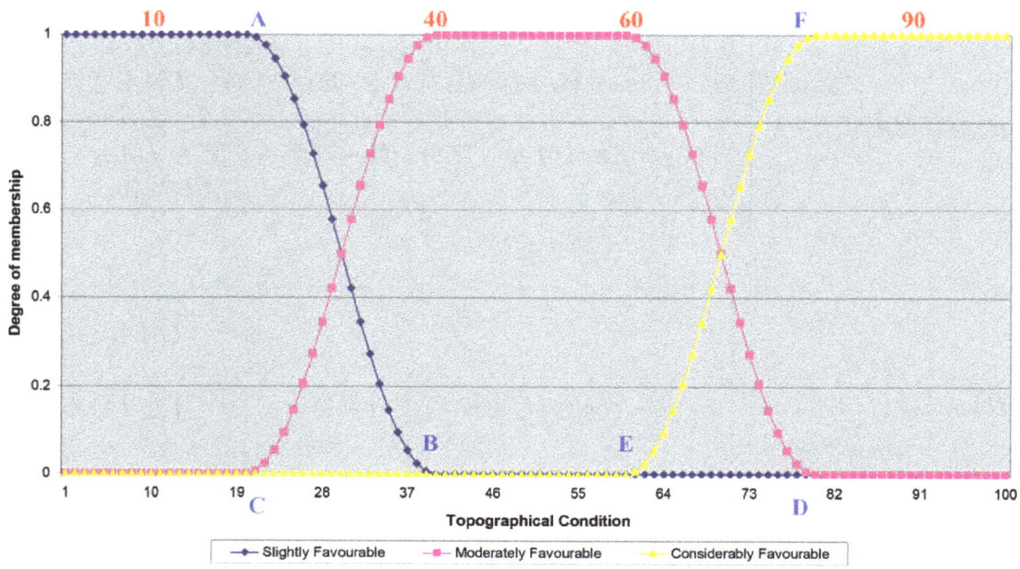

Figure 13. Membership functions of the TCF driving force.

favourable (CF). Each set describes a different condition of the TCF. **Figure 13** depicts the placement and shape of these three fuzzy sets (membership functions). Eqs. (20)–(22) provide the mathematical models for calculating these sets. According to **Figure 13**, the letters A–F are parameters which control the width of the fuzzy sets and are to be calibrated, while the numbers 10, 40, 60 and 90 are the fixed central values.

Linguistic Expression *Membership Function*

$$\mu_{Slightly\ Favorable}\ (TC_{ij}) = \begin{cases} 1 & TC_{ij} < A \\ ½\ (1 - \cos(\pi\ (TC_{ij} - A) / (B - A) & A \leq TC_{ij} \leq B \quad (20) \\ 0 & TC_{ij} > B \end{cases} \qquad (20)$$

$$\mu_{Moderately\ Favorable}\ (TC_{ij}) = \begin{cases} 0 & TC_{ij} < C \\ ½\ (1 + \cos(\pi\ (TC_{ij} - C) / (40 - C) & C \leq TC_{ij} \leq 40 \\ 1 & 40 < TC_{ij} < 60 \\ ½\ (1 - \cos(\pi\ (TC_{ij} - 60) / (D - 60) & 60 \leq TC_{ij} \leq D \\ 0 & TC_{ij\ j} > D \end{cases} \qquad (21)$$

$$\mu_{Considerably\ Favorable}\ (TC_{ij}) = \begin{cases} 0 & TC_{ij} < E \\ ½\ (1 - \cos(\pi\ (TC_{ij} - E) / (F - E) & E \leq TC_{ij} \leq F \\ 1 & TC_{ij} > F \end{cases} \qquad (22)$$

4.4.4. Planning policies and regulations factor (PPRF)

The PPRF is generated based on two *input variables:* (1) planning area (PA) and (2) excluded area (EA). These input variables are binary variables. The PPR is computed by combining the two variables as shown in Eq. (23).

$$PPR_{ij}^{t} = EA_{ij}^{t} * PA_{ij}^{t} \tag{23}$$

Linguistic Expression *Membership Function*

$$\mu_{Suitable}(PPR_{ij}) = \begin{cases} 0 & PPR_{ij} = 1 \\ 1 & PPR_{ij} = 0 \end{cases} \tag{24}$$

$$\mu_{Non\text{-}Suitable}(PPR_{ij}) = \begin{cases} 0 & PPR_{ij} = 0 \\ 1 & PPR_{ij} = 1 \end{cases} \tag{25}$$

26

PPRF is then transformed into linguistic variables comprising two fuzzy sets namely Suitable and Non-Suitable. Each set describes a different condition of the TCF. Eqs. (24) and (25) provide the mathematical models for calculating these sets.

The three fuzzy input variables, TSF, UAAF and TCF, which drive the urban growth in the FCUGM, consist of different membership functions. Al-Ahmadi et al. [27] showed and discussed the calibrated fuzzy input variable parameters using genetic algorithm (GA), parallel simulated annealing (PSA) and expert knowledge (EK). The calibrated fuzzy input parameters have similar patterns for each of the three algorithms (GA, PSA and EK), scenarios and over all periods with low variations. As an example, in relation to the TSF, the average of the calibrated A,B,C,D,E and F parameters for the three fuzzy sets weak (W), moderate (M) and strong (S) that form the TSF fuzzy input variable accounts for 15, 33, 18, 77, 74 and 81, respectively. **Figure 14** depicts the optimized shape and placement of the three fuzzy sets for TSF. These optimized values reflect the area that is covered by each fuzzy set, that is, the quantity of TSF that belongs to each set. The A-B values (15–33) for the weak set indicate that locations with TSF values between 1 and 15 have a weak transport support (crisp set of Weak, i.e. full membership function). Those locations above 15 and below 33 are partly weak, that is, they might gain both weak and moderate support according to their degree of membership (fuzzy set of weak). And those with TSF values above 33 do not belong to the weak set. As the range of the calibrated values between the B parameter of *weak* set and the C parameter of *moderate* set increase, the overlap between the two sets and the quantity of fuzzy values

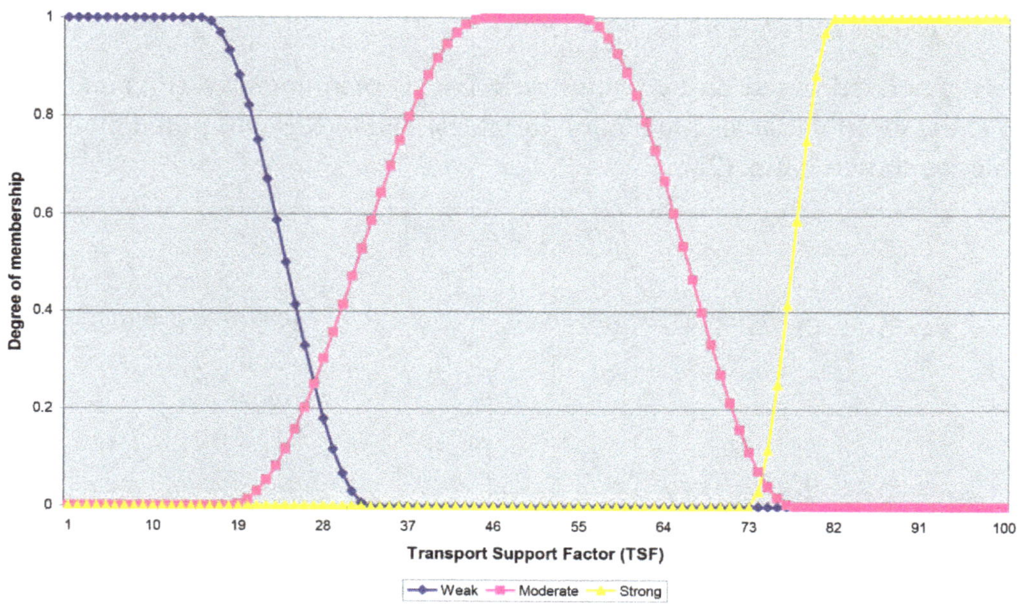

Figure 14. The mean optimized fuzzy sets for transportation support factor (TSF).

increases. Accordingly, as can be seen in **Figure 14**, the overlap between the weak and moderate sets (accounting for 15 as a range) is larger than the overlap between the moderate and strong sets (accounting for 3 as a range).

In terms of fuzzy modelling, it is better for the two adjacent sets to be intersected at 0.5 (crossover point) degree of membership, which provides a wider overlap area. In this sense, the larger the overlap areas between two sets, the better the results. This is because it allows different sets to use the same TSF values but with a different membership function according to the degree of ambiguity. The same interpretation can be applied to the other two factors UAAF and TCF.

5. Conclusions

This chapter presented an example of how fuzzy set theory can be applied to model urban growth factors and driving forces. The mathematical models for measuring, computing and addressing fuzzy input variables (accessibility and urban density) and fuzzy driving forces of the urban growth of Riyadh were discussed. The urban growth factors were established using fuzzy set theory, which quantified the effect of a distance decay using fuzziness. The driving forces of urban growth were designed for use with linguistics variables, which closely reflect human behaviours and attitudes regarding real-world phenomena. Such an approach provides a transparent method for interpreting the curve of a distance decay using linguistic variables.

The results indicate that fuzzy accessibility and fuzzy urban density factors are capable of mimicking and representing the uncertainty in the behaviour of the human decision-making process in land development in a very efficient manner. The behaviour of people was represented as one of three classes: 'full membership', 'partial membership' and 'no membership', corresponding to people who prefer to live 'near', 'fairly near' and 'far' from a certain feature, respectively. The driving forces of urban growth—the transportation support factor (TSF), urban agglomeration and attractiveness factor (UAAF) and topographical constraints factor (TCF)—were established by integrating fuzzy urban growth factors and were designed for use with linguistics variables, which closely reflect human behaviours and attitudes regarding real-world phenomena. These fuzzy driving forces were designed in an attempt to model the spatial patterns of urban growth using natural language statements, which made the modelling process more realistic and transparent than other approaches.

The author concluded that instead of defining the factors and driving forces of urban growth using deterministic equations, the FCUGM would apply fuzzy logic control to generate fuzzy transition rules. Such an approach is more transparent and more similar to the real processes of urban growth while also representing the uncertainty of various constraints and driving forces underlying urban growth. In conclusion, the chapter demonstrated that it was possible to treat the complexity and uncertainty inherent in decision-making processes in terms of site selection, which is usually a subjective process that follows individuals' preferences in which people generally use linguistic constructs for the evaluation of environmental or social situations.

Acknowledgements

The author acknowledges with gratitude King Abdulaziz City for Science and Technology for the accomplishment of this work.

Author details

Khalid Al-Ahmadi

Address all correspondence to: alahmadi@kacst.edu.sa

National Centre for Remote Sensing Technology, Space and Aeronautics Research Institute, King Abdulaziz City for Science and Technology, Riyadh, Saudi Arabia

References

[1] Kurtener D, Badenko V. Fuzzy algorithms to support planning. In Geertman S. and Stillwell J., (editors). Planning Support Systems in Practice. Berlin: Springer; 2003. pp. 245-269

[2] Wu F. Simulating urban encroachment on rural land with fuzzy-logic-controlled cellular automata in a geographical information system. Journal of Environmental Management. 1998;**53**:293-308

[3] Zadeh LA. Fuzzy sets. Information and Control. 1965;**8**:335-353

[4] Zimmermann HJ. Fuzzy Set Theory and its Applications. 4th ed. London: Springer; 2001

[5] Wu F. A linguistic cellular automata simulation approach for sustainable land development in a fast growing region. Computers, Environment, and Urban Systems. 1996;**20**: 367-387

[6] Altman D. Fuzzy set theoretic approaches for handling imprecision in spatial analysis. International Journal of Geographical Information Systems. 1994;**8**:271-289

[7] Wang F. Towards a natural language user interface: an approach of fuzzy query. International Journal of Geographical Information Systems. 1994;**8**:143-162

[8] Zadeh LA. The concept of a linguistic variable and its application to approximate reasoning. Memorandum ERL-M 411, Berkeley; 1973

[9] Ross T. Fuzzy Logic with Engineering Applications. West Sussex, UK: John Wiley and Sons; 1995

[10] Zadeh LA. From circuit theory to systems theory. IRE Proceedings. 1962;**50**:856-865

[11] Leung Y. Urban and regional planning with fuzzy information. In: Chatterjee L, Nijkamp P, editors. Urban and Regional Policy Analysis in Developing Countries. Aldershot: Gower; 1983. pp. 231-249

[12] Tao Y, Yang XM. Fuzzy comprehensive assessment, fuzzy clustering analysis and its application for urban traffic environment quality evaluation. Transportation Research Part D-Transportation and Environment. 1998;**3**:51-57

[13] Feng Z, Flowerdew R. Fuzzy geodemographics: A contribution from fuzzy clustering methods. In: Carver S, editor. Innovations in GIS 5. London: Taylor and Francis; 1998. pp. 119-127

[14] Davidson DA, Theocharopoulos SP, Bloksma RJ. A land evaluation project in Greece using GIS and based on Boolean and fuzzy set methodologies. International Journal of Geographical Information Science. 1994;**8**:369-384

[15] Hall G, Wang F, Subaryono. Comparison of Boolean and fuzzy classification methods in land suitability analysis by using geographical information systems. Environment and Planning A. 1992;**24**:497-516

[16] Leung Y, Leung KS. An intelllgent expert systems shell for knowledge-based geographical information systems 2: Some applications. International Journal of Geographical Information Science. 1993;**7**:201-213

[17] Wang F, Hall GB. Fuzzy representation of geographical boundaries in GIS. International Journal of Geographical Information Systems. 1996;**10**:573-590

[18] Banai R. Fuzziness in geographical information systems: Contributions from the analytic hierarch process. International Journal of Geographical Information Systems. 1993;**7**:315-329

[19] Kollias V, Viliotis A. Fuzzy reasoning in the development of geographical information systems FRSIS: A prototype soil information system with fuzzy retrieval capabilities. International Journal of Geographical Information Systems. 1991;**5**:209-223

[20] Tobler W. A computer movie simulating urban growth in the Detroit region. Economic Geography. 1970;**26**:234-240

[21] Batty M, Kim KS. Form follows function: Reformulating urban population density functions. Urban Studies. 1992;**29**:1043-1070

[22] Ward DP, Murray AT, Phinn SR. A stochastically constrained cellular model of urban growth. Computer, Environment and Urban Systems. 2000;**24**:539-558

[23] Cheng J, Masser I. Urban growth pattern modelling, a case study of Wuhan, P. R. China. Landscape and Urban Planning. 2003;**62**(4):199-217

[24] Cheng J, Masser I. Understanding spatial and temporal process of urban growth: Cellular automata modeling. Environment and Planning B: Planning and Design. 2004;**31**:167-194

[25] Torrens PM. How Cellular Models of Urban Systems Work. CASA Centre for Advanced Spatial Analysis, University Collage London, Working paper 28. 2000. URL: http://www.casa.ucl.ac.uk/working_papers/paper28.pdf. [Accessed: 2005]

[26] Vries J, Nijkamp P, Rietveld P. Exponential or Power Distance-Decay for Commuting? An Alternative Specification. Free University, Amsterdam. 2006. URL: http://www.tinbergen.nl/discussionpapers/04097.pdf. [Accessed: 2006]

[27] Al-Ahmadi K, Heppenstall AJ, Hogg J, See L. A fuzzy cellular automata urban growth model (FCAUGM) for the City of Riyadh, Saudi Arabia. Part 1: Model structure and validation. Applied Spatial Analysis. 2009;**2**(1):65-83

[28] Al-Ahmadi K, Heppenstall AJ, Hogg J, See L. A fuzzy cellular automata urban growth model (FCAUGM) for the City of Riyadh, Saudi Arabia. Part 2: Scenario analysis. Applied Spatial Analysis. 2009;**2**(2):85-105

[29] Al-Ahmadi K, See L, Heppenstall AJ, Hogg J. Calibration of a fuzzy cellular automata model of urban dynamics in Saudi Arabia. Ecological Complexity. 2009;**6**(2):80-101

[30] Al-Ahmadi K, See LM, Heppenstall AJ. Validating spatial patterns of urban growth using cellular automata. In: Salcido A, editor. Emerging Applications of Cellular Automata. Croatia, Rijeka: InTech; 2012. pp. 23-52

[31] Al-Ahmadi K, Alahmadi M, Alahmadi S. Spatial optimization of urban cellular automata model. In: Hung M, editor. Applications of Spatial Statistics. InTech; 2016. pp. 61-94

[32] Schenghe L, Sylvia P. Spatial patterns and dynamics mechanisms of Urban land use growth in China: Case study in Beijing and Shanghai. IR-02-005. Laxenburg, Austria: International Institute for Applied Systems Analysis; 2002

[33] Alonso W. Location and Land Use. Cambridge: Harvard University Press; 1964

[34] Van Thunen JH. Der Isolierte Staat im Beziehung auf Landwirtschaft und Nationalokonomie. Hamburg, The Author;1826

[35] Hansen W. How accessibility shapes land use. Journal of the American Institute of Planners. 1959;**25**:73-76

[36] Torrens PM. How Land-use Transport Models Work. Centre for Advanced Spatial Analysis, University Collage London, Working paper 20. 2000. URL: http://www.casa.ucl.ac.uk/publications/workingpapers.asp. [Accessed: 2005]

[37] Wegener M. Overview of land-use transport models. In: Henscher DA, Button K, (editors). Transport Geography and Spatial Systems. Kidlington: Pergamon and Elsevier Science; 2004

[38] Harvey J. Urban Land Economics. Houndsmills: Macmillan; 1996

[39] Clarke KC, Gaydos LJ. Loose-coupling a cellular automaton model and GIS: Long-term urban growth prediction for San Francisco and Washington/Baltimore. International Journal Geographical Information Sciences. 1998;**12**:699-714

[40] Wu F, Webster CJ. Simulation of land development through the integration of cellular automata and multi-criteria evaluation. Environment and Planning B: Planning and Design. 1998;**25**:103-126

[41] Wu F, Webster CJ. Simulating artificial cities in a GIS environment: Urban growth under alternative regulative regimes. International Journal of Geographical Information Science. 2000;**14**(7):625-648

[42] Yeh A, Li X. A constrained CA model for the simulation and planning of the sustainable urban forms by using GIS. Environment and Planning B: Planning and Design. 2001;**28**(5): 733-753

[43] Liu Y, Phinn S. Developing a cellular automaton model of urban growth incorporating fuzzy-set approaches. Computers, Environment and Urban Systems. 2003;**27**:637-658

[44] White R, Engelen G. High resolution integrated modelling of the spatial dynamics of urban and regional systems. Computers, Environment and Urban Systems. 2000;**24**:383-440

[45] Lau KH, Kam BH. A cellular automata model for urban land-use simulation. Environment and Planning B: Planning and Design. 2005;**32**:247-263

[46] Ziehr C. Fundamental of Geography. Education Course, Northeastern State University; 2005. URL: http://arapaho.nsuok.edu/~ziehr/courses/geog2243/Urban_Accessibility.htm, [Accessed: 2005]

[47] Arriyadh Development Authority (ADA) (2004) Arriyadh Metropolitan Strategy Plan: Part 2 State of the City, Background and Issues, Riyadh, Saudi Arabia

[48] Lowry IS. A Model of Metropolis. RM-4035-RC. Rand Corporation, Santa Monica, California; 1964

[49] Li X, Yeh A. Neural-network-based cellular automata for simulating multiple land use changes using GIS. International Journal of Geographical Information Science. 2002;**16**: 323-343

[50] Wegener M. Urban land-use transportation models. In: Maguire D, Batty M, Goodchild M, editors. GIS, Spatial Analysis, and Modelling. USA: ESRI Press; 2005. pp. 203-220

[51] Jang J, Sun C, Mizutani E. Neuro-Fuzzy and Soft Computing. New Jersey: Prentice Hall; 1997

[52] Dwinger P. Introduction to Boolean Algebras. Würzburg: Physica Verlag; 1971

[53] Zadeh LA. Towards a theory of fuzzy systems. In: Kalman RE, DeClaris N, editors. Aspects of Network and Systems Theory. Holt Rinehart, Winston; 1971

[54] Kainz W. Fuzzy Logic and GIS. Austria: Department of Geography and Regional Research, University of Vienna; 2005. URL: http://homepage.univie.ac.at/wolfgang.kainz/Lehrveranstaltungen/Fuzzy_Logic_and_GIS/Fuzzy_Logic_und_GIS_2_3spp.pdf. [Accessed: 2006]

[55] See L. Geographical Applications of Fuzzy Logic and Fuzzy Hybrid Techniques. Unpublished PhD thesis, School of Geography, University of Leeds, Leeds, UK;1999

[56] Cox E. The Fuzzy Systems Handbook: A Practitioner's Guide to Building, Using and Maintaining Fuzzy Systems. Cambridge, MA: AP Professional; 1994

[57] ESRI. ArcGIS 9.2 Desktop Help, Calculate Euclidian Distance. 2006. URL: http://webhelp.esri.com/arcgisdesktop/9.2/index.cfm?TopicName=eucdistance, [Accessed: 2006]

[58] Li X, Yeh A. Modelling sustainable urban development by the integration of constrained cellular automata and GIS. International Journal of Geographical Information Systems. 2000;**14**:131-1152

[59] Wu F. SimLand: A prototype to simulate land conversion through the integrated GIS and CA with AHP-derived transition rules. International Journal of Geographical Information Science. 1998;**12**:63-82

[60] ESRI. ArcGIS 9.2 Desktop Help, Calculating slope. 2006. URL: http://webhelp.esri.com/arcgisdesktop/9.2/index.cfm?TopicName=Calculating_slope. [Accessed: 2006]

2

Generating Reality with Geosimulation Models: An Agent-Based Social-Spatial Network Modelling Perspective

Andreas Koch

Abstract

Models in general and geosimulation in particular are epistemologically characterized by two principles: first, they produce reality through their existence and communication rather than simply representing it. Second, they reduce complexity in the process of mapping complexity. Since almost any current phenomenon is understood as complex without specifying how complex it is and in which sense, geosimulation models are important tools in solving this problem of specifying and representing complexity. This capability rests, among other things, upon its multilevel approach (bottom-up and top-down) and its ability to translate terms into numbers and thus into distinct singularities. A demonstration of such an understanding of models will be given by presenting a socio-spatial simulation approach in the domain of network analysis and social capital operationalization. Two Austrian regions serve as case studies using empirical and simulated data. The demonstration includes the endeavor to intertwine a place-based geography with a network-based geography.

Keywords: generative power of models, translating social capital into quantities, network mechanisms, semi-empirical model, network-based and place-based geography

1. Introduction

The mutual referencing of statements is a presupposition to comprehend the stated facts in a communication situation. Be it theories or methods, descriptions or explanations, texts or maps—relationships always create, explicitly or implicitly, these nexuses with which particularities are being contextualized. Contextualization generates meaning. However, neither the creation of relationships nor the creation of meaning is grounded in objective and independent criteria, neither in science nor in everyday life. They are, on the other hand, not completely arbitrary, but depend on particular social-cultural, temporal (historical) and spatial (geographical) contexts.

The referential contexts emphasize in addition how, why, and that complementarities are fundamental for comprehensive reasoning. Complementarities can be dialectic; the decisive point is that they simultaneously express a mutually excluding *and* complementing relationship, as is the case of the dualism between wave and particle or of impulse and location in physics. One characteristic of complementarity is that the so far unconnected (or differently connected) components do already exist. Every social construction of space, for instance, presupposes spatial rules, patterns, and conventions which are independent properties compared with the social rules, patterns, and conventions being applied when taking possession of social space. Among others, spatial rules ground in geometry, topology, or algorithmic computation. This is equally true for every statistical analysis. The result of a model run or a statistical analysis presumes the selected technique which itself is based on concrete procedural rules (e.g., that interactions among agents and between agents and their environment are taken into consideration). The process of computing a number is usually not reflected in the result, except it is made explicit. In this respect, a mathematical emergence of reality does not differ from a textual or (carto-) graphical emergence of reality and in all cases we refer to *a priori* objectives, reflections, and so on. What makes a difference is the nature of translation.

For modelling geographical phenomena from a social scientific perspective, an understanding of relations and complementarities is crucial because both spatial and social facts do rely heavily on their inner-temporal dynamics and develop in time differently [1]. Social-spatial network processes and mechanisms may obey common rules of homophily or of addressing roles and positions. The quality of the processes and mechanisms, however, together with its concrete manifestations of allocations of social, cultural, economic, and spatial capital is influenced by the idiosyncratic peculiarity of their local, temporal, and community compositions.

With these introductory remarks, we aim to highlight the underlying principles in the creation of meaning and intentionality derived from abstract information [2] and dedicated to be used in scientific, political and everyday-life communications. The notion of communication is used here as a means of providing us with the capability to deal with our perceived and interpreted environment in a meaningful way. This aim implies a closer look into the functions and purposes of models and quantifications, and is tied to the challenges that arise due to the referencing of complementarities, of numbers and words, of quantifications and qualifications.

2. The nature of models and quantifications and their impacts

2.1. The peculiarity of quantities and quantifications

In contrast to the common understanding of quantities which are assumed to devaluate the individual as an abstract number and negate its qualities, it is Lewin ([3], p. 150) who appreciates the peculiarity of quantification: "It is the increased desire, and also the increased ability, to comprehend *concrete particular cases*, and to comprehend them fully, which, together with the idea of the *homogeneity* of the physical world and that of the *continuity* of the properties of its objects, constituted the main impulse to the increasing quantification of physics." Numbers, quantifications and their emergence through quantitative methods are one possibility for raising a kind of uniqueness to facts, phenomena, and events which cannot be realized by the other approaches. "There is nothing more abstract and singular than numbers. Beyond numbers no further abstraction is thinkable, because abstracting from numbers would mean to disregard singularity. This in turn would mean to give up the Self, because the Self is only possible in contradiction to 'Another'" ([4], p. 4; translation A.K). It is worth noting that contemporary sociological and philosophical approaches to modern societies increasingly take the concept of singularity into account. [5–6]. Weiss [4] also points out that quantity (the single part) and quality (the whole) conflate; they can be distinguished analytically, but need to be mutually related in order to understand both sides of the coin—and the coin.

The complementarity of quantity and quality, of quantification and qualification, is realized through translation, and this procedure must be recalled when statements and numbers as well as their manifold representations are captured and utilized. Detractors of quantification tend to create an asymmetry between the two worlds. They argue that if a qualitative phenomenon is being tied to a number it will lose its eigenvalue—the only thing that counts henceforth is its numerical value. The processes and mechanisms of its creation, their cultural, geographical or temporal differences maintain obscured [7]. Though this problem definitely exists, it is not a problem of the quantification itself but a problem of exposing the cultural, political, social and scientific circumstances of its creation. Therefore, the use of numbers and words incorporates presuppositions about what is perceived as relevant and valuable.

2.2. Challenges of transformability of methods, numbers, and models

Measurements, quantifications and models result in one or many but always concrete single cases, as has been stated above. What they bear must be contextualized. From an epistemological perspective, it is the well-known threefold "context of discovery," "context of justification," and "context of utilization." Contextualization, in addition, must refer to other measurements and models in order to verify and validate model results, but also to utilize them comparatively in connection with other approaches [8]. What follows is not a competition between outcomes and paradigmatic settings but an abductive approaching to the explicit problem(s) at hand.

This may sound trivial. Indeed, the claim for *trans*formability of methodological approaches, of modelling aims and types among and between modelling paradigms, scientific and epistemological explanations has been uttered many times over the past six or so decades. One such utterance, apart from that of Thomas Kuhn [9], was made by Ludwik Fleck, a Polish physician who refused to accept an "absolute truth" as an epistemological counterpart of scientific exploration. He insisted on three social factors which inherently determine scientific reasoning ([1], p. xxii):

1. The "weigh of education": knowledge foremost consists of learnt items, being, however, then subtly transformed by learning and communication.

2. The "burden of tradition": new recognition is primarily coined by already existing recognition.
3. The "effect of the recognition sequence": what has been once conceptualized theoretically and/or methodologically restricts the opportunities of new concepts.

Incorporating these social factors into the evaluation of research results may help to relax the researcher from unobtainable truths and to focus more strictly on the processes and purposes of model production.

2.3. Utilization of knowledge, numbers and models

Taking the challenges of contextualization, translation and transformation explicitly into account would also mean having to deal with knowledge and models in a different way. The qualitative experience with social phenomena and problems feeds mainly on very subjective and local living conditions; it is, however, simultaneously an amalgamation of aggregated knowledge as "socialized subjectivity" [10]. This kind of knowledge which refers to local and subjective-social living conditions can be termed "local knowledge." Local knowledge, though derived from a different epistemology and perspective (ethnography and natives in developing countries) [11], can be transferred to contemporary (post)modern, globalized *and* localized living conditions as well. Local knowledge of people having, for example, a different social or economic status, living in impoverished or wealthy regions is, apart from scaling, important for social-spatial network analysis because it incorporates another mode of observation and perception. While knowledge about socioeconomic inequality or disparate living conditions in scientific and political contexts is mostly a perspective of people who are not affected by these disparities (a second order observation), local knowledge of socially discriminated and excluded people offers a first order observation.

In addition to mutual relationships between different modes of recognition, ordering, and understanding, it is important to account for the relationships within the quantitative methodology. It can be confirmed that "calculating is existing equality of opportunity" ([12], p. 43) because societal problems such as discrimination and inequality can be—and must be—made comparable through exact differentiation and objectivation which can then be used for opinion-forming in the political arena. As long as statistical analyses are considered as an instrument among others, one would be able to agree with the above quote. If not, the "equality of opportunity" turns out to be misused as ideology, representing an interest- and power-driven inequality of opportunity, since there is no longer explicit advice for a reduction of complexity.

A reduction of complexity is another and inevitable step in perceiving and grasping our world. It is, however, simultaneously necessary to deal with this reduction of complexity in an explicit and deliberate way. Explicit means to not only publish the results of an analysis but also the process of its realization. Deliberate means taking a critical attitude with regard to the chosen method(s) (see, among others, [13] for a detailed description of the creation of the "risk of poverty threshold" and its implications, and [14] with delineation of measuring poverty in London in the nineteenth and twentieth century).

Another problem of the internal confirmatory referencing of quantitative methods is the danger of not only making the measured values absolute but also the models and techniques applied. This problem can be phrased as "to explain the real through the impossible." Economics can serve as an example here: "Economics creates mathematical models which could never be built in reality but are used nevertheless to compute and reduce complex economic processes to a few numbers. Here too, they try to describe the real through the impossible. [...] Neoclassical economics assumes a kind of market harmony. If markets are left to their own resources then everything develops well. Dummy arguments are used to confirm this opinion by misusing mathematics in order to circulate ideology" ([15], p. 112f, translation A.K). The critical point is not that quantitative methods deal with artificial experimental settings or models, but that they immediately equate models with reality. Instead, models— through their construction and focus—*create a reality* through their use in science and everyday life, which will be explained in more detail in the next section. In addition, the methodological and technological progress must also be thought of explicitly.

2.4. The purpose and meaning of models

For an appropriate appreciation of quantification and modelling, a different kind of concluding translation between method and epistemology seems to be necessary. Statistics often assumes an absence of ideologies and normative values which is not true when considering the modes of acquiring, producing, and publishing data and results, respectively. The basic instruments with which we observe, describe, explain, and interpret the world are models—there is no immediate access to our spatial and social environment.

This must be briefly explained. I agree with Schurz ([16], p. 56) who argues for a hypothetical-constructivist realism approach. According to this realism, our perception and imagination of reality is not *a priori* given, but constructed and conditioned through active cognition (which is referred to as "epistemic" constructivism). Contrary to an ontological constructivism which conclusively claims that reality too is not *a priori* given, the hypothetical-constructivist realism does not link perception and reality so tightly. Instead, it assumes a structural correspondence which transforms information between perception and reality, which is neither complete nor unambiguous. Taking such a corresponding linkage into consideration, it leads to an understanding of models that does not claim a straightforward coincidence of models with *the* reality, *the* nature, or *the* world. Models are not simply simplified representations of reality; they are images (imaginations) which we make from our environment. We constantly construct and reconstruct our environmental entry by using many and different instruments and tools proactively and interactively. These instruments and tools in turn influence our ways of (re-) construction. This making of is not always an act of creation, we also (and maybe mainly) use images (imaginations) made by others, be it maps, news articles, social network blogs, novels, theater plays, and the like. The application of images differs with respect to experience, social roles and positions, and many more contexts.

If this assumption of how reality can be accessed is true, then the role and meaning of models changes significantly. Models, then, are created, developed, and applied to generate reality, not to represent it (although representation remains one, but only one important characteristic).

In so doing, they untie from an assumed objective, true or total reality which has to be imitated by applying certain sets of rules. They establish a kind of independence. Models, moreover, reduce complexity in order to make the subject matter concrete. Hence, it is not the unimaginable, opaque complexity addressed theoretically to "reality" which is under investigation in models and computationally translated to quantities. Complexity of reality is a metaphor which will act as a counterpart to models, theories and quantifications. To compare it with system theory, we can conceive of characteristics of systems because they emerge as distinct objects structurally and functionally. Initially, we are not able to talk about a system's environment—it is just the unassigned other side of the system.

If the function of a model is not to represent or imitate the complexity of an unknown reality, then a model inhabits inherently its own justification by explicitly expressing its purpose, assumptions, and ways of reducing complexity. This is quite similar to map-making by applying rules of generalization or to statistical analysis by deliberately selecting variables and techniques in order to achieve a certain result.

The aim of this line of thought is to avoid absolute external reality as a reference in modelling, but instead putting the inherent purposes, assumptions, framing conditions, parameter settings, and so on in the foreground. In so doing, it is a realization of a modelling epistemology that has been claimed by several researchers, among others Epstein [17], by arguing for a "generative social science" which accounts for letting the phenomenon of interest grow in a simulation model of interacting agents, or Küppers et al. [18] who equate simulation models epistemologically with a "pragmatic construction of reality," whereby reality emerges inside the models (further hints are given in [19]).

It would therefore be more appropriate to refer to "originals" instead of "reality" which are being generated and represented by models. Originals refer to other originals and in doing so there is no need to refer to an absolute truth/reality. This is in line with Stachowiak's [20] General Model Theory, whereby models are defined through three characteristics: (1) a model is always a representation of a natural or artificial original, and the original can itself be an original; (2) a model does not encompass all attributes of an original, but only those which seem to be relevant for the model purpose; (3) a model does not conflate with the original inherently, but depends on the purpose, thus on the questions 'whereto,' 'for what and whom,' 'when,' and 'where.'

2.5. The risk of instrumentalizing models

Without relationships and associations, models and numbers tend to be used as ideology or sheer instruments. They become stylized facts, and the images (imaginations) they produce induce a kind of a factual inevitability. A social capital value derived from network analysis, a poverty threshold, or a correlation between voter turnout and social status is equated with reality that not even one of the parameters would represent as a single case accidentally. A related problem is given with the use of the notion of "optimization"—model optimization is often equalized with empirical optimization, misappropriating the purpose(s) and premise(s) of the model.

Complementary to the ideologization *of* models, we have to take the ideologization *through* models into account. Members of a social community or milieu potentially identify with scientific and political model results or analyses, be it confirmatory or deprecatory or something in between. Mechanisms of self- and external exclusion are one of the most negative and sustainable effects of stigmatized identification.

3. Social capital as a reference in social-spatial network modelling

The facts and problems that have been addressed so far, in order to adequately deal with models in general and geosimulation models in particular, will now be put into the context of social-spatial network modelling. The contextualization of information and the mutual transformation of complementary reference units in social network analyses are, *inter alia,* challenged by the difficulty of translating terms into numbers—in other words, of quantifying notions of social capital, solidarity or trust. Before introducing a geosimulation network modelling approach that attempts to explicitly struggle with this difficulty methodologically we begin with a brief delineation of the theoretical problem.

"Social interactions in collective human relations are commonly understood as a fundamental condition for human beings to live a satisfactory life. Establishing a personal identity and autonomy on the one hand, and trust, solidarity and commitment on the other hand require social-normative rules which function as a glue that helps agents to connect to one another. While the nature of social interaction is seen as an unquestioned fact its valuation in operational terms is much harder to achieve. One apparently and generally accepted approach to operationalize the value of social interactions is given by the theory of social capital. A huge body of theoretical reflections as well as empirical studies [21, 22] consider transferring the concept of economic capital into the social realm as a suitable methodology for coping with social relations" [23].

"The size and composition of social networks rely to a large part on the characteristics of the actors involved, i.e. their capabilities, needs and aspirations to collaborate with others. They also depend on the reasons for cooperation, the kinds of problems, and the (in)tangible infrastructure necessary to communicate via different channels. An important additional issue, however, comes with the nature of network mechanisms themselves – the process of how relations between actors emerge structurally and which determinants are thought to be relevant" [23]. There are several different approaches which relate the analysis of network mechanisms to social facts, one of which is dedicated to social capital.

Bourdieu ([24], p. 248) defines social capital as "the aggregate of the actual or potential resources which are linked to possession of a durable network of more or less institutionalized relationships of mutual acquaintance and recognition." This definition is in favor of an actor-centered and utilitarian perspective that not only presupposes an affiliation to one or many social networks as a constituent characteristic, but also assumes knowledge about the structure and function(s) of these. Networks vary significantly in size and complexity; they can be concrete and manageable as, for instance, families and cliques as well as abstract and opaque, as is the case for associations, organizations, labor markets, or electoral rolls. [25]

Furthermore, it is remarkable that the unit of value, the currency, is implicitly given by solidarity, reciprocity, and trust, expressed explicitly as network connections, however. "The volume of the social capital possessed by a given agent thus depends on the size of the network of connections he can effectively mobilize and on the volume of the capital (economic, cultural or symbolic) possessed in his own right by each of those to whom he is connected" [25]. This quantitative relation—the more connections one has the higher his/her social capital is—sounds odd because it makes no qualitative difference in terms of network structure (density of relations) or the nature and value of relations (positions and roles of agents, weights of directions). "Resources" within the social capital context are understood as the availability of network relationships, which are, however, not specified—neither quantitatively nor qualitatively. A functional approach to social capital differs from an actor-centered one and accounts more suitably for assigning value to social relation: "Social capital is defined by its function. It is not a single entity, but a variety of different entities having two characteristics in common: They all consist of some aspect of a social structure, and they facilitate certain actions of individuals who are within the structure" ([26], p. 302).

An understanding of social capital is commonly embedded into a broader conceptualization of capitalizing social relations. While this idea is tempting with economic capital, it is less so with social capital. "The three functions of capital and money, respectively, are (i) a means of exchange, (ii) a means of value hedge and preservation, and (iii) a unit of calculation." Accumulation may play a major role in all of these functions, but is not restricted to them. The worth of capital also stretches to the functions of distribution and circulation within a commodity and service economy [27]. Taxation and social security contributions are other examples of how capital is used. In addition, economic capital accumulation implies accumulation of both surplus and debt. The execution of all these functions requires not only a general usability of capital but also a (statistical) scale of measurement that enables comparisons and evaluations across a diverse field of objectives, and of spatial and temporal scales. All this is not given—or not sufficiently so—with social capital. It is possible to weigh and to qualify the direction of network relations, but still no adequate measures are available which account for the latent variables associated with social capital. In other words, it makes no sense to say "we must double solidarity or trust by doing X." It would, however, be a scientifically fruitful endeavor to develop indexes of social capital—based on trust, solidarity, network relations, etc.—with the aim of investigating social networks comparatively." [25]

In so doing, a strategy of obscuring or fuzziness is not constructive as can sometimes be detected in qualitative empirical social network research. For example, Dill ([28], p. 85) claims for his own approach that a definition should be a clear and simple one: "Social capital is the sum of intangible merits and goods within a community" (translation A.K.). In fact, this definition is anything but clear and simple. What can be accepted as "intangible merits" in communities and by whom? Who or what defines communities, and why are we dealing now with communities instead of networks? Are they identical? Who is producing and who is consuming intangible goods, based on which distribution rules? How does one sum up intangible goods? This definition is different from the accumulation and the "unit of calculation" concept because it provides no clues about distributional rules (it is a simple sum), the nature of intangible goods, and the nature of communities (which is much harder to specify than talking about social networks). The distinctive property of social capital in comparison with human or economic capital is, according to Dill [28], totally idealistic nature.

The following chapter presents one attempt to operationalize the qualities that are associated with social capital in quantitative terms in the context of a social-spatial network. It seeks to address the problem of team assembly mechanisms (the topological space) with its potential impacts on the spatial distribution of these teams (the chorological space) in terms of the success experienced in conducting their projects.

4. An agent-based geosimulation model of a semi-empirical social network

4.1. Introductory remarks

The basic idea of the approach to translate the qualitative characteristics of social capital into quantitative properties is given with the number of face-to-face meeting opportunities. Though we are aware of the fact that the creation and continuation of social relations today are significantly influenced, or even determined, by social media technologies occupying virtual spaces, we are convinced that meeting opportunities of physically present people also have a strong impact on the process of tying relationships between them. The purpose of investigating networking mechanisms is, therefore, threefold: first, the geosimulation model simulates processes that represent the evolution of a large(r) network derived from a finite number of small(er) team networks. This evolution is thought to represent an increase of network connections, that is, an increase of social capital. Second, the model seeks to carve out the statistical meaning of our core variable *number of face-to-face meeting opportunities* as a latent variable that points to one condition of creating and allocating social capital. Third, since face-to-face communication takes place in physical space, the geosimulation model is used to analyze potential theoretical relationships between a network-based and a place-based geography.

With respect to our interest of studying networking mechanisms, we argue that the evolution toward a large connected network contributes to overcoming isolated and fragmented efforts toward a common goal. Achieving synergies through effective trans-local knowledge transfer is a goal that has been investigated in (social) network analysis for some time [29–31]. In this respect, agent-based simulation is an appropriate method to explore the underlying processes that lead to these networking mechanisms [32].

As a first step of this investigation, we selected and then adapted to our own needs a theoretical reference model developed by Guimera et al. [33] and made available as a NetLogo model by Bakshy and Wilensky [34]. The question raised ([33], p. 699) remains evident to some degree in our context too: "Is there a large connected cluster comprising most of the agents or is the network composed of numerous smaller clusters?" The ideal size does not necessarily correspond simply to exactly one all-encompassing network: "Successful teams evolve toward a size that is large enough to enable specialization and effective division of labor among teammates but small enough to avoid overwhelming costs of group coordination" ([33], p. 697). Where our empirical research is concerned, we are less interested in economic costs than in the creation of sustainable social interactions.

These assumptions are taken as relevant premises for networking mechanisms. In so doing, our focus shifts to the parameters that influence or determine the construction of large(r) social networks that are initially small and more or less isolated due to the design of the social project. These mechanisms are understood as operators of translating social capital into quantitative units. The parameters used can be divided into three general domains: (i) actor-based parameters, (ii) linkage-based parameters and (iii) place-based parameters.

4.2. The empirical case studies

The empirical case studies for which an agent-based simulation model was constructed, one which relies theoretically on the prototype model, derives from an Austrian social project entitled "Keep the Ball Rolling." This project aims to enhance social well-being at a regional level by encouraging the local population to put into practice ideas that they are convinced are relevant. Individuals or small teams are called to submit project proposals that help reduce social injustice and promote social cohesion. Successful teams are awarded a grant to fund their projects, and in addition receive organizational support. The project started in the Lungau region (Salzburg) in 2011 and was taken up by Steirische Eisenstrasse (Styria), Mühlviertler Alm (Upper Austria), and Mostviertel Mitte (Lower Austria) regions [35–36]. The empirical data used for the following simulation models are from the Styrian and Upper Austrian regions.

The project proposals are presented by the teams at three jury meetings. The successful candidates must implement their projects within 18 months. Every jury session is followed by a public celebration where the successful projects are presented. Because the implementation of a project is supported scientifically by a team of researchers for the duration of the project, several further meetings, ranging from small informal meetings to larger stakeholder workshops, are offered. Every team leader is invited to participate in a semi-standardized questionnaire designed to obtain knowledge about the project team in terms of its size and composition.

The analysis is concerned with how the number of such meetings (1) determines the team assembly mechanisms, (2) correlates with other determinants and (3) correlates with a place-based geography.

4.3. The model design

In order to analyze the development of collaborative socio-spatial network structures of initially small(er) and unconnected teams in two Austrian regions, a simulation model was constructed that includes some of the general ideas of the original model [33]. However, there are a number of major differences between our model and the original one: (1) The number of teams is initialized at three time steps and not stepwise; with this adaptation, we represent the selection process of successful teams. (2) Teams can vary in size. (3) Teams can also vary in network structure, while the original model only allows the implementation of teams of three actors that are fully connected, the "Keep-The-Ball-Rolling" project does not have such restrictions. (4) Individual agents (and not only teams of three agents) are inserted as new potential collaborators; this feature will play a crucial role in the future modelling process when data sets are available that will be collected approximately 2 years after the official end of the social project (data for the Mühlviertler Alm region are currently missing, while data for the Steirische Eisenstrasse are not yet fully edited). Therefore, we do not refer to this characteristic of the model in this paper. (5) The original teams can be merged, either by a team leader (i.e., the leader of a project team) or by a team member (which is not possible in the reference model) who is selected stochastically for a new connection. This adaptation was implemented due to the nature of events that are realized during the different types of project meetings which can consist of stakeholder workshops, project presentations and informal meetings, as well as *ad hoc* assistance from the staff of the regional offices. Although we do not consider these different event properties explicitly, they remain important to justify the "connecting role" of both agent types, that is, team leaders and team members. The original model takes only team leaders as connectors into account.

The construction of the two regional socio-spatial network models is based on standardized questionnaires that were conducted at the beginning of the project in each region. Every team leader of a project who was awarded a grant to participate with her/his project and who took part in the survey was asked about her/his team collaborators. We also know the team leader's home address (but not those of the collaborators, which affects the analysis). The numbers of actors and network ties are given in **Table 1**.

The two models are initialized with these settings of nodes and edges. In the style of the reference model, a six-dimensional parameter space is used to analyze the networking mechanisms among team leaders, team members and between them. The aim of this procedure is to detect common patterns of relevance of and relationship between the six parameters that seem to determine the behavior of social networks in terms of their structure and dynamic. **Table 2** gives an overview of the parameters and the ranges of values within which the further simulation results have been analyzed.

The parameter *selAgents* determines the likelihood that agents attending an event are willing to collaborate; for example, *selAgents* = 50% means that 50% of all attendees of a workshop are willing to collaborate. The parameter *selAgentType* specifies which agent type—the team leader or the team member—is more likely to be willing to collaborate. If *selAgentType* = 50%, both agent types are equally likely to be willing to collaborate; if *selAgentType* < 50%, more team leaders are likely to be willing to collaborate. The interval of both parameters has a large range in order to detect network differences because of asymmetric probabilities. The maximum number of agents who are likely to collaborate, *maxSelAgents*, is a conservative estimate based on our experience of events we have organized in both regions. These three parameters affect the behavior of the agents, while the following two affect the relationship

Number of ...	Team leaders	Team members	Network ties
Styrian region	79	396	441
Upper Austrian region	59	418	655

Table 1. Number of registered team leaders, team members, and network ties among and between them in the Steirische Eisenstrasse (Styrian region) and the Mühlviertler Alm (upper Austrian region).

between them. The *conTeamLeader* and *conTeamMember* parameters determine stochastically how agents connect among themselves. If, for example, *conTeamLeader* = 10%, then in 10% of all cases team leaders connect with other team leaders, and in 90% of all cases they connect with team members. The same rule applies to *conTeamMember*.

The last parameter of **Table 2**, *numEvents*, determines the number of events that take place during a simulation run. This is our core parameter for implementing social capital as a quantitative approximation. Events are defined here as face-to-face meetings and hence exclude Internet-based communications. The range of values varies between 1 event and 160 events. Taking into account that one simulation run takes 160 steps, which represents a time period of 160 weeks or almost 3 years (i.e., the project time of approx. 80 weeks and the post-project time of another 80 weeks when *a posteriori* questionnaires are conducted to evaluate the sustainability of the projects), the number of events varies between just one meeting during or after the social festival, and meetings on a weekly basis.

The social network simulation models were created using NetLogo 6.0 [37]. The analysis of the modelling results has two stages: first, a statistical analysis composed of a multivariate linear regression analysis and a cluster analysis was conducted. Second, a simulation analysis consisting of representative simulation runs investigated the behavior of the social networks by considering the network parameters 'closeness centrality' and 'betweenness centrality.' **Figure 1** represents the simulation process graphically. The model used here is available as an updated version at OpenABM (https://www.openabm.org/model/5583/version/1).

4.4. Some selected model results

All possible combinations of values (that is, 3888 in this case) within the intervals of the six variables were computed, which led to 3888 mean values across all variations. These mean values were then used to compute average bivariate correlations and measurements of

Parameter	Description	Interval of analysis
selAgents	Probability of selecting agents per event who are willing to collaborate	[20, (30)[1], 80] %
selAgentType	Probability of selecting team leaders or team members per event who are willing to collaborate	[20, (30), 80] %
maxSelAgents	Potential maximum number of agents per event who are willing to collaborate	[4, (4), 12] abs.
conTeamLeader	Probability of team leaders actually connecting with other agents	[5, (15), 35] %
conTeamMember	Probability of team members actually connecting with other agents	[0, (10), 30] %
numEvents	Number of events that enable the establishment of new ties over a complete simulation run	[1, 2, 4, 8, 16, 32, 40, 80, 160] abs.

[1]Values in round brackets indicate the increment value; for example, the parameter values for *selAgents* are 20, 50 and 80.

Table 2. Description of parameters used in the simulation models.

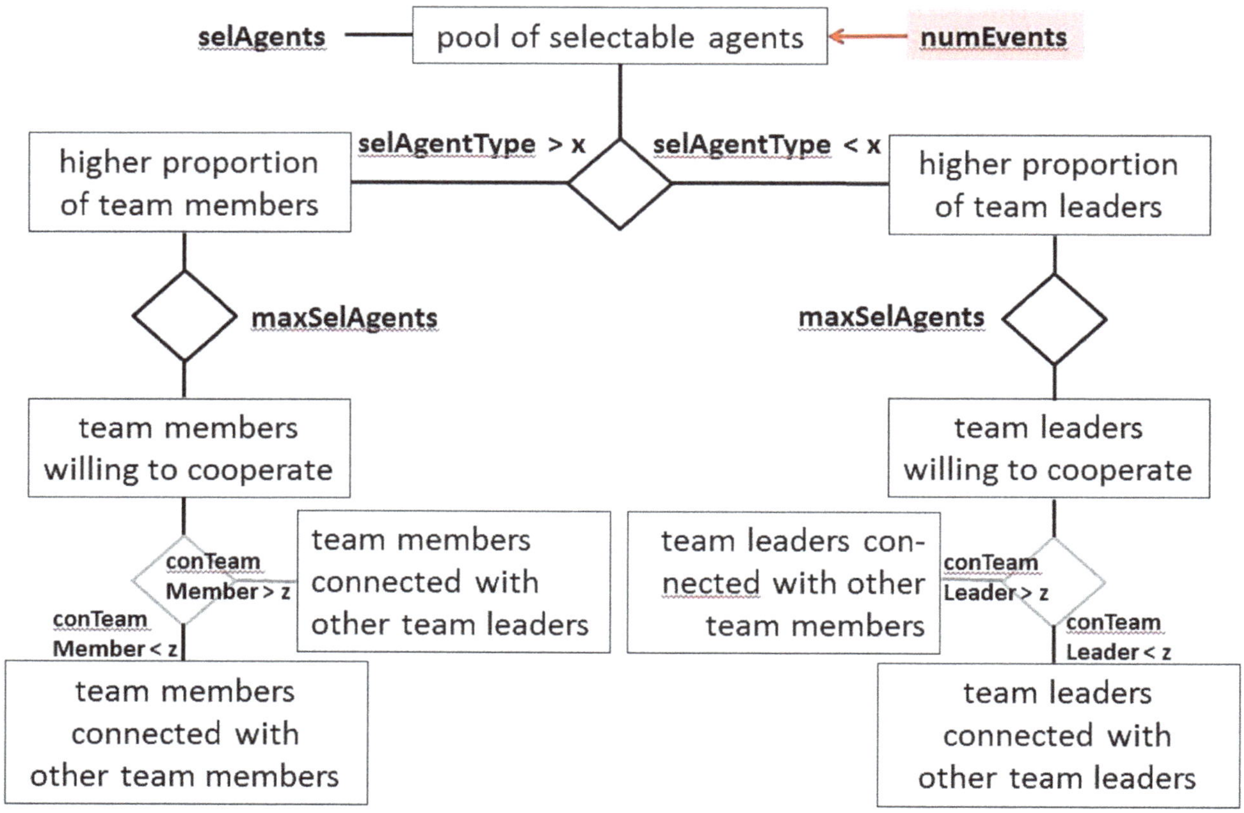

Figure 1. Flow diagram of the simulation model.

determination. The aim was to determine the strength and direction of relationships of the six independent variables by which the versatile network structures can be explained. The emergence of these different network structures consists of both the initial network relations of the original teams (empirical data) and the network relations created during the simulation (modeled data).

The method used to create multiple regressions was "stepwise selection," which avoids multicollinearity to some degree. **Table 3** reveals that our core parameter *numEvents* is the most relevant independent variable for all three dependent variables representing the linkages between nodes. The measurement of determination ($\mathbf{R}^2$) confirms this statement. The parameter *numEvents* is, however, more relevant to explain the variation of the distribution of the links among team leaders than among team members. On average, team leaders benefit more than team members from an increase in the number of workshops and meetings. This is due in part to an implicit bias, occurring because there are fewer team leaders than team members, which quickly leads to a higher number of linkages. This assessment is confirmed by the contrary fact of there being only a few events (1–8): under this condition, team leaders are far more involved than team members in creating a large connected social network. This fact has to be taken into account to avoid lock-in effects of a well-informed stakeholder group. The statistical influence of the remaining five parameters is significantly less relevant in explaining the variation of ties among and between the two groups when compared with *numEvents*, as is illustrated in the last column of **Table 3**.

Dependent variable	Most relevant independent variable	R^2 of most relevant independent variable	R^2 of all included independent variables
Number of connections among team leaders (Styria)	Number of events r = 0.785	0.617	0.759
Number of connections among team leaders (Upper Austria)	Number of events r = 0.779	0.606	0.751
Number of connections among team members (Styria)	Number of events r = 0.633	0.401	0.525
Number of connections among team members (Upper Austria)	Number of events r = 0.623	0.388	0.505
Number of connections among all actors (Styria)	Number of events r = 0.756	0.572	0.687
Number of connections among all actors (Upper Austria)	Number of events r = 0.735	0.540	0.652

Table 3. Regression patterns of the three dependent edge-related variables for the Steirische Eisenstrasse (Styria) and the Mühlviertler Alm (Upper Austria).

The cluster analysis aims to explore further structures that have been unknown so far. The cluster algorithm used here is the "Ward method," which yields more or less evenly distributed clusters. A variation of 4–6 clusters was applied, and the solution with five clusters provided good results with respect to a good discrimination of the values and interpretation of the results. As **Tables 4** and **5** illustrate for both regions in a very similar way, the highest numbers of connected agents (cluster 5 in both cases) are achieved when *numEvents* is the highest (which is not surprising), the likelihood of selected agents is relatively high, the proportion of team leaders is higher than that of team members, the maximum number of potentially selectable agents is relatively high, and the likelihood that agents will establish ties is high. If more team members are likely to be selected (*selAgentType* > 50%), then a considerable decrease of realized linkages follows. The least relevant parameters are *conTeamLeader* and *conTeamMember* (although they are responsible for the relevant discrimination between cluster 4 and cluster 5 in the upper Austrian case study).

Although *numEvents* appears to be a highly relevant determinant in the development of large network compositions, it is also highlighted in the cluster analysis results that team members benefit significantly from an increase in the number of events. A comparison of cluster 1 and cluster 2 for the Steirische Eisenstrasse reveals that almost doubling *numEvents* leads to a considerably higher increase of tied team members (approx. 2100 compared with 800) than of tied team leaders (approx. 1500 compared with 700).

In addition to the statistical analysis, a simulation analysis was performed in order to investigate the process of the network creation. For this purpose, two common centrality measures were used, namely the closeness centrality and the betweenness centrality. Both centrality measures characterize an agent's position or role in the entire network. In NetLogo, closeness

Dependent variables	Cluster 1	Cluster 2	Cluster 3	Cluster 4	Cluster 5
number of ties among team leaders	700	1041	156	1263	1415
number of ties among team members	805	2075	83	3402	9074
number of all ties	2132	4248	497	7824	10,423
Independent variables					
numEvents	73	118	18	141	160
selAgents	58	58	47	70	73
selAgentType	52	50	51	40	32
maxSelAgents	7	7	6	8	8
conTeamLeader/conTeamMember	20/15	20/15	20/15	19/13	20/25

Table 4. Cluster analysis results for the Styrian case study. Values represent mean values.

Dependent variables	Cluster 1	Cluster 2	Cluster 3	Cluster 4	Cluster 5
Number of ties among team leaders	184	932	1004	1299	1364
Number of ties among team members	141	1458	5293	2682	10,459
Number of all ties	817	4056	7229	12,152	12,844
Independent variables					
numEvents	22	104	133	155	160
selAgents	47	60	66	70	72
selAgentType	51	51	38	34	34
maxSelAgents	6	7	7	9	9
conTeamLeader/conTeamMember	20/15	20/13	20/22	20/4	21/25

Table 5. Cluster analysis results for the upper Austrian case study. Values represent mean values.

centrality is defined as "[…] the inverse of the average of an [agent's] distances to all other [agents]" [38]. Distances are defined as shortest paths. Betweenness centrality, by contrast, refers to the mediator function of an agent (for example, mediating communication flows). To calculate the betweenness centrality of an agent, "[…] you take every other possible pairs of [agents] and, for each pair, you calculate the proportion of shortest paths between members of the pair that passes through the current [agent]. The betweenness centrality of an [agent] is the sum of these" [38].

Based on typical and representative simulation runs of the abovementioned behavior space analysis in NetLogo, a data subset with 32 cases has been extracted. Extraction here means that extreme values of parameters have been excluded; for example, *numEvents* was set to 80 in one case and to 8 in another. **Figure 2** illustrates the results of agents that have a high closeness centrality ("high" defined as above the threshold value of 0.5) for the Steirische Eisenstrasse region. The two maps differ in the number of events during the simulation run.

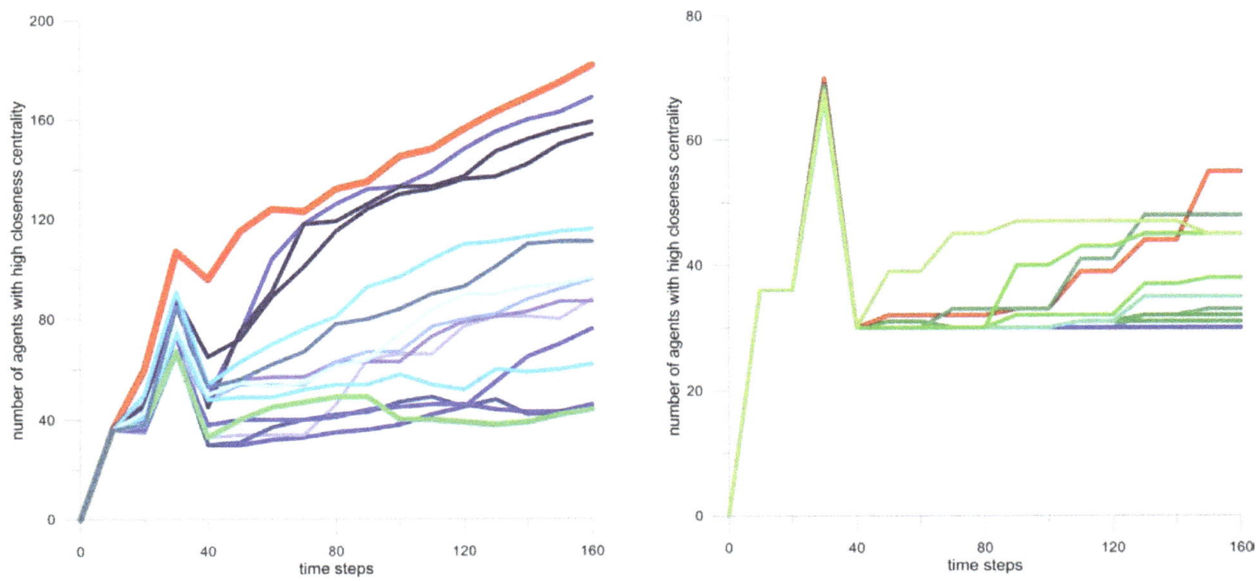

Figure 2. Distribution of high closeness centrality agents for *numEvents* = 80 (left) and *numEvents* = 8 (right) in the Styrian study area.

Sixteen out of the 32 simulation runs were executed using a high number of meeting events (left-hand side), while the other 16 used a low number of events (right-hand side). When we compare the two graphs of **Figure 2**, the most obvious fact is that the variation of results with *numEvents* = 80 is considerably larger than with *numEvents* = 8. The case representing the lowest number of agents gaining a high closeness centrality (green graph in the left diagram) is characterized by a high number of selected team leaders (because *selAgents* is high and *selAgentType* is significantly below the 50% threshold value) and by high linkage percentages for both agent types. The case with the highest number of agents gaining a high closeness centrality (red graph in the left diagram) differs in the relationship of *selAgents* and *selAgentType* which is now exactly the opposite. These results highlight the important contribution of team leaders in two ways: (1) they are important as central nodes within the social network; (2) they also act as multipliers for team members to achieve a central position in the network.

Fluctuations for cases with *numEvents* = *8* (right diagram of **Figure 2**) are far less significant. Here, values range from 30 (blue graph) to 55 (red graph) (there are six different combinations of parameter values that lead to 30 agents gaining high closeness centrality). The most relevant determinants to explain the differences are *selAgents* and *conTeamMember* – one node-related and one linkage-related parameter. A comparison of the two diagrams of **Figure 2** convincingly shows that even a high number of events supplied do not guarantee a sufficiently high number of agents who are tightly linked together if the other parameter values do not foster successful collaboration efforts.

4.5. The geographical dimension of the social network geosimulation model

The statistical and simulation analyses reveal that the construction and (sustainable) consolidation of social networks are influenced by a high number of factors whose inter-relationships are quite complicated in terms of generating a large(r) connected network. Offering a high number of events does not automatically ensure that a high proportion of agents will gain high centrality in order to provide for efficient knowledge transfer within the social network. In fact, even only a few events can result in a reliable number of durable linkages among agents.

However, one must take the specific sequence of the three jury meetings into account. Due to the inclusion of new team leaders and team members at predefined time steps, a temporary decline in the number of agents with high betweenness and/or closeness centrality can arise. Fluctuations are large(r) if the number of events provided is high because meetings are likely to take place between jury sessions, too.

Therefore, the organization of events to foster collaboration is a challenging undertaking, as social network analysis has shown. The supply of opportunities to meet each other in order to exchange knowledge and experience has, in addition to its qualitative component (obligatory stakeholder meeting vs. informal team meeting), a quantitative tendency. Setting aside all the network-based determinants discussed so far, one should not forget the geographical domain, that is, the geospatial distribution of the relevant actors. In other words, a translation from a space-of-flows geography to a space-of-places geography seems appropriate.

Figures 3 and **4** are representative extracts of the spatial distribution of agents with high betweenness centrality (yellow) and high closeness centrality (violet), and remaining team leaders (red) for a high and low numbers of events (*numEvents* = 80 and *numEvents* = 8 respectively). Linkages are hidden and the scale of resolution of the agents' locations, which is based on the questionnaire, is the municipal level. Distribution within the municipalities is for visualization purposes only.

One important conclusion that can be drawn from the distribution patterns is that the vertical structures of the social networks do not completely coincide with the structure of the places where the team leaders reside and where the projects were implemented. The initial network pattern is characterized by many small and largely unconnected teams each with one leader and a couple of members who are, of necessity, linked to the leader. This vertical structure was then reiterated as team leaders connected to one another or to other team members. Agents with high closeness and/or betweenness centrality are disproportionately more often located in municipalities with comparatively fewer implemented projects (light green colored areas). This is true both for situations with a high number of events and those with a low number. If a high number of meetings were to be offered (maps at the left-hand side), then the distribution of agents with high betweenness centrality (= important communicators) would be more even than the distribution of the projects. The result is true for the Styrian as well as the Upper Austrian study area. In fact, this statement can be extended to the situation of a low number of offered meetings if the closeness centrality (=strong ties between agents) is taken into consideration, as can be seen in the maps at the right-hand side of **Figures 3** and **4**. The peculiar relationship between the two geographies immediately prompts the conclusion that the decisions about adequate venues for meetings should be made by taking the whole project region into account and not concentrating mostly on the region's larger towns.

Another conclusion that can be drawn is that with a more even spatial distribution of highly centralized agents, a proper coverage of network geography and place-based geography can be achieved in terms of communications (space of flows) and localized decisions (space of places).

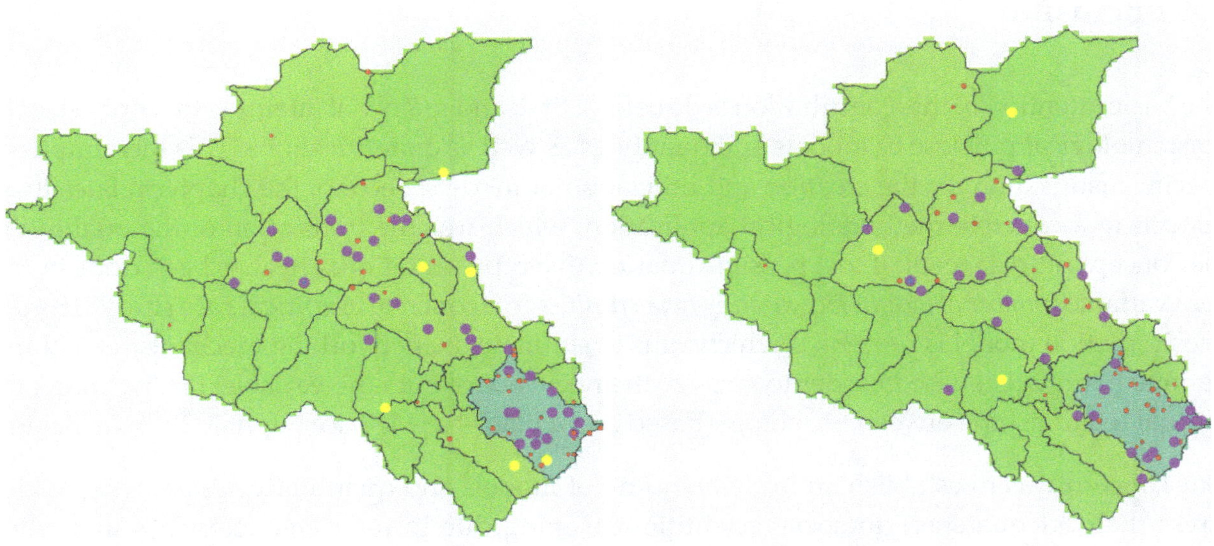

Figure 3. Distribution of agents with high betweenness centrality (yellow) and high closeness centrality (violet) for *numEvents* = 80 (left) and for *numEvents* = 8 (right) in the Steirische Eisenstrasse. Colors of municipalities indicate less than 2% of all projects (light green); between 2 and 20% (green); more than 20% (dark green).

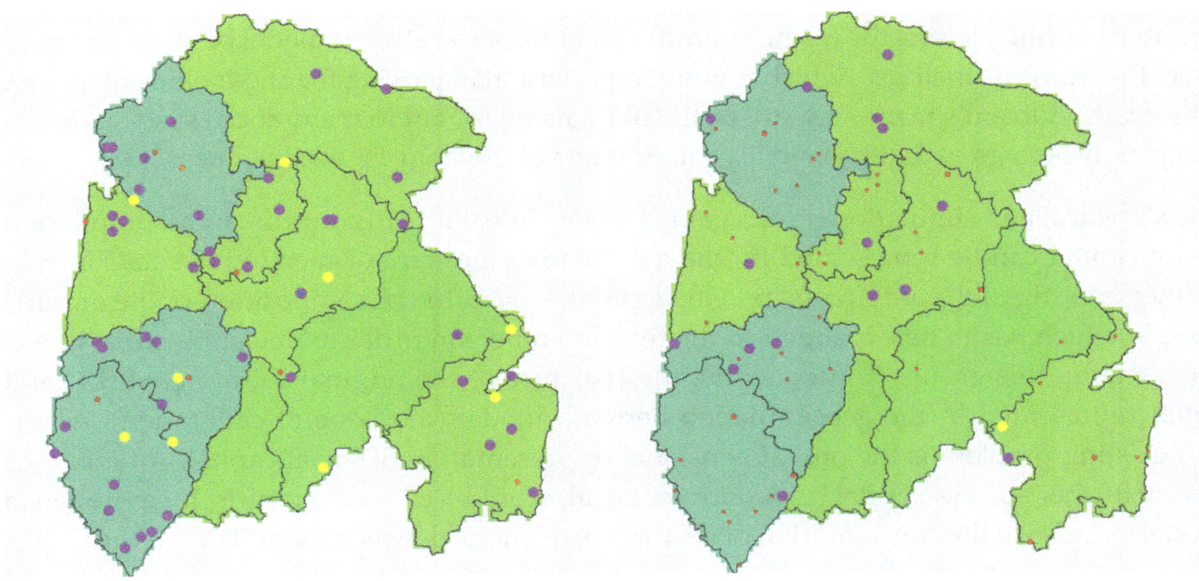

Figure 4. Distribution of agents with high betweenness centrality (yellow) and high closeness centrality (violet) for *numEvents* = 80 (left) and for *numEvents* = 8 (right) in the Mühlviertler Alm region. Colors of municipalities indicate: Less than 5% of all projects (light green); between 5 and 15% (green); more than 15% (dark green).

If this is true, then local projects can benefit from each other thanks to this type of knowledge dissemination. Ultimately, personal engagement in one's own local social environment also needs to be appreciated by rotating meeting locations across the entire region, because then "peripherally located" agents can act as hosts and can proudly present their project work in immediately visible form.

5. Conclusion

The basic intention of this contribution is threefold: first, and mainly, it attempts to emphasize the epistemological purpose of models and quantities as well as quantifications. This then implies a specific justification for the creation and utilization of models. Second, the chapter relates these reasons to a concrete methodological application, which was the translation of the qualities of the concept of social capital to a possible quantitative representation. This has been done by the core variable *number of face-to-face meeting opportunities* in a concrete empirical case study. Third, a geosimulation model is presented which aims to simulate the networking mechanisms within a definite parameter space in order to analyze the relevance of our core variable. Furthermore, the mutual relationship between a topology-based and a place-based space ought to be investigated.

Our concern was to establish an understanding of models and quantitative approaches which stress their adequateness in social scientific reasoning due to their characteristics in dealing with the subjects that matter. It has been argued that models do not refer to an absolute truth or reality that they do not represent or imitate reality but create their own reality and impact by their application within scientific communities. The hypothetical-constructivist realism appears to be a proper theoretical foundation to this argument. In fact, models and numbers attempt to make phenomena we observe or deduce theoretically tractable, graspable and visible. Since many social processes designed by models cannot be perceived straightforwardly or are unobservable in principle, it is the (communicative) relationship between model builders and model users that rewards their use, which is more important than looking for and looking at an intangible truth. "Occam's razor may still be the ultimate quest, but in many social systems, evident complexity is so great that plausibility rather than validity may be the real quest" [38].

The statistical and simulation results gained some plausibility to confirm our assumption that social capital can be represented by the quantitative measure of meeting events. However, further investigations are necessary. One issue is related to the composition of the empirical sample which has to be expanded to the team members in order to detect potential connections among this group. Consequently, the structure of the original social networks can be better represented in the geosimulation network model (the issue of calibration). Another still existing problem is the one-dimensional re-presentation of social capital. An incorporation of further sorts of capital (economic, cultural) would lead to an empirically more reliable model because of the consideration of mutual correlations between them. This, in turn, would improve the model results which influence our understanding of the empirical reality (the issue of verification and validation). Finally, the geographical context—in its two dimensions of topology and chorology—ought to be investigated more deeply. What interdependencies between networks and places in terms of spatial proximity and center-periphery relationships might have an impact on the size and composition of the network, and the distribution of actors on the allocation of social capital, should be further discussed. Though an integration of these issues will definitely improve our understanding and imagination of social networks which take social capital explicitly into account, the geosimulation model presented here justifies the continuation of research on this scientific path.

Acknowledgements

Chapters 2 and 4 are excerpts which were taken from my previous publications, cited as [23] and [39]. These excerpts have been extended and aligned to fit into the structure of this

Author details

Andreas Koch

Address all correspondence to: andreas.koch@sbg.ac.at

University of Salzburg, Salzburg, Austria

References

[1] Schäfer L, Schnelle T, Fleck L, editors. Entstehung und Entwicklung einer wissenschaftli chen Tatsache. 9th ed. Frankfurt a.M.: Suhrkamp; 2012

-

[2] Couclelis H. Ontologies of geographic information. International Journal of Geographical Information Science. 1785-1809;**24**:12. DOI: 10.1080/13658816.2010.484392

[3] Lewin K. The conflict between Aristotelian and Galileian modes of thought in contem porary psychology. Journal of General Psychology. 2002;**5**:141-177 -

[4] Weiss W. Gödels Unvollständigkeitssatz, die Vernunft, Metawahrheit(en) und Everetts Vielweltentheorie. Vienna: Wissenschaftliche Nachrichten des Bundesministeriums für Unterricht, Kunst und Kultur; 2010. pp. 3-12

[5] Reckwitz A. Die Gesellschaft der Singularitäten. Berlin: Suhrkamp; 2017

[6] Rosanvallon P. Die Gesellschaft der Gleichen. Hamburg: Hamburger Edition; 2013

[7] Mau S. Das metrische Wir. Über die Quantifizierung des Sozialen. Berlin: Suhrkamp; 2017

[8] Christie M, Cliffe A, Dawid P, Senn S. Simplicity, Complexity and Modelling. Chichester: Wiley & Sons; 2011

[9] Kuhn T. The Structure of Scientific Revolution. Chicago: University of Chicago Press; 2012

[10] Bourdieu P, Wacquant L. Reflexive Anthropologie. Suhrkamp: Frankfurt a.M.; 1996

[11] Geertz C. Local Knowledge. London: Harper Collins Publishers; 1993

[12] Lotter W. Wir rechnen mit allem. 13th ed. brand eins; 2011. pp. 40-54; Heft 11

[13] Cremer G. Armut in Deutschland. C. H. Beck: München; 2016

[14] Armut LP. Ursachen, Formen, Auswege. C. H. Beck: München; 2017

[15] Link O. Die Welt lässt sich nicht berechnen. 13th ed. brand eins; 2011. pp. 110-115; Heft 11

[16] Schurz G. Einführung in die Wissenschaftstheorie. Darmstadt: WBG; 2008

[17] Epstein JM. Generative Social Science. Studies in Agent-Based Computational Modeling. Princeton and Oxford: Princeton University Press; 2006

[18] Küppers G, Lenhard J, Shinn T. Computer simulation: Practice, epistemology, and social dynamics. In: Lenhard J, Küppers G, Shinn T, editors. Simulation. Pragmatic Construction of Reality. Dordrecht: Spinger; 2006. pp. 3-22

[19] Edmonds B, Meyer R, editors. Simulating Social Complexity. Heidelberg, New York, Dordrecht, London: Springer; 2013

[20] Stachowiak W. Allgemeine Modelltheorie. Wien New York: Springer; 1973 (Reprint)

[21] Field J. Social Capital. 3rd ed. New York: Routledge; 2016

[22] Halpern D. Social Capital. Cambridge: Polity Press; 2005

[23] Koch A. Determinants of social networking mechanisms and their potential effects on a place-based geography: An agent-based simulation approach. GI_Forum. 2017;**1**:369-382. DOI: 10.1553/giscience2017_01_s369

[24] Bourdieu P. The forms of capital. In: Richardson J, editor. Handbook of Theory and Research for the Sociology of sEducation (Translated by Richard Nice). New York; 1986. pp. 241-258. Retrieved from https://www.marxists.org/reference/subject/philosophy/works/fr/bourdieu-forms-capital.htm (2018-03-19)

[25] Koch A. Capital, the social, and the institution – Bourdieu's theory of social capital revisited. In: Kapferer E, Gstach I, Koch A, Sedmak C, editors. Rethinking Social Capital. Newcastle upon Tyne: Cambridge Scholars Publishing; 2016. pp. 3-17

[26] Coleman JS. Foundations of Social Theory. Cambridge: Harvard University Press; 1990

[27] Hopkins R. The Power of Just Doing Stuff. Cambridge: UIT/Green Books; 2013

[28] Dill A. Gemeinsam sind wir reich. Wie Gemeinschaften ohne Geld Werte schaffen. München: Oekom Verlag; 2012

[29] Batagelj V, Doreian P, Ferligoj A, Kejžar N. Understanding Large Temporal Networks and Spatial Networks. Chichester: John Wiley & Sons; 2014

[30] de Nooy W, Mrvar A, Batagelj V. Exploratory Social Network Analysis with Pajek. Cambridge: Cambridge University Press; 2005

[31] Robins G, Pattison P. Interdependencies and social processes: Dependence graphs and generalized dependence structures. In: Carrington PJ, Scott J, Wasserman S, editors. Models and Methods in Social Network Analysis. Cambridge University Press: Cambridge; 2005

[32] Namatame A, Chen S-H. Agent-Based Modeling and Network Dynamics. Oxford: Oxford University Press; 2016

[33] Guimera R, Uzzi B, Spiro J, Amaral LAN. Team assembly mechanisms determine collaboration network structure and team performance. Science. 29 April 2005;**308**:697-702

[34] Bakshy, E, Wilensky U. NetLogo Team Assembly model. http://ccl.northwestern.edu/netlogo/models/TeamAssembly. Center for Connected Learning and Computer-Based Modeling. Evanston, IL: Northwestern University; 2007

[35] Gstach I, Kapferer E, Koch A, Sedmak C, editors. Sozialatlas Steirische Eisenstraße. Mandelbaum Verlag: Wien; 2013

[36] Gstach I, Kapferer E, Koch A, Sedmak C, editors. Sozialatlas Mühlviertler Alm. Mandelbaum Verlag: Wien; 2015

[37] Wilensky U. NetLogo. http://ccl.northwestern.edu/netlogo/. Center for Connected Learning and Computer-Based Modeling. Evanston, IL: Northwestern University; 1999

[38] Batty M. A generic framework for computational spatial modelling. In: Heppenstall AJ, Crooks AT, See LM, Batty M, editors. Agent-Based Models of Geographical Systems. Dordrecht, Heidelberg, London, New York: Springer; 2012. pp. 19-50

[39] Koch A. Quod erat expectandum: The modus operandi of models and quantifications. GI_Forum. 2018;**1** (in print)

3

Quantification and Prediction of Land Consumption and its Climate Effects in the Rhineland Metropolitan Area Based on Multispectral Satellite Data and Land-Use Modelling 1975–2030

Andreas Rienow, Nora Jennifer Schneevoigt and Frank Thonfeld

Abstract

Land use and soil sealing are particularly high in metropolitan regions. They bring about conflicts of use: the demand for housing, business and economy is enormous, but at the same time, quality of life depends on a network of green spaces. With the aid of remote sensing, the change of urban areas can be observed and quantified over time. This study investigates the change dynamics of land cover and land use in North Rhine-Westphalia (NRW) with multispectral satellite data, focussing on imperviousness. Landsat data is used to monitor and analyse half a century of landscape development. In addition, recent trends in land surface temperature (LST) are estimated from MODIS data. Changes to the LST are caused by land cover and land use changes amongst other factors. Accordingly, a link can be shown between the medium-term LST changes and the hotspots of landscape transformation in NRW. Due to global climate change, land consumption is increasingly affecting the densely populated urban areas, which calls for measures to increase their resilience. The results of the study can be used by decision makers to assess the environmental impact of land use, the loss of agricultural land or the resulting effects of climate change.

Keywords: land consumption, land use and land cover change, land use modelling, climate adaptation, land surface temperature trends

1. Introduction

Anthropogenic climate change is widely accepted as a major challenge to human society. Climate change is a worldwide phenomenon that is also described in detail for Germany [1]. The mitigation of climate change impacts is one strategy to deal with the consequences of a changing climate. Climate change-related events are increased flooding frequency, increased frequency of storm events, heat waves or droughts, amongst other things. Nonetheless, most of the coming consequences of climate change have been caused in the past, fostered by the release of greenhouse gases [2]. Even if climate mitigation strategies do succeed, this will not prevent us from increasing average air temperatures and the hazards mentioned above [2]. The means we are taking today are supposed to limit the expected temperature rise to a certain level, e.g. the two degrees′ goal compared to pre-industrial levels as declared by the United Nations Framework Convention on Climate Change (UNFCCC). In other words, temperatures will increase in the next decades even if the agenda of the UNFCCC is implemented successfully. Therefore, climate mitigation will not be sufficient to prepare the world's inhabitants for future conditions [2].

It is well documented that climate change impacts differ widely in a global perspective [1, 2]. However, even on a regional or local scale, there are pronounced differences. Vegetation in general and forests in particular function as a buffer against warming, for example, whereas artificial surfaces such as urban fabric often have contrasting characteristics and heat up easily. This leads to an increased vulnerability of urban areas against climate change. Many places of the world have experienced dramatic land cover and land use changes over the past decades, usually characterised by the conversion of natural environments into artificial landscapes such as industrial cropland and urban areas. Although some experts argue that land use change has an impact on river runoff [3], it is not yet well understood how climate change relates to land use change. At the same time, many administrative units and urban decision makers have identified the necessity to take action against climate change. Besides climate mitigation programmes, they have started implementing climate adaptation strategies for sustainable urban life. Moreover, urban areas host the majority of people in most parts of the world, hence their particular importance.

In industrial societies, cities often experience money scarcity and need to cooperate with one another on a broader scale to bridge the financial gaps. German metropolitan areas are consortia offering approaches of integrated strategies for regional development. They integrate the strengths of economy, science and the public sector, link urban and rural areas and implement regional projects towards sustainable development. In particular, in conurbations, the need for land leads to conflicts of use since the demand for both housing space and land for economic development is enormous. At the same time, it is important to ensure and improve the livability of the region by providing and sustaining green areas. Rural and urban areas are supposed to complement each other to satisfy all interests. There are, however, big challenges regarding infrastructure planning, energy supply and organisation of our living environment. The demand for recreational sites, housing, traffic and commercial space must be balanced and translated into purposeful planning structures.

In 2002, the federal government of Germany presented a national sustainability strategy, "Perspektiven für Deutschland" ("Outlooks for Germany"), in which they defined the so-called 30-ha goal. Its purpose is to reduce Germany's land consumption to a maximum of 30 ha per day until 2020—meanwhile the time limit has been extended to 2030 [4]. Land consumption is commonly defined as the conversion of agricultural and silvicultural areas into built-up areas [5], where built-up areas include also non-impervious surfaces like parks or golf courses. Land consumption in Germany currently amounts to 66 ha per day [4]. While daily land consumption has decreased over the past years, it was still at 10 ha per day in the German federal state of North Rhine-Westphalia (NRW) in 2011 alone [6]. The motivations for the 30-ha goal are of both qualitative and quantitative natures: soil, the essential component for food production, is a finite resource. Landscape structure is a considerable factor for fauna and flora, which again is significant for the existence of individual species. The consumption of contiguous areas leads to a dissection of associated habitats and has various impacts on several biocenoses. Furthermore, the cultural landscape, grown over centuries, might lose its outstanding characteristics. Considering the ageing society and population decline, an increase of built-up areas also involves growing maintenance costs for supply lines, sewer systems, roads, etc. [7].

Considering this background, we assess land use and land cover change (LULCC) in NRW over the past four decades with a focus on the Rhineland Metropolitan area. We demonstrate the relationship between LULCC and land surface temperature, a parameter linked to climate. Finally, we develop different land use scenarios to offer options for future development.

2. Study site: The Rhineland Metropolitan area

The Rhineland Metropolitan area is located in the western part of NRW (**Figure 1**). It is a federation of 35 districts and cities in the realm of the district governments of Cologne and Duesseldorf. **Figure 1** shows the mean normalised difference vegetation index (NDVI) [8] of NRW, calculated from a series of Landsat-8 OLI summer time images taken between 2014 and 2017. The index is sensitive to vegetation condition and cover. Green tones indicate high vegetation cover; red tones indicate non-vegetated areas. Hence, red areas in **Figure 1** can be interpreted as mainly urban areas. They are concentrated along the Rhine River and in the Ruhr area, a former coal-mining district and industrial hotspot of Central Europe. Other areas characterised by low vegetation cover (red areas) are formed by the Rhenish open pit mining area in the western part of NRW. The central southern part of NRW and its eastern part are characterised by high NDVI values due to high forest cover. North of the Ruhr area, NRW is characterised by agricultural use.

Besides large industrial districts, the Rhineland Metropolitan area is characterised by intensive agriculture supported by fertile soils and favourable climatic conditions. It is one of the most populous areas in Europe with 690 inhabitants per square kilometre. As most parts of the world, NRW is experiencing urbanisation with a population increase in larger agglomerations with strong economies and a population decrease in rural areas. Migration takes place

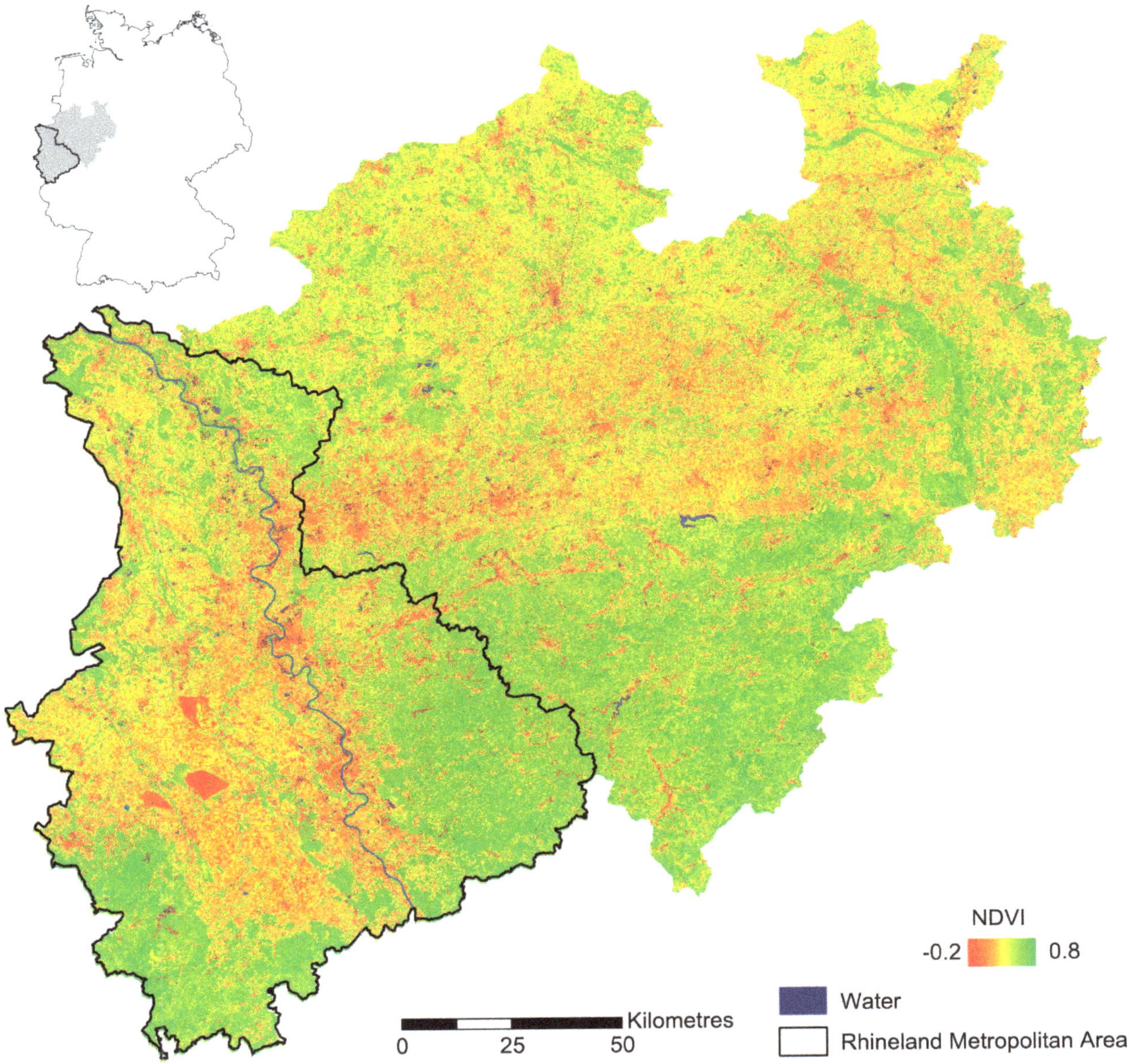

Figure 1. Rhineland Metropolitan area (black outline) in North Rhine-Westphalia (NRW), western Germany. The mean NDVI image shows urban areas and the Rhenish open pit mining area in the western part of NRW in red tones; vegetated areas are depicted in green tones. Yellowish tones can be attributed to agricultural areas.

between rural and urban areas, but there is also migration from outside NRW. **Figure 2** shows the population dynamics in NRW between 2005 and 2015. In larger urban areas such as Bonn, Cologne, Duesseldorf or Muenster, the numbers of inhabitants are increasing, whereas the greater Ruhr area suffers from a loss of population. The Ruhr area is characterised by a huge economic transformation after the breakdown of the coal and steel industry. In this context, the town of Gelsenkirchen exemplifies current population decrease (**Figure 2**) within a generally densely populated urban area. However, declining population numbers pose also a rural challenge in NRW.

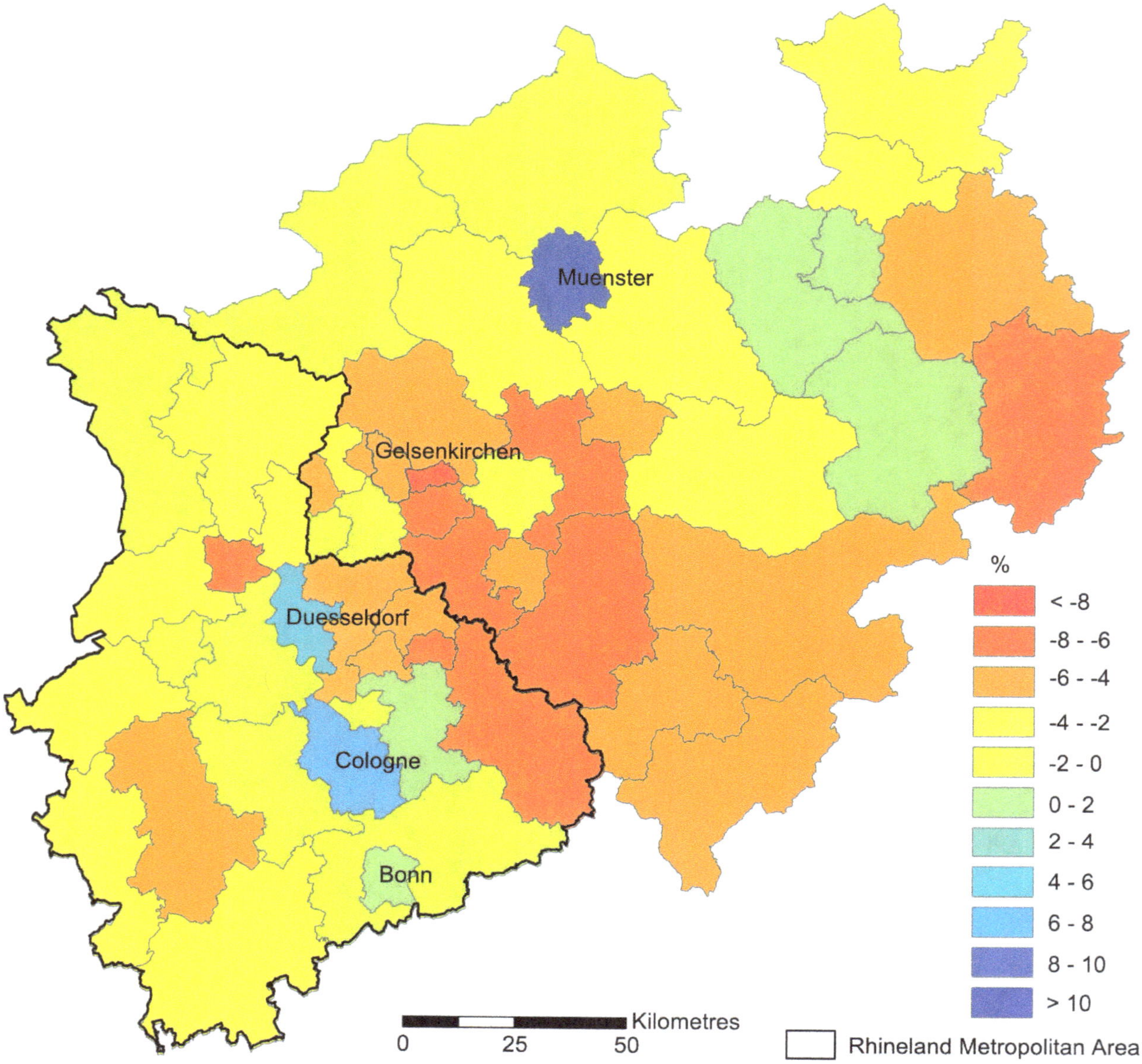

Figure 2. Population dynamics in NRW between 2005 and 2015 (data source: © IT.NRW, Duesseldorf, 2018).

3. Methods and data

3.1. Long-term LULCC assessment

Landscapes can be characterised by either land cover or land use. The former refers to the different materials on the Earth's surface such as vegetation, bare soil or water bodies. Land use, on the other hand, describes how these surfaces are used by humans, e.g. as forests, parks or commercial areas. Information on land cover and land use is usually provided by agencies in the form of statistical figures often relating to administrative units. Therefore, small-scale heterogeneities are often not captured. Remote sensing products based on satellite data or aerial images can help detect structures, processes and dynamics at various spatial and temporal scales. Thanks to long-term satellite programmes like the Landsat programme by NASA, there are image archives that cover a period of over 45 years. Since 2015, this valuable long-term archive is complemented by data from the Sentinel satellites within the European Copernicus programme (http://www.copernicus.eu/). The Satellite image archives comprise spectral information, which can be visualised as coloured images for various purposes but are mostly used for scientific analyses. Currently, the Landsat-7 and Landsat-8 satellites are collecting data of the same area every 16 days. The scanning track is around 180 km wide and has a spatial resolution of 30 m [9]. Sentinel-2 collects data in similar spectral bands in the visible, near-infrared and short-wave infrared spectrum. Its spatial resolution is 10 m, 20 m and 60 m, respectively. The scanning track is 290 km wide, and repetition time is up to 5 days with Sentinel-2A and B operating in one constellation [10]. The calibrated data can be compiled into time series, which allow an insight into the Earth's dynamics.

Although man-made land consumption propagates rather slowly compared to the effects of most natural hazards such as storms or floods, its constant behaviour and its widespread occurrence make it a serious threat to any well-balanced environment. Monitoring at regular intervals is required to accurately quantify recent land consumption rates. In this example, we classify satellite images of the Rhineland metropolitan region from 1975, 1984, 2001 and 2015 using the Random Forests classifier [11].

3.2. Land surface temperature (LST) trends

The satellite sensor Moderate Resolution Imaging Spectroradiometer (MODIS), developed to observe the dynamics of the Earth's surface, is recording data since 2000. Many products can be derived from its data, including datasets on the condition of the atmosphere, land surfaces as well as oceans and ice/snow expanses [12]. One of the land products is the dataset MOD11A2, which holds information on land surface temperature and emissivity. The data has a spatial resolution of 1 km and is provided as a composite at a repetition rate of 8 days [13, 14]. This data is available free of charge and can be explored using cloud-based applications like the Google Earth Engine platform (https://earthengine.google.com/), which is currently free of charge for scientific purposes. By calculating trends, it is possible to visualise regions of heating or cooling in a specific period. Further analyses are needed to reveal the causes for these trends, though.

Here, we calculate LST trends with different methods as there is no agreement on an optimal algorithm. First, we compute a simple linear trend per pixel only considering data from July and August between 2000 and 2015. In doing so, we can assess long-term effects of summer LST. Second, we make use of the Annual-Aggregated-Trends (AAT) method implemented in the "greenbrown" package [15, 16] using the free statistical software R [17]. This method calculates trends on annual aggregated time series using the annual mean and an ordinary least squares (OLS) regression. The significance of the trend is estimated using the Mann-Kendall test. The AAT method accounts for seasonal effects and considers all observations, thereby assessing annual effects rather than seasonal effects.

3.3. Land use and land cover modelling

Land consumption and the ecological problems associated with it are a huge unresolved challenge for Germany. Even in shrinking regions, urban sprawl can be observed. Land use and land cover modelling can shed light on processes, causes and effects of urban growth—socially and ecologically. There are several types of land use and land cover models. Spatially explicit models often simulate the future urban structure based on classified satellite data. Cellular automata (CA) are typical "bottom-up" models, which depict cities as systems and try to comprehend their complexity. They combine spatial units and model, based on spatial structures in a cellular environment, neighbourhood relations and simple transformation rules.

To model the land consumption in the Rhineland Metropolitan area, the cellular automaton SLEUTH is applied. The model was developed at the University of California in Santa Barbara

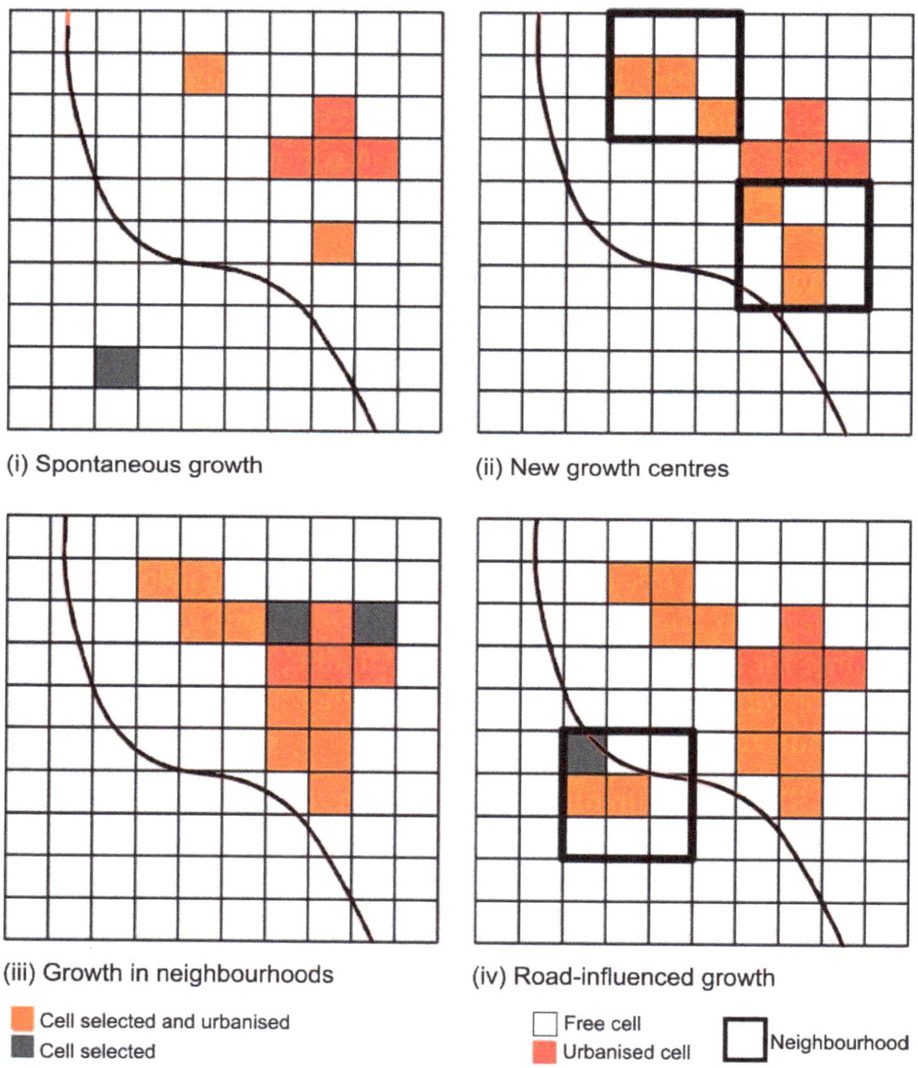

Figure 3. Rules for growth of SLEUTH's simulation cycle (adapted from [21, 22]).

(USA) back in 1996 and was used to answer various issues in many different cities [18, 19]. The land use maps serve as the main data basis; classifications from 2010 and 2015 are used to calibrate the CA. For SLEUTH, the binary version of the data with a spatial resolution of 100 m is sufficient to distinguish impervious from non-impervious surfaces. In its core, SLEUTH is an urban growth model. Its name is an acronym that consists of its main input data: slope, land use, exclusion, urban extent, transportation and hillshade. "Exclusion" includes all those areas where urban growth is restricted (e.g. nature reserves). Five growth coefficients control the four rules of the CA. These include spontaneous growth, which simulates the random development of settlements in their distinct neighbourhood, the growth of city centres, extensive growth in the direct neighbourhood and specific road-influenced growth (**Figure 3**).

The calibration of the model is necessary to synchronise the individual coefficients and to calculate the volume of the settlement areas to be distributed. It is possible to implement different scenarios for land consumption after calibration. In this study, the base year is 2015. One decade after the deadline of the 30-ha goal, the year 2030 is defined as the final year of the simulation, to demonstrate how the Rhineland Metropolitan area will look like both if the goal is achieved and, more importantly, if the goal cannot be reached [20]. For the simulation of land consumption, three scenarios are defined:

- Scenario A: "sustainable development". The 30-ha goal has been adopted by municipal planning, economy and the general public. Inner-urban development takes precedence over urban expansion. The focus lies on space-saving designs; development of land-intensive housing is reduced. Residents use public transportation rather than individual transport.
- Scenario B: "business as usual". Land consumption continues at the pace of the past. Caused by economic developments, land consumption is on the rise again, after a short-term decrease in the early 2000s. Cities are expanding, and the process of densification is advancing as well. Most of the settlement area is used for single-family homes. The construction of apartment buildings only accounts for 10% of the settlement area.
- Scenario C: "desire to live in a green area". Land consumption is increasing. The main driver is the increased construction of single-family homes. Cities are expanding rather than densifying inner-city areas. The "green meadow" is the preferred location for the construction of commercial areas and logistic centres. The parish-pump politics of the municipalities lead to an increasing designation of commercial areas in the periphery. The ageing population causes remnant effects, while society is oriented towards single households; the demand for housing rises.

4. Results

4.1. Remote sensing in support of the 30-ha goal

The maps resulting from the LULCC classification differentiate between water, arable land, grassland, deciduous forest, coniferous forest, mixed forest, open-pit mines as well as low, medium and high levels of imperviousness. A low level of imperviousness is defined as containing less than 40% impervious surfaces, medium level as 40 to 80% impervious surfaces and high level as more than 80% impervious surfaces per pixel. **Figure 4** shows how land consumption has altered the region's settlement structure. The impervious surfaces are constantly increasing since 1975, while arable land and grassland are decreasing. The analysis of the region's satellite images shows that impervious surfaces comprised about 126,800 ha in 1975. 40 years later, impervious surfaces have almost doubled, amounting to 245,600 ha in 2015. Between 2010 and 2015, land consumption was as high as 8 ha per day in the Rhineland Metropolitan area alone. It should be kept in mind that the goal for 2030 is to stay below 30 ha per day—nationwide. Increasing land consumption as well as the densification of urban areas not only affects our immediate living environment; they also have measurable impacts on the Earth's radiation budget and therefore on micro- and mesoclimate. **Figure 4** shows that some hotspots of land use change pop up in the LST trend map as places of heating and cooling.

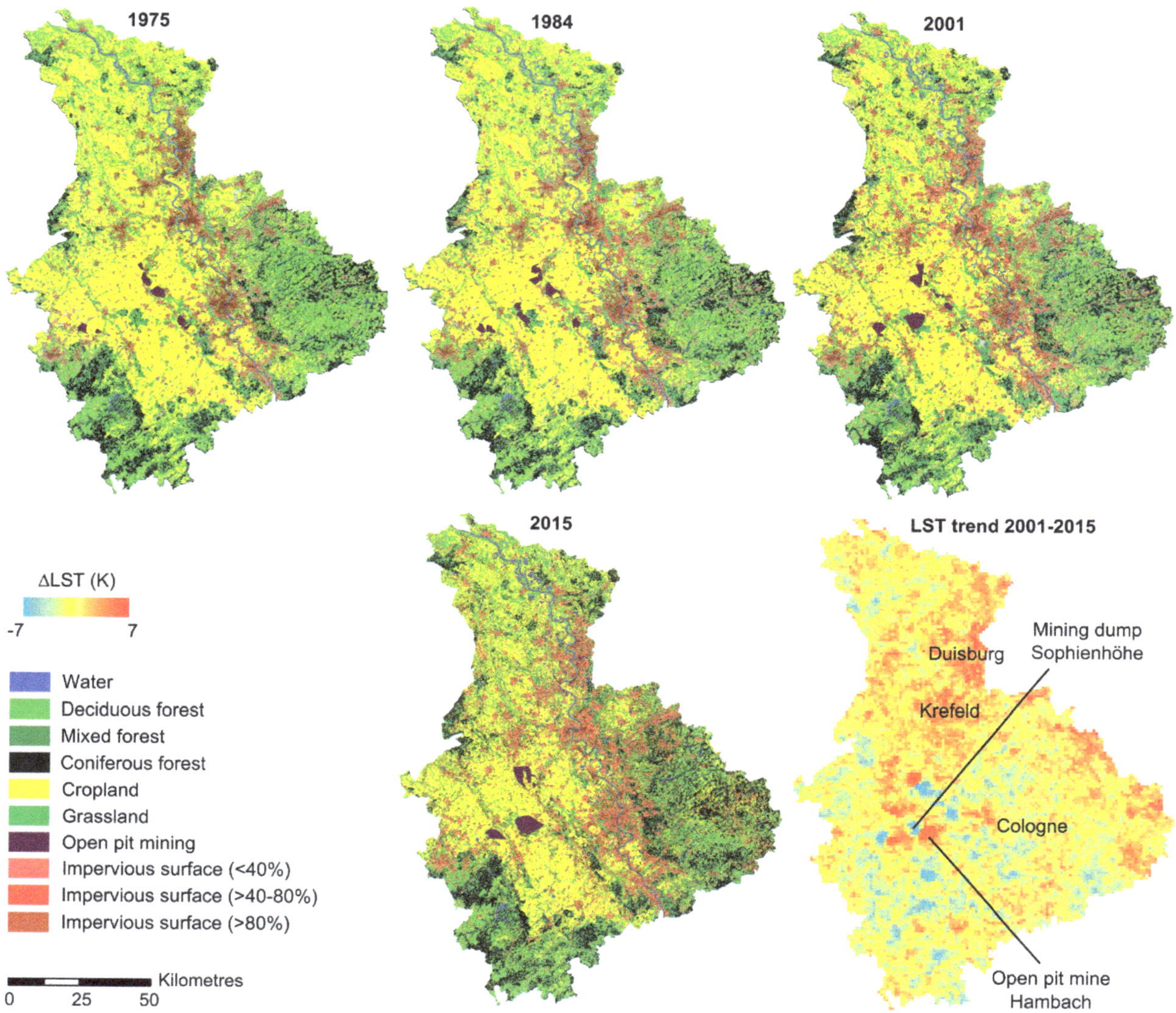

Figure 4. LULCC maps of 1975, 1984, 2001 and 2015 as well as LST linear trends of the Rhineland Metropolitan area (adapted from [23]).

Most prominent and shown in red indicating heating are the open pit mines of the Rhenish lignite mining area. The respective dump sites—meanwhile recultivated with vegetation—show cooling trends, indicated by blue colours. Other areas of LST warming trends are urban areas related to the cities of Cologne, Krefeld, Duesseldorf and Duisburg.

Figure 5 depicts trends in land surface temperature for the entire NRW, where LST trends from July and August between 2000 and 2015 are shown in **Figure 5a**. This figure complements the LST trend map in **Figure 4** but extends further east and covers areas beyond the Rhineland Metropolitan area. Accordingly, it shows some additional eastern hotspots of warming (red) and a few areas of cooling (blue). Areas that show no trend are depicted in yellow. Statistically irrelevant areas have not been masked out in **Figure 4** and **Figure 5a**. The hotspots are primarily urban areas; the cities of Krefeld and Duisburg as well as the western part of Cologne are easy to spot. Most of the named regions have an increased level of imperviousness and experience urban sprawl as a consequence of population growth (**Figure 2**). Usually, such alterations happen at the expense of green cover and agricultural land. Agricultural and vegetated areas usually display a cooling effect as a consequence of the vegetation's evaporation.

Densified urban areas on the other hand are commonly characterised by excessive heating compared to their surroundings, since runoff is accelerated and man-made materials heat up considerably. The newly developed open-cast mining areas in the Rhenish lignite mining region are also areas of significant heating, mainly caused by the removal of vegetation and the exposure of soil, sand and sediments. In addition to that, groundwater has to be drained in the excavation area, which leads to a shortfall of water and a diminished cooling effect of evapotranspiration. The mining process also exposes sand and coal seams to the surface, which heat up more than the cultivated land and the forests that covered the region before. Other areas that have shown

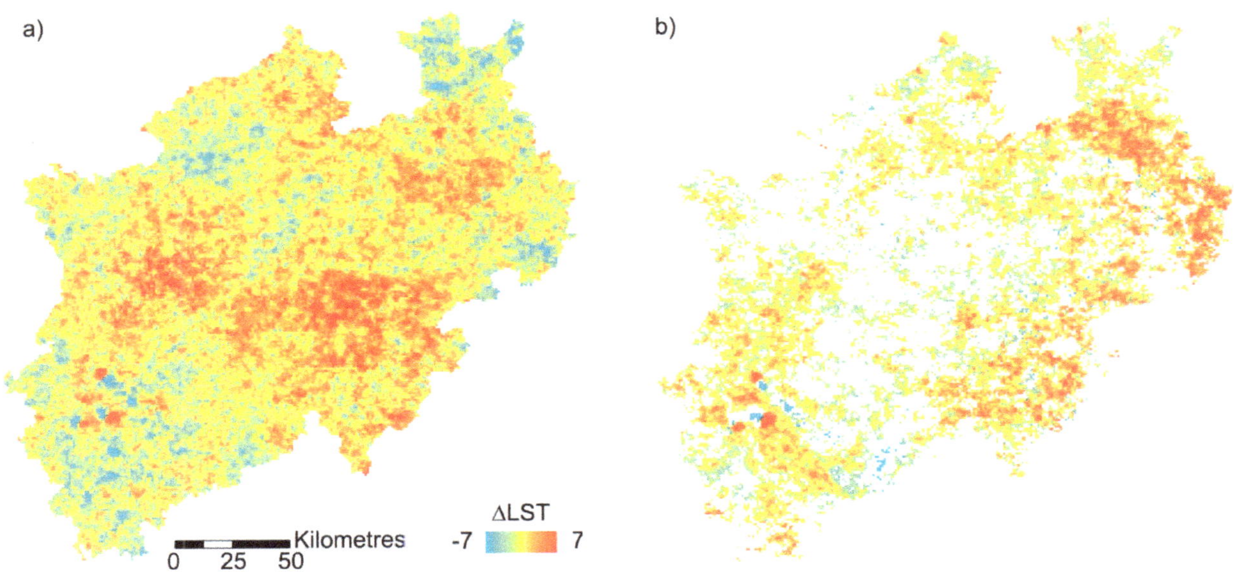

Figure 5. Trends of LST: (a) linear trend of July and august observations between 2000 and 2015 and (b) significant trends of all observations between 2000 and 2015 based on AAT.

an increase in LST are further east in vegetated areas (see **Figure 2**). The warming here is probably the result of forest damage and degradation caused by storms like Kyrill in 2007 [23].

Areas of summer cooling trends are primarily surfaces that have been classified as arable land or grassland in 2001 as well as in 2015 but also recultivated opencast mining sites (**Figure 4**). A possible explanation for this cooling is that the harvest date of the crops may have been delayed from year to year, so that the fields were vegetated for a longer period, hence the prolonged vegetative cooling effect. Another possible cause could be the cultivation of crops like corn and sugar beets that are harvested in autumn as opposed to the common winter cereals like wheat and barley, which are harvested in July and August. The impact of land use change on the surface temperature has not yet been fully investigated and shows different patterns for different time periods. In contrast to **Figure 5a**, in which only the LST trend of the summer months between 2000 and 2015 is shown, **Figure 5b** displays the statistically significant trends ($p < 0.05$) for all measured values based on AAT, i.e. the whole course of the year has been considered. The patterns are identical to some extent, as seen in the areas of opencast mining, but for some parts, they differ considerably.

Figure 5b makes two things abundantly clear: on the one hand, areas with statistically significant trends are predominantly detected in urban spaces and to some extent in agricultural land and forests. On the other hand, the trends indicate a rise in LST nearly everywhere—up to 3 K in urban spaces and agricultural land and even more in forests. Only a few recultivated opencast mines show a striking decrease in temperature. The temperature increase in urban spaces is caused by the expansion and concentration of urban structures. The temperature increase in agricultural land could be caused by increasing periods of drought in spring, fall and winter, when the fields lack vegetation regulating temperature. If and to what extent crop rotations or other causes can influence the change in LST has not been investigated yet. However, our findings are in line with other studies that show correlations between land use change, vegetation decrease and LST increase [24].

The cooling effect of recultivated opencast mines is juxtaposed to the long-term warming of newly developed mining areas in **Figure 6**. Cooling already sets in when a population of pioneer plants develops on formerly barren land, regulating water and radiation budget. During vegetation growth and succession, biomass, the level of transpiration and finally the cooling effect increase. The lines showing a linear trend in **Figure 6** have been added to emphasise the land surface temperature evolution since 2000. The warming might actually be an abrupt change from one state into another. **Figure 6** shows that in the area of the recultivated dump site "Sophienhöhe", temperature has decreased by about 3 K between 2000 and 2015, while the temperature of the newly developed mining area of the "Hambach" open pit mine has increased by about 3 K.

4.2. Perspectives on future land consumption in the Rhineland Metropolitan area in 2030

Figure 7 shows the results of the modelling for 2030. The quantity of newly urbanised and occupied areas oscillates between 29,200 ha and 79,000 ha, depending on the scenario

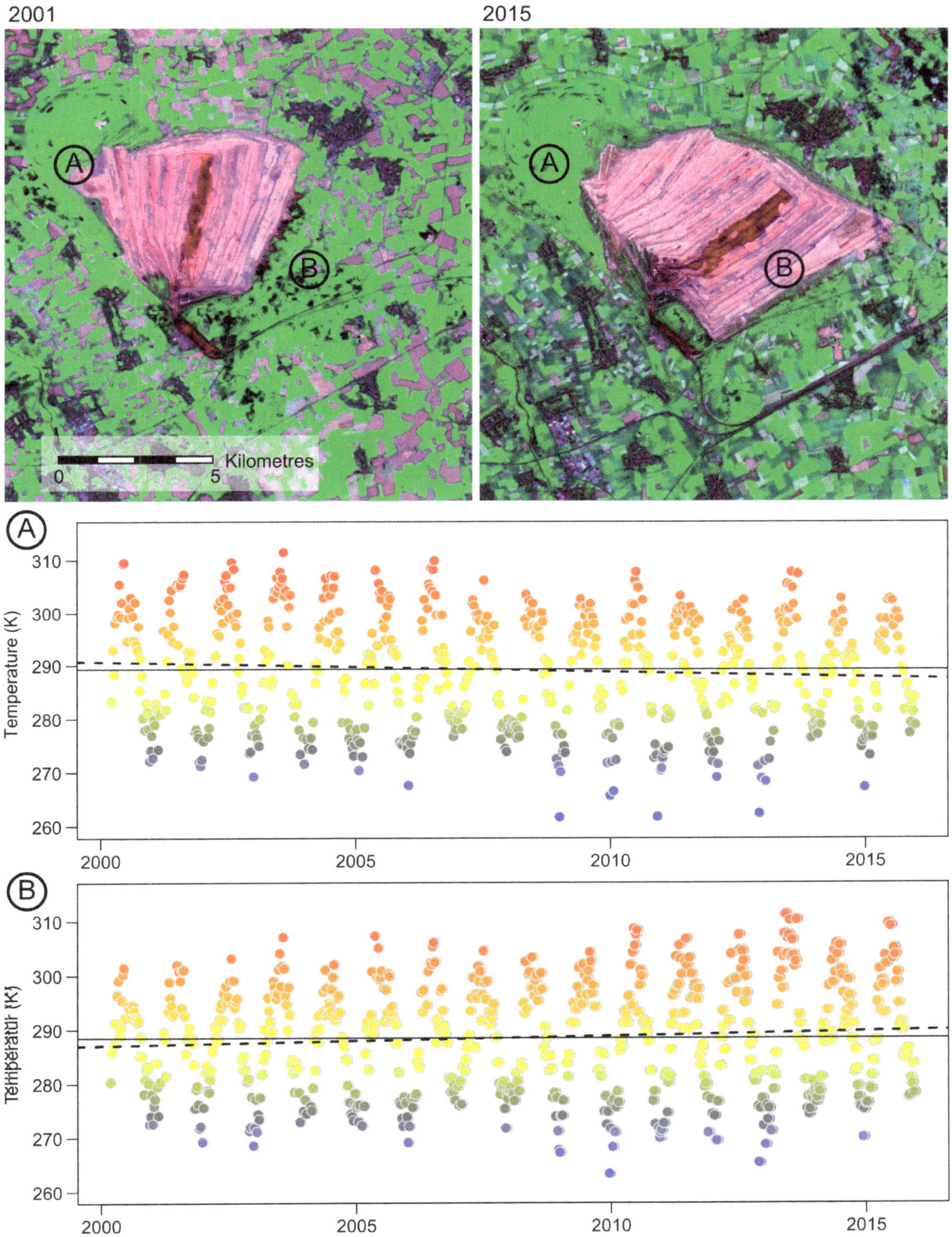

Figure 6. LULCC in Rhenish open pit mining between 2001 and 2015. (A) Indicates the location of "Sophienhöhe" mining dump and (B) the migration of "Hambach" open pit mine towards southeast. The respective time series of a (cooling trend) and B (warming trend) of all observations of LST between 2000 and 2015. Red colours indicate warmer LST during summer; blue colours indicate cold winter LST.

(**Figure 8**). This would represent a daily consumption of 5 ha, 10 ha and 15 ha, respectively. The "business as usual" scenario would already exceed the 30 ha benchmark. In the scenario of "sustainable development", the land consumption in the Rhineland Metropolitan area would reduce by 3 ha per day. In the opposite scenario "desire to live in a green area", land consumption would almost double (**Figure 8**). All areas that are consumed in scenario A (red) are consumed in the other scenarios, too. All areas that are consumed in scenario B (yellow) would be added to those of scenario A (red + yellow). In scenario C, most areas would be consumed (red + yellow + blue). Note that cellular automata only issue an outlook on what could happen and not necessarily what will. The outcome visualises a basic trend, which occurs in all three scenarios: the settlement area increase firstly leads to a densification of present settlements. Looking at the distance that lies between newly developed settlement areas and present ones, this statement is underlined. In addition to that, the development of new urban areas is taking place in the suburban and exurban space, which leads to disperse structures. The next step could be to link the land consumption modelling to a prediction of the development of LST. This could emphasise the importance of a sustainable management of land resources and add to the debate on climate change. Densification and infill development are to be preferred over expansion and the development of new building areas, since forests and agricultural areas can be preserved that way. However, increasing densification also holds challenges, especially concerning logistics and urban climate.

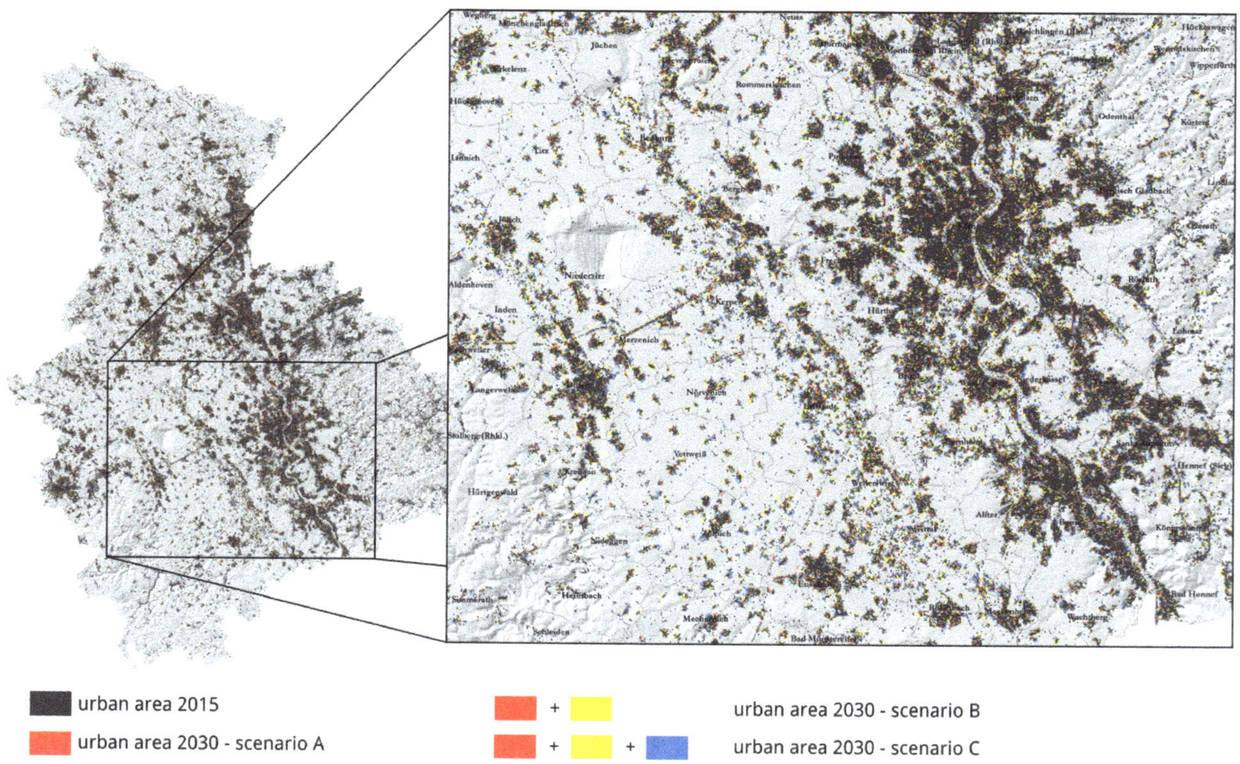

Figure 7. Modelling of land consumption in the Rhineland Metropolitan area in 2030 according to the scenarios "sustainable development" (A), "business as usual" (B) and "desire to live in a green area" (C) (figure modified after [25]).

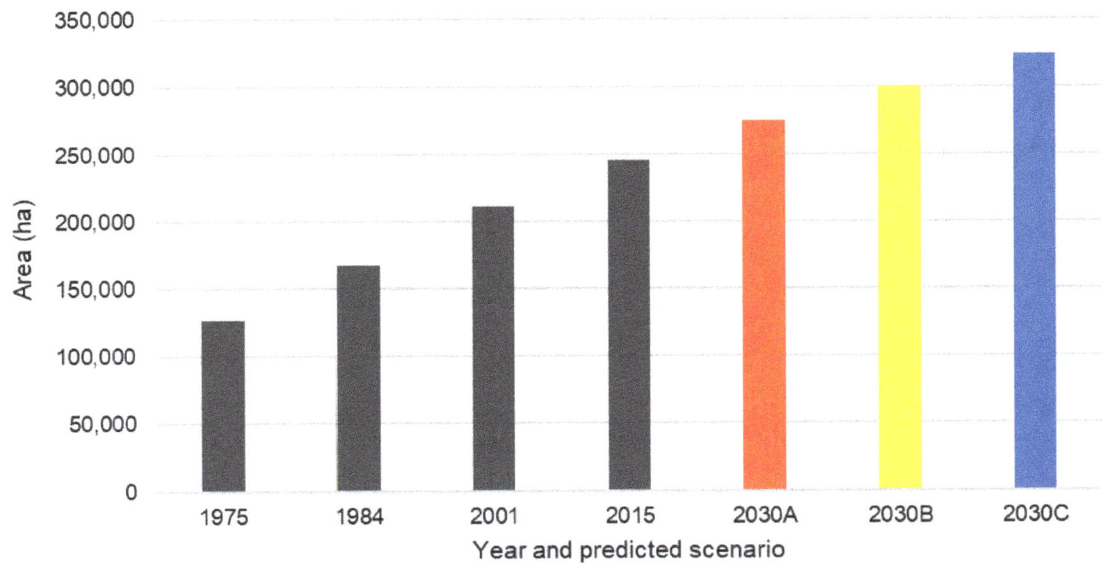

Figure 8. Urban development in the Rhineland Metropolitan area between 1975 and 2030 from an analysis of the region's satellite images (1975–2015) and from modelling (2030) according to the scenarios "sustainable development" (A), "business as usual" (B) and "desire to live in a green area" (C).

5. Conclusion and outlook: Land consumption and climate adaptation: A glance beyond the metropolitan area

Due to current climate change, the topic of land consumption and change of landscape structure particularly concerns the densely populated areas and requires measures to increase their resilience. The classified satellite data (**Figure 4**) illustrate how settlement structure, building density and green structure of the Rhineland Metropolitan area have changed over the past decades. This can be seen when looking at the settlement development along the Rhine, in particular between Dusseldorf and Cologne, where open spaces are constantly receding. Looking at the individual neighbourhoods and roads, more links between land use and microclimate become clear—ventilation, heat development and percolation ability, for example. Areas with a high degree of imperviousness are at a higher risk to develop urban heat island effects than green spaces [25]. This is primarily caused by the heat storage of asphalted surfaces, the missing cooling effect of vegetation, limited air exhaustion due to dense construction and, of course, anthropogenic emissions. On some days, it might be nice when it is a little warmer and calmer in the street canyon than in the open field, but the development of such heat islands causes many problems in summer, be it for health, water budget or vegetation growing conditions [1, 2].

The link between the land cover maps and LST trends increases the validity of the results and establishes new courses of action. That way, it is possible to detect areas that have the same degree of imperviousness but still differ with regard to their respective microclimate. The next step would be to take a closer look at the current situation and derive solutions. The orientation, the height, or the design of the different building structures might all imply reasons for differences in the respective LULCC and LST results.

In order to bring those aspects in accordance with requirements for future development and to make upcoming plans as climate friendly as possible, many cities already run map services showing urban climate, which not only show the status quo but also predictions. Although these maps are a valuable basis for planning, they are not strictly necessary to take initial steps towards resilience. Many measures can be taken without detailed climate prognoses. Besides politics and administration, local parties like homeowners and housing associations have to contribute, too.

To fulfil the potential at hand and to inspire willingness to act, it is useful to raise awareness of climate change and the associated need for adaption. To do that, there are no measurements needed; observations can be sufficient. Satellite images—supplemented by climate data—can be very helpful in doing so. Other examples, besides the already mentioned heat islands, are rivers and creeks, which, over the last years, have been subject to floods caused by climate change. While many things can cause a river to overflow, flooded creeks are much very suitable to identify causal climatic relationships and to raise awareness.

The Rhineland Metropolitan area is just one of many examples. The observations made here can be applied to numerous regions in NRW and Germany. It is a promising strategy to consider both short- and long-term approaches as well as local and regional approaches. The development of small-scale solutions on how to handle climate change can help to raise awareness and build a basis for political action. This is first and foremost the responsibility of urban development planning, as can be seen in many local climate change adaption concepts [5]. Politics and administration can make suggestions and requirements for newly planned construction projects, which are based on the insights gained by remote sensing data. Especially the urban sprawl over the past decades becomes alarmingly clear when looking at the classified satellite images of 1975 and 2015 (**Figure 4**). Even though many municipalities are required to favour inner development over outer development, land consumption is still at a high level—especially when considering the consumption per capita.

Urban development planning is not the only determining factor for climate adjustment, though. Modifications of portfolio properties can also contribute to the task at hand. The use of sustainable materials and the greening of roofs and facades of private residences, logistics centres, company premises, commercial and industrial parks and communal buildings can also move things forward. Numerous models and pilot projects show the potential and the willingness of companies and associations to contribute to a climate-friendly remodelling [1, 5]. The topics selected as examples here reveal how important it is to combine sectoral-based strategies. The link could be created by regular sustainability reporting on a municipal level. Surveys done by the statewide Agenda 21 workgroup "LAG 21 NRW" (Landesarbeitsgemeinschaft Agenda 21 NRW) show that only a few municipalities are taking advantage of this opportunity so far [5]. However, those who do can record short- and long-term success. Communities which are willing to address these problems find assistance in a special training session offered by the Education Centre for the Supply and Disposal Economy (Bildungszentrum für die Ver- und Entsorgungswirtschaft, BEW). The Ministry for Environment, Agriculture, Conservation and Consumer Protection of the State of NRW supports this approach and awards municipalities that successfully finish the course with a certificate [5].

Climate adaption is a very complex topic, which is hard to impart. Small-scale possibilities of change like the renaturation of a creek are a well-suited starting point to demonstrate individual and local courses of action. The inclusion of the public also creates the opportunity for creative potential, which no city should pass up. Cities in the Rhineland Metropolitan area have already taken action. In association with the Universities of Bonn and Bochum, the Bonn Science Shop (Wissenschaftsladen Bonn, WILA) and the cities of Bonn and Gelsenkirchen are developing practical climate adaptation measures and guidelines in the project called "Town and Country in the Flow – Network for the Creation of a sustainable Climate Landscape (KlimNet)" to include parties at all levels as well as citizens (www.klimalandschaften-nrw.de).

Acknowledgements

The project KlimNet is funded by the Project Management Acency Jülich (Projektträger Jülich, PTJ) with means of the Federal Ministry for the Environment, Nature Conservation and Nuclear Safety following a decision of the German Bundestag (funding code 03DAS098B) within the German Strategy for Adaptation to Climate Change (Deutsche Anpassungsstrategie an den Klimawandel, DAS). The authors would like to cordially thank their colleagues at the Bonn Science Shop (Wissenschaftsladen Bonn, WILA), namely, Dr. Anke Valentin, for their constant and productive collaboration on all levels. Javier Muro kindly gave MODIS data processing support at the Center for Remote Sensing of Land Surfaces (Zentrum für Fernerkundung der Landoberfläche, ZFL), University of Bonn. Special thanks go to the former members of the "NRWpro "project team, especially to Dr. Roland Goetzke, Dr. Birte Schöttker and Prof. Dr. Gunter Menz (†). Their results from more than a decade ago formed the basis of new research ideas and projects, one of which is currently KlimNet.

Conflict of interest

The authors declare no conflict of interest.

Author details

Andreas Rienow[1*], Nora Jennifer Schneevoigt[1] and Frank Thonfeld[2]

*Address all correspondence to: andreas.rienow@rub.de

1 Geomatics Research Group, Department of Geography, Ruhr-University Bochum, Germany

2 Remote Sensing Research Group, Department of Geography, University of Bonn, Germany

References

[1] Brasseur GP, Jacob D, Schuck-Zöller S. Klimawandel in Deutschland: Entwicklung, Folgen, Risiken und Perspektiven. Springer-Verlag; 2017. 352 p

[2] Stocker TF, Qin D, Plattner G-K, Tignor M, Allen SK, Boschung J. Herausgeber. Climate Change 2013: The Physical Science Basis. Contribution of Working Group I to the Fifth Assessment Report of the Intergovernmental Panel on Climate Change [Internet]. Cambridge & New York: Cambridge University Press; 2013 [last access 11 May 2018]. 1535 p. (IPCC). Available from: https://www.ipcc.ch/report/ar5/wg1/

[3] Piao S, Friedlingstein P, Ciais P, de Noblet-Ducoudré N, Labat D, Zaehle S. Changes in climate and land use have a larger direct impact than rising CO_2 on global river runoff trends. Proceedings of the National Academy of Sciences of the United States of America. 2007;**104**(39):15242-15247

[4] BMU (Bundesministeriums für Umwelt, Naturschutz und nukleare Sicherheit). Flächenverbrauch – Worum geht es? [Internet]. Reduzierung des Flächenverbrauchs. 2018 [last access 11 May 2018]. Available from: https://www.bmu.de/themen/nachhaltigkeit-internationales/nachhaltige-entwicklung/strategie-und-umsetzung/reduzierung-des-flaechenverbrauchs/

[5] Reuter K, Breyer K, Herausgeber. Flächenmanagement als partizipativer Prozess einer nachhaltigen Stadtentwicklung. Dokumentation [Internet]. Schwerte: LAG 21 NRW (Landesarbeitsgemeinschaft Agenda 21 NRW e.V.); 2008. 48 p. Available from: http://www.flaechenportal.nrw.de/fileadmin/user_upload/2008_LAG21NRW_Flaechenmanagement_Modellprojekt_I_Dokumentation.pdf

[6] Hoymann J, Goetzke R. Die Zukunft der Landnutzung in Deutschland – Darstellung eines methodischen Frameworks. Raumforschung und Raumordnung. 2014;**72**(3):211-225

[7] Hauger G. Ökologische Bewertung der Flächeninanspruchnahme durch Verkehrsinfrastruktur. In: Petz KC, Herausgeber. Versiegelt Österreich? Der Flächenverbrauch und seine Eignung als Indikator für Umweltbeeinträchtigungen. Wien: Umweltbundesamt; 2001. p. 53-62

[8] Tucker CJ. Red and photographic infrared linear combinations for monitoring vegetation. Remote Sensing of Environment. 1979;**8**(2):127-150

[9] Irons JR, Dwyer JL, Barsi JA. The next Landsat satellite: The Landsat data continuity mission. Remote Sensing of Environment. 2012;**122**:11-21

[10] Drusch M, Del Bello U, Carlier S, Colin O, Fernandez V, Gascon F. Sentinel-2: ESA's optical high-resolution mission for GMES operational services. Remote Sensing of Environment. 2012;**120**:25-36

[11] Breiman L. Random forests. Machine Learning. 2001;**45**(1):5-32

[12] NASA LP DAAC (Land Process Distributed Active Archive Center). MODIS Products Table [Internet]. 2016 [last access 11 May 2018]. Available from: https://lpdaac.usgs.gov/dataset_discovery/modis/modis_products_table

[13] Wan Z, Hook S, Hulley G. MOD11A2 MODIS/Terra Land Surface Temperature/Emissivity 8-Day L3 Global 1 km SIN Grid, Version 6 [Internet]. NASA EOSDIS Land Processes DAAC, USGS; 2015. Available from: https://lpdaac.usgs.gov/node/820

[14] NASA LP DAAC (Land Process Distributed Active Archive Center). NASA Land Data Products and Services [Internet]. 2017 [last access 11 May 2018]. Available from: https://lpdaac.usgs.gov/

[15] Forkel M, Carvalhais N, Verbesselt J, Mahecha MD, Neigh CSR, Reichstein M. Trend change detection in NDVI time series: Effects of inter-annual variability and methodology. Remote Sensing. 2013;**5**(5):2113-2144

[16] Forkel M, Migliavacca M, Thonicke K, Reichstein M, Schaphoff S, Weber U. Codominant water control on global interannual variability and trends in land surface phenology and greenness. Global Change Biology. 2015;**21**(9):3414-3435

[17] R Core Team. R: A language and environment for statistical computing. R foundation for statistical computing [Internet]. Vienna, Austria; 2018. Available from: https://www.R-project.org/

[18] Jat MK, Choudhary M, Saxena A. Application of geo-spatial techniques and cellular automata for modelling urban growth of a heterogeneous urban fringe. Egyptian Journal of Remote Sensing and Space Science. 2017;**20**(2):223-241

[19] Chaudhuri G, Clarke K. The SLEUTH land use change model: A review. Environmental Resource Research. 2013;**1**(1):88-105

[20] Baron M, Dross M. Fläche sparen trotz Wohnungsnot: Geht das? – Das 30-Hektar-Ziel der Bundesregierung steht unter Druck. Vol. 62016. pp. 2-3

[21] Clarke KC, Hoppen S, Gaydos L. A self-modifying cellular automaton model of historical urbanization in the San Francisco Bay Area. Environment and Planning. B, Planning & Design. 1997;**24**(2):247-261

[22] Rienow A, Thonfeld F, Valentin A. Flächenverbrauch in der Metropolregion Rheinland 1975-2030 [Internet]. Wiesbaden: Springer Fachmedien Wiesbaden; 2018. (essentials). Available from: http://link.springer.com/10.1007/978-3-658-20399-3

[23] Fink AH, Brücher T, Ermert V, Krüger A, Pinto JG. The European storm Kyrill in January 2007: Synoptic evolution, meteorological impacts and some considerations with respect to climate change. Natural Hazards and Earth System Sciences. 2009;**9**(2):405-423

[24] Muro J, Strauch A, Heinemann S, Steinbach S, Thonfeld F, Waske B. Land surface temperature trends as indicator of land use changes in wetlands. International Journal of Applied Earth Observation and Geoinformation. 2018;**70**:62-71

4

Spatial Analysis, Modelling, and Planning

José António Tenedório and Jorge Rocha

1. Introduction

It can be difficult to separate spatial analysis from other fields of interest such as geography, location analysis, geographic information science, etc. Yet, its beginnings are to some extent easy to identify. It started with both the "spatial thinking" paradigm and the geography quantitative revolution in the 1950s–1960s [1, 2].

The first promoters of this paradigm shift (Brian Berry, Waldo Tobler, Art Getis, etc.) had a geography background and, as such, conducted their work in a multidisciplinary crossroad approach, allowing crossing ideas and spatial analysis approaches from substantially different disciplines (e.g., statistics and computer science). One of the most marking add-ins to spatial analysis was Peter Haggett's [3] work, which remains as a reference for spatial analysis researchers and scholars.

Spatial analysis stands over the principle that there is some spatial component—absolute, relative, or both—in data. Indeed, in the beginning of the twentieth century, 80% of all data have already some kind of spatial explanation [4]. Spatial analysis comprehends numerous representational models of reality based on the spatial properties of the data features [5].

The importance gathered by spatial analysis within geography, its achievement in getting into the analytical framework of several sciences (e.g., natural, social, and physical), and its prominence as a pillar of geographic information science [6] reflects that geographic space does matter and greatly influences the way natural, social, and physical processes evolve.

Spatial patterns and processes have idiosyncratic properties [7] that establish the core of the spatial analysis paradigm. One is spatial dependence, which postulates that the spatially located semantic information gives some insights about the existent information in nearby locations. This is known as spatial autocorrelation (i.e., a kind of statistical dependence relationship) and when it applies to univariate analysis is often understood as some kind of spatial expansion process. Here it is impossible not to mention Tobler's (1970) First Law of Geography, stipulating that all things are related, but near things are more related than distant things [8].

Another basic principle of spatial analysis is known as spatial heterogeneity. Here univariate/multivariate analysis is possibly not static throughout the geographic space, that is, anisotropy. Thus, one may find local hot and cold spots [9], because the parameter calibration of these models may vary athwart the study area, mirroring local variations of the global model adjusted for the study area as an all [10].

In a narrow view, one can consider that spatial data is special [11]. Yet, a rigid interpretation has often resulted in the postulate that the geographic space exists objectively and independently of the social and natural processes that operate across the spatial extension and in the conceptual and operational separation of spatial and semantic information in spatial models. Most recently this idea has been considered dogmatic and detached from the authenticity of the geographic space [12].

Geography has a history of a, sometimes crispy, relation between law-seeking (nomothetic) and description-seeking (idiographic) knowledge [13]. Wisely, physical geographers get away from these debates, but the nomothetic-idiographic tension keeps on in human geography [13–15]. Possibly without surprise, geography has been censured for invalidated theories, results that cannot be reproduced, and a division among practice and science [16].

Goodchild and Li [17] debate that old synthesis process, which is usually hidden from general view and not easily related to the final result, will be more explicit and of serious importance in the new big data era. Much of the geographic knowledge is made of formal theories, models, and equations that need to be processed in an informal manner. By the contrary, data mining techniques require explicit representations, for example, rules and hierarchies, with straight access deprived of processing [18].

One can state that multiple regression is undoubtedly the most widely used statistical approach in geography. This model assumes that the model that is being used in the analysis is the most correct [19]. Regrettably, the background theories are hardly ever satisfactorily developed to include even the most pertinent variables. Also, often many researchers do not even know what can be those missing variables. Hence, one often finds us testing a limited set of variables, and the incapacity to include crucial mislaid variables can have severe implications in the model's accuracy and thus in the conclusions drawn from the data [20].

In opposition to this traditional methodology, the data analytic approach trusts on multiple models or a group of models. Instead of selecting the only best model and accepting that it properly defines the data generation process, a group of models analyze all the possible models to be resultant from the existing variables set and combine the results through a multiplicity of techniques, for example, bootstrap aggregation, bagging, boosting, support vector machines, neural networks, genetic algorithms, and Bayesian model averaging [21, 22]. The subsequent group of models constantly achieves better results than the designated best model making higher accurate predictions across a wide group of domains and techniques [23, 24].

However, one should note that the new advanced data analytic techniques do not always outperform the more traditional techniques [25].

This book is a gathering of original research contributions focusing on recent developments in spatial analysis and modelling with implications to—spatial—planning. The book is organized in three parts that make use of spatial analytic approaches in a progressively integrated and systemic way. It pretends to show how computational methods of spatial analysis and modelling in a geographic information system (GIS) environment can be applied on systems comprehension and allow a more informed spatial planning and, thus, theoretically improved and more effective. The 12 topics comprise new types of data, analysis to distinguish the importance of data in structures, functions and processes, and the use of approaches to backing decision-making.

2. Spatial analysis

The emergence of critical geography (mainly physical), critical GIS, and radical approaches to quantitative geography fostered the idea that geographers are well prepared to combine quantitative methods with technical practice and critical analysis [26]. This proved to be not quite true, but presently big data opens, specially through data mining, new possibilities for spatial analysis research [27] and can extend the limits of quantitative approaches to a wide array of problems usually addressed qualitatively [27, 28].

Despite that big data puts challenges to conventional concepts and practices of "hard" sciences, where geographic information science is included [29, 30], the predominance of big data will undoubtedly lead to a new quantitative turn in geography [31]. This is clearly a new paradigm shift in geography research methodologies: a fourth—data-intensive—paradigm [32].

The alleged spatially integrated social sciences intend to influence GIS in order to analyze the enormous amounts of available geocoded data [33]. Making sense of these data requires both computationally based analysis methods and the ability to situate the results [34] and brings together the risk of plunging traditional interpretative approaches [35]. The big data era calls for new capacities of synthesis and synergies between qualitative and quantitative approaches [36].

This paradox alliance between "poets and geeks" [37] can be a unique opportunity for geography, stimulating wider efforts to create a bridge over the qualitative-quantitative crater [15] and enabling smart combinations of quantitative and qualitative methodologies [38–40].

It is a similar case to the rebirth of social network theory and analysis where due to the growing availability of relational datasets covering human interactions and relationships, network researchers manage to implement a new set of theoretical techniques and concepts [41].

Surveys are an example of this new paradigm. This methodology is at a crisis because of the decline of response rates, sampling frames, and the narrow ability to record certain variables that are the core of geographical analysis, for example, accurate geographical location [42].

Gradually, self-reported surveys quantifying human motivations and behaviors are being studied and compared with more "biological" data sources [43].

These limitations are still more pronounced if one considers two additional features: (i) the majority of social survey data is cross-sectionally deprived of a longitudinal temporal facet [44] and (ii) most social datasets are rough clusters of variables due to the limitations of what can be asked in self-reported approaches.

Big data is leading to advances on both aspects, shifting from static snapshots to dynamic recounting and from rough aggregations to high resolution, spatiotemporal, data. Here, what matters the most is the likelihood of an increased emphasis of geography on processes rather than structures. Again, network analysis works as a good example as the availability of longitudinal relational data generated the latest procedural and theoretic advances on network dynamics [41].

Big data and its influence on geographic research have to be interpreted in the context of the computational and algorithmic shift that is progressively influencing geography research methods. To fully understand such shift, one can make the distinction between two modelling approaches [45]: (i) the data modelling approach which assumes a stochastic data model and (ii) the algorithmic modelling approach that considers the data as complex and unknown. The first evaluates the parameter values from the data and then uses the model for information and/or prediction, and in the second, there is a move from data models to algorithms properties.

This is precisely the type of data created from immense complex systems simulations [46], but a big percentage of it is provided by sensors and/or software that collect a wide range of social and environmental patterns and processes [47, 48]. The geographic sources of this spatial and temporal data embrace location-aware tools such as mobile phones, airborne (e.g., unmanned aerial vehicles) and satellite remote sensors, other sensors attached to infrastructures or vehicles, and georeferenced social media, among others [18, 49, 50].

There is in big data an enormous potential for innovative statistics [51]. Perhaps the upmost importance is the necessity for a distinct mind-set because big data points toward a paradigm shift, comprising an increased and improved use of modelling practices [52, 53]. Taking in consideration the growing importance of location, it is fundamental for geographers to stop just questioning "where?" but also start to enquire "why?" and "how?" [47].

Spatial analysis is defined as a way of looking at the geographical patterns of data and analyzes the relationships between the entities. In spatial analysis, the tendency in the direction of local statistics, for example, geographically weighted regression [54] and (local) indicators of spatial association [9], characterizes a concession where the main rules of nomothetic geography can evolve in their own way across the geographic space. Goodchild [55] sees GIS as a mix of both the nomothetic and idiographic characteristics, retained, respectively, on the software and algorithms, and within the databases.

Hence, spatial analysis is some sort of modelling procedure that relates data features over a geographic space (2D), across several spaces (3D), and along time dimension (4D).

3. Spatial modelling

What is a model? Well, in a broad sense, a model is a simplification of the reality: thus, all models are wrong [56]. As one can understand, it is impractical or even functionally impossible to collect cartographic information using an exact match between the representation and the real objects; the elements generated would be a replica of the studied area and not a model. The acquisition of information is therefore a numerical relationship between reality and the cartographic representation and, therefore, requires a semantic transfer, inseparable from the graphic and thematic generalization processes.

Lewis Carroll, the world-renowned writer for his book *Alice in Wonderland*, in his poetic tale *The Hunting of the Snark (An Agony in 8 Fits)* [57], presents a very particular vision of the relation: greater abstraction versus less information versus more extensive understanding, by proposing an empty map (the Bellman's Map, Figure 1). This blank sheet of paper, with suggestions for navigation (North, South, etc.) and very mysterious, can represent the total ignorance of humans in relation to their location but at the same time was a map that everyone understood. The point is only the simplification/selection, since in the middle of the ocean, this map can be quite accurate, if there is nothing else to consider than water itself.

Chorley and Haggett [58] mention that one of the approaches to model building can start with the simplification of a system to its essential and then start building an increasingly complex structure, by induction, a priori reasoning, and so on. Hardly there may be a standard procedure for the construction of a system model never before modeled, but the suggestion of ways to address the problem given by the authors can help in a first approach to the problem. The original thought processes are difficult to understand and explain, and the solutions of the problems auto-suggest in strange shapes and times. It is not expected that two researchers working on the same subject address two models in the same way. What is expected is that they start with a topic of interest and then try to model it their own way.

All information is gathered at a certain range. This can be set, in a somewhat crude manner, as the number of real-world metrics units that correspond to a same unit in the spatial model. As one reduces the operating scale, the level of detail decreases according to the implicit generalization. However, before doing it, this option should be weighted because, in practice, it is not always possible to reduce and then enlarge a map, without such procedures will lead to a loss of information.

According to a story collected by Jorge Luís Borges from "Travels of Praiseworthy Men" written by Suárez Miranda (1658), and published in the chapter "Of Exactitude in Science" of the book named *A Universal History of Infamy* [59], this would not have been the understanding of a group of cartographers who, perhaps compelled by the thirst for the power of an empire, intended to make a map of their country. More driven by the greatness than by the desire to better understand that territory, these cartographers endeavored to design or, rather, copy the shape of their territory in increasing scales—1:10,000, 1:1000, 1:100, and 1:10—until 1 day they reached what they considered the perfect representation, that is, a map with a 1:1 scale.

Inevitably, "less attentive to the Study of Cartography, succeeding generations came to judge a map of such magnitude cumbersome, and, not without irreverence, they abandoned it to the rigors of sun and rain" [59].

The Empire's cartographers had copied the territory in an obsessive way as if it were a text. It is possible that, except if the Empire comes to decline, the next step would have been to represent each of the transformations of any details of that territory to the extent that it would be impossible to distinguish the importance between the representation and the object represented. These cartographers though believed achieving an increasingly better representation, through a perfect copy of the geometry of place, distorted, in inverse proportion, the ability of these maps to explain the territory of the Empire. Today we can associate modernity, to which Marc Augé refers [60], to the excess of information that submerges us with spatial data; but at the same time reduces the distances due to information and communication technologies (ICT).

Nowadays, spatial modelling and in a broad sense geography have shifted from a data-scarce to a data-rich environment. Contrary to the generalized idea, the critical change is not about the data volume, but relatively to the variety and the velocity at which georeferenced data can be taken. Data-driven geography is (re)emerging due to a massive georeferenced dataflow coming from sensors and people.

The notion of data-driven science defends that the generation of hypothesis and theory creation is up to date by an iterative process where data is used inductively. Hence, it is possible to name a new category of big data research that leads to the creation of new knowledge [61]. One should note that the inductive process should not start in a theory-less void. Preexisting knowledge is used to outline the analytic engine in order to inform the knowledge discovery process, to originate valuable conclusions instead of detecting any-and-all possible relations [62].

Data-driven geography raises some issues that in fact have been long-lasting problems debated within the geographic community. Just to name a few, one can point dealing with large data volumes the problem of samples versus populations, the data fuzziness, and the frictions between idiographic and nomothetic approaches. Yet, the conviction that location matters (i.e., spatial context) is intrinsic to geography and acts as a strong motivation to approaches such as spatial statistics, time geography, and geographic information science as an all.

Models can have very distinct applications, from the conception of suitability, vulnerability, or risk indicators, to simulation to the assessment of planning scenarios. In a GIS framework, modelling can provide insights about the way real systems work with enough precision and accuracy to permit prediction and assertive decision-making.

Nowadays, two distinct cultures of modelling coexist [45, 63]. By one side, one can start imaging a stochastic data model in what can be called a data modelling culture. The other one, the algorithmic modelling culture, assumes that the core of the model is complex and unidentified. The former uses the model for both information and prediction after retrieving the parameter values from the data. In the latter, a shift exists from the data models to the algorithms properties.

Putka and Oswald [64] indicate how geography could benefit by implementing the data algorithmic philosophy. They claim that the actual data modelling philosophy prevents the ability to predict results more accurately, generates models that do not integrate a phenomenon's key drivers, and cannot incorporate models' uncertainty and complexity in a satisfactory manner.

4. Modelling and planning

The history of territories reveals cycles, both of progress and decline, if we consider only the opposites. Each cycle mirrors, in scales, dimensions, and variable rhythms, the importance of political decisions. Planning the territory constitutes an instituted praxis from which the models of the desired evolution are derived. As a general rule, the models are drawn up on the basis of essentially qualitative assumptions. They establish themselves as models that transpose the dominant ideas resulting from the interpretation of the spirit of the laws and regulations, from the debate of the technical solutions, and from public participation.

However, in the light of the recent theories on the territorial dynamics, there is the possibility to resort to quantitative models that reveal the self-organizing systems of the territories (e.g., cellular automata, multi-agent systems, fractal analysis, etc.). These models use intensively spatial modelling in a GIS environment and future scenarios simulation based on historical information of geographical changes. When, for instance, one quantifies the land use/cover changes and relates them with what is predicted in the plans, the conditions for quantitative modelling are created, favoring the dialectic between models. Therefore, it is not a question of using the confrontation between qualitative and quantitative models to nullify the relevance and/or excessive valorization of one of them. On the contrary, it is to evaluate the potential of each other and make use of it to improve the technical efficiency in the moment of preparation, monitoring, and evaluation of the territorial management instruments.

The legal systems and regulations of each country can be an opportunity to use geosimulation models, of quantitative root, to enrich the political and technical debate, about the planning of the territories in the future. It was in this context that we captured the questions relating to the analysis and spatial modelling as fundamentals of urban planning and regional planning, which, as we know, are complex processes of geographic space organization.

5. Conclusions

The difficulties to interpret and understand the territory, particularly with regard to the mixing of subsystems, inevitably require using the notion of complexity. Thus, it is essential to provide tools that could address complexity, linking both spatial organizations and the system of actors who make them evolve. Therefore, the systems approach presents itself as a paradigm capable of guiding the use and understanding of complex systems and as a prerequisite for more advanced modelling approaches.

Understanding social complexity requires the use of a large variety of computational approaches. For instance, the multiscale nature of social clusters comprises a countless diversity of organizational, temporal, and spatial dimensions, occasionally at once. Moreover, computation denotes several computer-based tools, as well as essential concepts and theories, varying from information extraction algorithms to simulation models [65, 66].

Location analysis and modelling as an integrating part of spatial analysis [67] come out from Weber's industrial location theory. Location models might embrace a descriptive methodology, but they can also be very operative as normative environments. Hence, spatial analysis overlaps typical data analytic methods such as statistics, network analysis, and several data science viewpoints, such as data mining and machine learning.

Whereas there is an interesting discussion between statistics and machine learning researchers about the advantages and disadvantages of each method, it is unmistakable that the huge mainstream of quantitative analytical methods falls inside the concept of data modelling culture. This enables a profounder knowledge about the importance of spatial values in shaping the geographic space.

The spatial analysis overlapping with numerous fields of application leads to the coin of the designation spatial science [68], which seems to better represent its singularities. In addition to geography, spatial analysis has a clear linkage to regional science.

Ever since its beginning, regional science has dealt with knowledge discovery adopting a neopositivism approach. It embraces the emerging architype of geospatial data integration rooted in geographic information science [69–71] to analyze the complex systems and the spatiotemporal processes that make them. It also extended the procedural boundary of spatial analysis, through both exploratory spatial data analysis [72] and confirmatory spatial data analysis [73].

Thus, spatial analysis and modelling is an interesting area of application within geographic information science, directing analysis, modelling, and improving the comprehension of spatiotemporal processes. It comprises a group of narrowly connected subareas, for example, geographic knowledge discovery, data analytics, spatiotemporal statistics, social network analysis, spatiotemporal modelling, and agent-based simulation.

Author details

José António Tenedório[1] and Jorge Rocha[2*]

*Address all correspondence to: jorge.rocha@campus.ul.pt

1 Interdisciplinary Centre of Social Sciences (CICS.NOVA), Faculty of Social Sciences and Humanities (NOVA FCSH), Universidade NOVA de Lisboa, Lisbon, Portugal

2 Universidade de Lisboa, Lisbon, Portugal

References

[1] Fotheringham AS, Rogerson P. Introduction. In: Fotheringham AS, Rogerson P, editors. The SAGE Handbook of Spatial Analysis. Los Angeles: SAGE Publications; 2009. pp. 1-4

[2] Johnston R. Location analysis. In: Kitchin R, Thrift N, editors. International Encyclopedia of Human Geography. Oxford: Elsevier Inc.; 2009. pp. 384-395

[3] Haggett P. Location Analysis in Human Geography. London: Edward Arnold; 1965. 339 p

[4] Franklin C. An introduction to geographic information systems: Linking maps to databases. Database. 1992;**15**(2):12-21

[5] Longley PA. Spatial analysis and modeling. In: Madden M, editor. Manual of Geographic Information Systems. Bethesda: American Society for Photogrammetry and Remote Sensing; 2009. pp. 587-590

[6] Longley PA, Goodchild ME, Maguire DJ, Rhind DW. Geographic information science. In: Madden M, editor. Manual of Geographic Information Systems. Bethesda: American Society for Photogrammetry and Remote Sensing; 2009. pp. 19-24

[7] Haining R. The special nature of spatial data. In: Fotheringham AS, Rogerson P, editors. The SAGE Handbook of Spatial Analysis. Los Angeles: SAGE Publications; 2009. pp. 5-23

[8] Tobler W. A computer movie simulating urban growth in the detroit region. Economic Geography. 1970;**46**(Supplement):234-240

[9] Anselin L. Local indicators of spatial association—LISA. Geographical Analysis. 1995; **27**(2):93-115

[10] Fotheringham AS, Brunsdon C, Charlton M. Quantitative Geography: Perspectives on Spatial Data Analysis. London, UK: SAGE Publications; 2000. 270 p. (Online access: Sage Sage Research Methods Online)

[11] Anselin L. What is special about spatial data? Alternative perspectives on spatial data analysis. In: Griffith DA, editor. Spatial Statistics: Past, Present, and Future. Monograph Ser., No. 12 (Book 12). Institute of Mathematical Geography; 1990. pp. 63-77

[12] Thill J-C. Is spatial really that special? A tale of spaces. In: Popovich V, Claramunt C, Devogele T, Schrenk M, Korolenko K, editors. Information Fusion and Geographic Information Systems: Towards the Digital Ocean. Berlin Heidelberg: Springer-Verlag; 2011. pp. 3-12

[13] Cresswell T. Geographic Thought: A Critical Introduction. Chichester, England: Wiley-Blackwell; 2013. 298 p

[14] DeLyser D, Sui D. Crossing the qualitative-quantitative divide II: Inventive approaches to big data, mobile methods, and rhythm analysis. Progress in Human Geography. 2012;**37**(2):293-305

[15] Sui D, DeLyser D. Crossing the qualitative-quantitative chasm I: Hybrid geographies, the spatial turn, and volunteered geographic information (VGI). Progress in Human Geography. 2011;**36**(1):111-124

[16] Landis RS, Cortina JM. Is ours a hard science (and do we care)? In: Lance CE, Vandenberg RJ, editors. More Statistical and Methodological Myths and Urban Legends. New York, NY, USA: Routledge (Taylor & Francis); 2015. pp. 9-35

[17] Goodchild MF, Li L. Assuring the quality of volunteered geographic information. Spatial Statistics. 2012;**1**(Supplement C):110-120

[18] Miller HJ. The data avalanche is here. Shouldn't we be digging? Journal of Regional Science. 2010;**50**(1):181-201

[19] Tonidandel S, King EB, Cortina JM. Big data methods: Leveraging modern data analytic techniques to build organizational science. Organizational Research Methods. 2016;**21**(3):525-547. DOI: 1094428116677299

[20] Antonakis J, Bendahan S, Jacquart P, Lalive R. On making causal claims: A review and recommendations. The Leadership Quarterly. 2010;**21**(6):1086-1120

[21] Rokach L. Ensemble-based classifiers. Artificial Intelligence Review. 2010;**33**(1):1-39

[22] Seni G, Elder JF. Ensemble methods in data mining: Improving accuracy through combining predictions. Synthesis Lectures on Data Mining and Knowledge Discovery. 2010;**2**(1):1-126

[23] Kaplan D, Chen J. Bayesian model averaging for propensity score analysis. Multivariate Behavioral Research. 2014;**49**(6):505-517

[24] Markon KE, Chmielewski M. The effect of response model misspecification and uncertainty on the psychometric properties of estimates. In: Millsap RE, van der Ark LA, Bolt DM, Woods CM, editors. New Developments in Quantitative Psychology: Presentations from the 77th Annual Psychometric Society Meeting. New York, NY: Springer New York; 2013. pp. 85-114

[25] Schmidt-Atzert L, Krumm S, Lubbe D. Toward stable predictions of apprentices' training success. Journal of Personnel Psychology. 2011;**10**(1):34-42

[26] Lave R, Wilson MW, Barron ES, Biermann C, Carey MA, Duvall CS, et al. Intervention: Critical physical geography. The Canadian Geographer/Le Géographe Canadien. 2014; **58**(1):1-10

[27] Michel J-B, Shen YK, Aiden AP, Veres A, Gray MK, Pickett JP, et al. Quantitative analysis of culture using millions of digitized books. Science. 2011;**331**(6014):176-182

[28] Lieberman-Aiden E, Michel J-B. Culturomics: Quantitative analysis of culture using millions of digitized books. In: Digital Humanities. Stanford, CA, USA: Conference Abstracts, Stanford University; June 19-22 2011

[29] Goodchild MF. The quality of big (geo)data. Dialogues in Human Geography. 2013;**3**(3): 280-284

[30] Gorman SP. The danger of a big data episteme and the need to evolve geographic information systems. Dialogues in Human Geography. 2013;**3**(3):285-291

[31] Ruppert E. Rethinking empirical social sciences. Dialogues in Human Geography. 2013; **3**(3):268-273

[32] Nielsen M. Reinventing Discovery: The New Era of Networked Science. Princeton, New Jersey: Princeton University Press; 2011. 272 p

[33] Gibson C, Brennan-Horley C, Warren A. Geographic information technologies for cultural research: Cultural mapping and the prospects of colliding epistemologies. Cultural Trends. 2010;**19**(4):325-348

[34] Berry DM, editor. Understanding Digital Humanities. 1st ed. Basingstoke: Palgrave Macmillan UK; 2012. 318 p

[35] Gold MK, editor. Debates in the Digital Humanities. Minneapolis, MN: University of Minnesota Press; 2012. 504 p

[36] Sieber RE, Wellen CC, Jin Y. Spatial cyberinfrastructures, ontologies, and the humanities. Proceedings of the National Academy of Sciences of the United States of America. 2011;**108**(14):5504-5509

[37] Cohen P. Humanities 2.0: Digital keys for unlocking humanities' riches [Internet]. The New York Times. 2010 [cited 2017 Dec 22]. Available from: http://www.nytimes.com/2010/11/17/arts/17digital.html

[38] Bodenhamer DJ, Corrigan J, Harris TM, editors. The Spatial Humanities: GIS and the Future of Humanities Scholarship, The Spatial Humanities Series. Indianapolis, IN, USA: Indiana University Press; 2010. 222 p

[39] Daniels S, De Lyser D, Entrikin JN, Richardson D, editors. Envisioning Landscapes, Making Worlds: Geography and the Humanities. Abingdon: Routledge (Taylor & Francis); 2011. 336 p

[40] Dear M, Ketchum J, Luria S, Richardson D, editors. GeoHumanities: Art, History, Text at the Edge of Place. Abingdon: Routledge (Taylor & Francis); 2011. 344 p

[41] Barabási A-L, Pósfai M. Network science. 1st ed. Cambridge: Cambridge University Press; 2016. 456 p

[42] Burrows R, Savage M. After the crisis? Big data and the methodological challenges of empirical sociology. Big Data & Society. 2014;**1**(1):2053951714540280

[43] Curti M, Iacus SM, Porro G, Siletti E. Measuring social well being in the big data era: Asking or listening? Computing Research Repository (CoRR); 2015. 40 p. arXiv:1512.07271

[44] Veltri GA. Big data is not only about data: The two cultures of modelling. Big Data & Society. 2017;**4**(1):1-6

[45] Gentle JE, Härdle WK, Mori Y, editors. Handbook of Computational Statistics: Concepts and Methods. Berlin: Springer-Verlag Berlin Heidelberg; 2012. 1192 p

[46] Miller HJ, Goodchild MF. Data-driven geography. GeoJournal. 2015;**80**(4):449-461

[47] Graham M, Shelton T. Geography and the future of big data, big data and the future of geography. Dialogues in Human Geography. 2013;**3**(3):255-261

[48] Kitchin R. Big data and human geography: Opportunities, challenges and risks. Dialogues in Human Geography. 2013;**3**(3):262-267

[49] Sui D, Goodchild M. The convergence of GIS and social media: Challenges for GIScience. International Journal of Geographical Information Science. 2011;**25**(11):1737-1748

[50] Townsend AM. Smart Cities: Big Data, Civic Hackers, and the Quest for a New Utopia. New York, NY, USA: W. W. Norton & Company; 2013. 400 p

[51] Daas PJ, Marco PJ, Buelens B, van den Hurk PAM. Big data as a source for official statistics. Journal of Official Statistics. 2015;**31**:249

[52] Struijs P, Daas P. Big Data, big impact? In: Conference of European Sattisticians—Seminar on Statistical Data Collection—Topic (v): Integration and management of new data sources. Geneva, Switzerland: United Nations: Economica Commission for Europe; 2013. p. 9

[53] Daas PJH, Puts MJH. Big Data as a Source of Statistical Information. The Survey Statistician. The Hague, The Netherlands: International Association of Survey Statisticians; 2014. pp. 22-31

[54] Fotheringham AS, Brunsdon C, Charlton ME. Geographically Weighted Regression: The Analysis of Spatially Varying Relationships. Chichester, England: Wiley; 2002. 284 p

[55] Goodchild MF. GIScience, geography, form, and process. Annals of the Association of American Geographers. 2004;**94**(4):709-714

[56] Box GEP. Science and statistics. Journal of the American Statistical Association. 1976; **71**(356):791-799

[57] Carroll L. The Hunting of the Snark: An Agony, in Eight Fits. London, UK: Macmillan. p. 1898

[58] Chorley RJ, Haggett P, editors. Models in Geography. London: Methuen & Company, Limited; 1967. 816 p. (Madingley lectures, 2d)

[59] Borges JL. A Universal History of Infamy (Historia universal de la infamia). 1st ed. 1954. Translated by Norman Thomas di Giovanni, editor. Middlesex, England: Penguin Books; 1975. 137 p

[60] Augé M. Non-lieux: Introduction à une anthropologie de la surmodernité. Seuil: La Librairie du XXIe siècle; 1992. 160 p

[61] Bakshy E, Eckles D, Bernstein MS. Designing and deploying online field experiments. In: Proceedings of the 23rd International Conference on World Wide Web. New York, NY, USA: ACM; 2014. pp. 283-292. (WWW '14)

[62] Kitchin R. Big data, new epistemologies and paradigm shifts. Big Data & Society. 2014; **1**(1):2053951714528481

[63] Breiman L. Statistical modeling: The two cultures (with comments and a rejoinder by the author). Statistical Science. 2001;**16**(3):199-231

[64] Putka DJ, Oswald FL. Implications of the big data movement for the advancement of IO science and practice. In: Tonidande S, King E, Cortina J, editors. Big Data at Work: The Data Science Revolution and Organizational Psychology. New York, NY, USA: Routledge (Taylor & Francis); 2015. pp. 181-212

[65] Alvarez RM. Computational Social Science: Discovery and Prediction. Analytical Methods for Social Research. Cambridge: Cambridge University Press; 2016. 327 p

[66] Cioffi-Revilla C. Introduction to Computational Social Science: Principles and Applications. London, UK: Springer-Verlag; 2014. 320 p

[67] Horner MW. Location analysis. In: Kitchin R, Thrift N, editors. International Encyclopedia of Human Geography. Oxford, UK: Elsevier; 2009. pp. 263-269

[68] Johnston R. Spatial science. In: Kitchin R, Thrift N, editors. International Encyclopedia of Human Geography. Oxford, UK: Elsevier; 2009. pp. 263-269

[69] Fischer MM, Nijkamp P. Geographic information systems and spatial analysis. The Annals of Regional Science. 1992;**26**(1):3-17

[70] Yeh AGO, Batty M. Applications of geographic information systems in urban and regional planning. Environment and Planning. B, Planning & Design. 1990;**17**(4):369-374

[71] Thill J-C. Regional science. In: Richardson D, Castree N, Goodchild MF, Kobayashi A, Liu W, Marston RA, editors. International Encyclopedia of Geography: People, the Earth, Environment and Technology. New York: Wiley-Blackwell; 2017. p. 8464

[72] Anselin L. Exploratory spatial data analysis in a geocomputational environment. In: Longley PA, Brooks SM, McDonnell R, MacMillan B, editors. Geocomputation: A Primer. 1st ed. New York: Wiley; 1998. pp. 77-94

[73] Anselin L, Getis A. Spatial statistical analysis and geographic information systems. The Annals of Regional Science. 1992;**26**(1):19-33

5

Fusion Study of Geography and Environmental Engineering

Toshiaki Ichinose

Abstract

The author focuses on urban area and shows case studies on urban energy planning and urban climate induced by concentrated human activities like energy use based on geographic information like land use data and numerical climate model linked to geographic information system (GIS). (1) In finding the candidate places to settle district heat supply facilities that make effective reuse of heat obtained from sewage, the usage of GIS gives the reasonable solution in scientific view point. (2) The spatiotemporal data in high resolution on human activity like energy consumption in urban area enable us to evaluate the impact of anthropogenic heat to urban climate by a numerical simulation model of the local climate. (3) The recent precise geographic information of urban vegetation cover derived from remote sensing (RS) data enables us to evaluate the contribution of urban green to cool our community with higher accuracy. (4) The author's approach is available not only to evaluate the current urban environment but also to evaluate the past one. If a certain scenario of future's urban land use and urban activity, this approach will be available for the future prediction. Thus, these outputs will give some sustainable urban design in viewpoint of climate change (mitigation and adaptation).

Keywords: urban environment, urban climate, energy system, GIS, sustainability

1. Introduction

Geography can be used to clarify the spatiotemporal distributions of various phenomena like climate, flora, fauna, human activities on the Earth, as well as their underlying mechanisms. In contrast, environmental engineering emphasizes the connections among individual elements such as air, water, soil, and living organisms, and attempts to link research findings to policy-making and technical development.

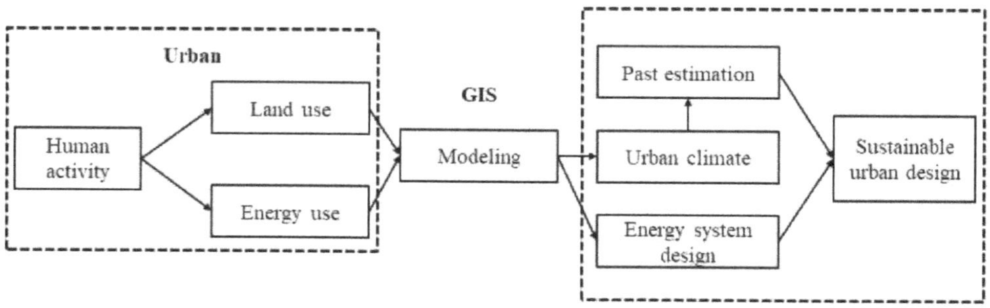

Figure 1. Conceptual structure of this chapter.

The author wants to promote environmental engineering studies to achieve urban design that incorporates environmental factors and fosters the diversity of natural ecosystems, with a focus on Asian cities, which are experiencing the fastest urbanization rate in the world. The author will show how urban physical forms can be designed to take advantage of surrounding natural environments, while also considering specific political and social systems. Another aim is to contribute to global urban design by helping to shift the paradigm of urban planning through innovations that will make low-carbon cities a reality.

Urban areas receive energy, water, products, and raw materials from the outside and emit other products, wastes, sewage, exhaust gases, and waste heat. These waste products are often emitted directly into the environment. This process is called "urban metabolism" in reference to its similarity to the metabolism of a living organism [1, 2]. Mitigating the environmental burden of urban activities requires optimizing the structure of urban metabolism. For example, incineration and heat pumps make it possible to recover heat from wastes and sewage, and the effective use of such untapped energy sources will not only reduce the environmental burden by cycling the energy flow within urban areas, but will also more effectively use energy and resources.

In this chapter, the author focuses on urban area and shows case studies on urban energy planning and urban climate induced by concentrated human activities like energy use based on geographic information like land use data and numerical climate model linked to GIS (**Figure 1**). These outputs will give some sustainable urban design in viewpoint of climate change (mitigation and adaptation).

2. Urban energy planning

Waste incineration plants supply far more concentrated heat than can sewage plants that produce energy, but incineration plants tend to be isolated so they can naturally serve only limited areas. With sewage, heat pumps make it possible to recover heat by taking advantage of the temperature difference between air and sewage. Tokyo (meaning the 23 special wards of the city proper) is covered by a sewerage network [3] (**Figure 2**); although the sewage in this network provides heat distributed widely in a low quality, heat recovery and supply are expected to be possible in broad areas. For such heat utilization to work, there must be quantitative congruence between heat demand and the use of waste heat in heat supply

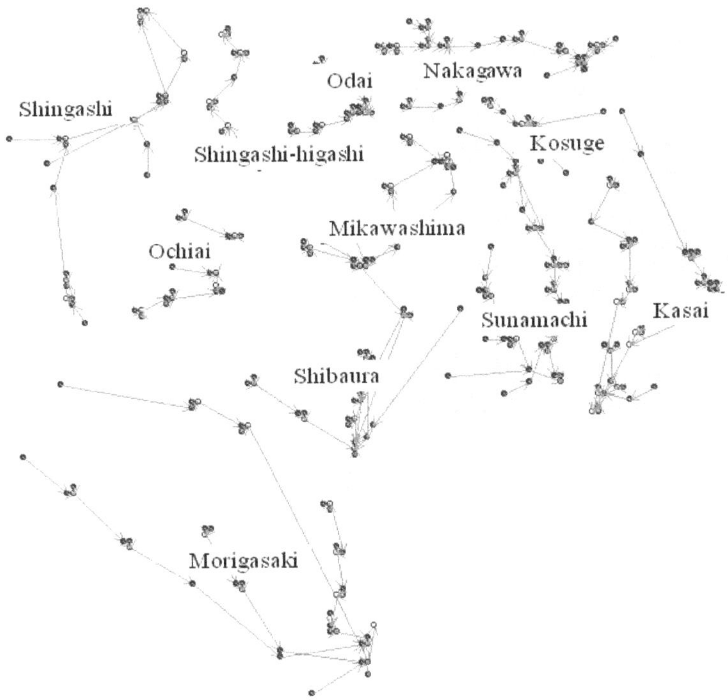

Figure 2. Extracted sewerage lines and sewerage treatment districts in the 23 special wards of Tokyo Metropolis [4]. (N-S: 32 km, E-W: 33 km).

operations. It would seem to be useful to search for such areas, and using a GIS may facilitate this process.

Connolly et al. [5] and Manfren et al. [6] reviewed computer tools that can be used to analyze the integration of renewable energy, including the currently untapped (and diffuse) sources of city energy. Some of the tools are GIS-based and applicable at the local community scale [7–10], but their target is mainly electricity (solar, wind, and biomass burning). Gils et al. [11] introduced a methodology for a GIS-based analysis of district heating potential. They assessed the energy demand for space heating and hot water in residential and commercial sectors in U.S. cities. The author's interest is the spatial conformity between energy demand and supply when referring to community-scale urban structure. This point is the reason why the author think GIS analysis is applicable for urban energy planning.

Accordingly, in relation to established plans of district heat supply facilities that make effective reuse of heat obtained from sewage, the author developed the GIS software to perform analyses of the spatial congruence between heat demand and the use of waste heat in heat supply operations [4]. The temporal and spatial distribution structure of energy consumption [12] was used as the basic data for the GIS, as was the distribution information for land-use types, height of buildings, and other relevant items used in the approximations.

The sewer system simulation model (SSSM) uses raster data on land-use types and building height (number of floors), and polygon data of sewage collection areas defined by sewerage lines, as inputs for performing calculations and mapping [4]. Main data entered into the model included floor space by type of building use for each 250 × 250 m grid cell [13], energy consumption per unit floor space according to industry type [12], and heat demand data according to type of building used [12, 14]. This model can calculate heat demand in each grid cell (**Figure 3**), calculate the sewage flow rate at sewerage line nodes (**Figure 2**), and calculate the amount of recoverable sewage heat at the nodes and surrounding grid cells (**Figure 4**).

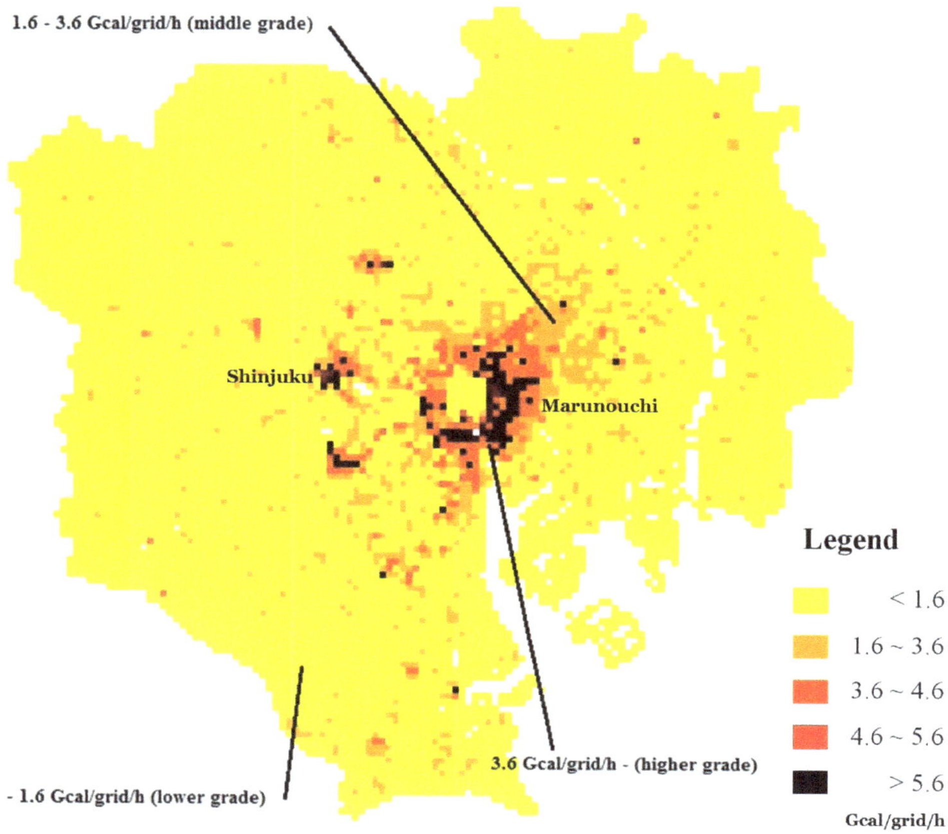

Figure 3. Distribution of heat demand density in Tokyo (daytime, winter, grid size: 250 m) [4].

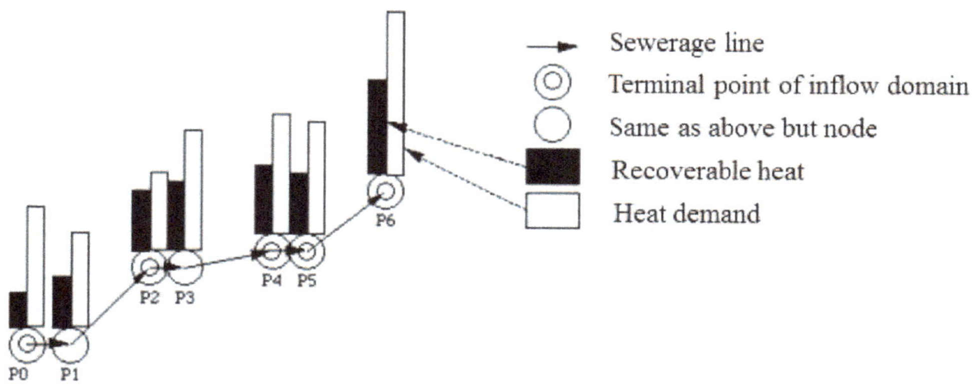

Figure 4. A schematic diagram of theoretical calculation results for recoverable heat at the points along a sewerage line [4].

In the areas with a large amount of usable heat, there are long sewerage lines and many possible locations of heat recovery in general. Owing to the many possible locations of recovery, each with large heat demand in areas such as business districts, the amount of usable heat per heat pump is large. There were also a few cases where the flow rate was a limiting condition. When sewerage lines run through residential areas in main, the heat demand along the sewerage lines is small [4]. In finding the candidate places to settle district heat supply facilities that make effective reuse of heat obtained from sewage, the usage of GIS gives the reasonable solution in scientific view point.

3. Assessment of urban climate

3.1. Urban energy use and its impact

Expansion of urban areas and increase in urban activity have several impacts on urban environments. These impacts are represented by air pollution, water pollution, thermal pollution, caused by anthropogenic heat emission, and so on. Anthropogenic heat affects local climate in urban areas producing phenomena such as heat islands [12]. This process has been explored in a numerical simulation [12, 15]. It was reported that urbanization has both positive and negative impacts on thermal environment, as an increase in urban energy use and a decrease in incoming solar radiation [16]. On the other hand, evaluation of the impact of urban activity through land use change and anthropogenic heat on the urban thermal environment has been necessary in view of urban planning. Quantitative analysis of how human activities and urban structures should be changed to reduce heat island phenomena requires detailed data on anthropogenic heat sources, one of the surface boundary conditions, for use in numerical simulations of such impacts. Such analysis needs evaluation of the influence of spatial and temporal structures of anthropogenic heat on urban thermal environment. Field surveys of energy consumption over wide areas are difficult to carry out. The data for energy consumption as a source of anthropogenic heat must be in the form of grid cell data to be referenced geographically. Some studies which mapped the anthropogenic heat in Tokyo have been published [12, 15]. A time-series analysis of the distribution of urban energy consumption is necessary to evaluate its impact on the thermal environment. The author attempted a detailed analysis of the temporal variability of the distribution of energy consumption in the 23 special wards of the Tokyo Metropolis (Tokyo) [12]. He used a digital geographic land use data set containing the number of floors of building at each grid point to tabulate energy consumption (**Figure 5**) on an areal basis. The numerical simulation of urban climate was consequently performed to determine the effect of anthropogenic heat on urban climate in Tokyo. Annual and diurnal variability of energy consumption (**Figure 6**) is necessary for computer mapping of urban anthropogenic heat.

The author drew detailed maps of anthropogenic heat in Tokyo with data from energy statistics and a detailed digital geographic land use data set including the number of floors of building at each grid point (**Figure 7**) [12]. Computer graphics animation of the diurnal and annual variability in anthropogenic heat of Tokyo was also prepared with the same data sources. These outputs characterize scenarios of anthropogenic heat emission and can be applied to a numerical simulation model of the urban climate [12]. The anthropogenic heat flux in central Tokyo exceeded 400 W m^{-2} in daytime, and the maximum value was 1590 W m^{-2} in winter. The hot water supply in offices and hotels contributed 51% of this 1590 W m^{-2}. The anthropogenic heat flux from the household sector in the suburbs reached

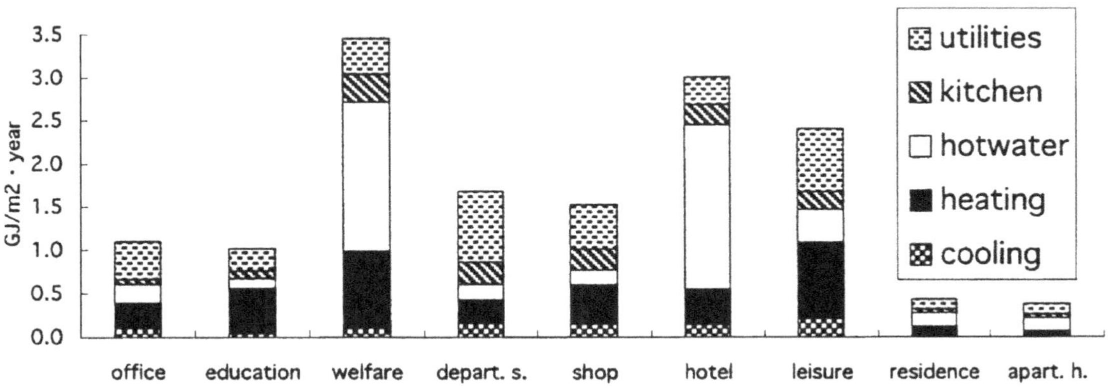

Figure 5. Areal energy consumption for each category on business or land use in Tokyo in 1989 [12]. It excludes industry, motor vehicle and trains. "Depart. S." means department store. "Apart. H." means apartment house.

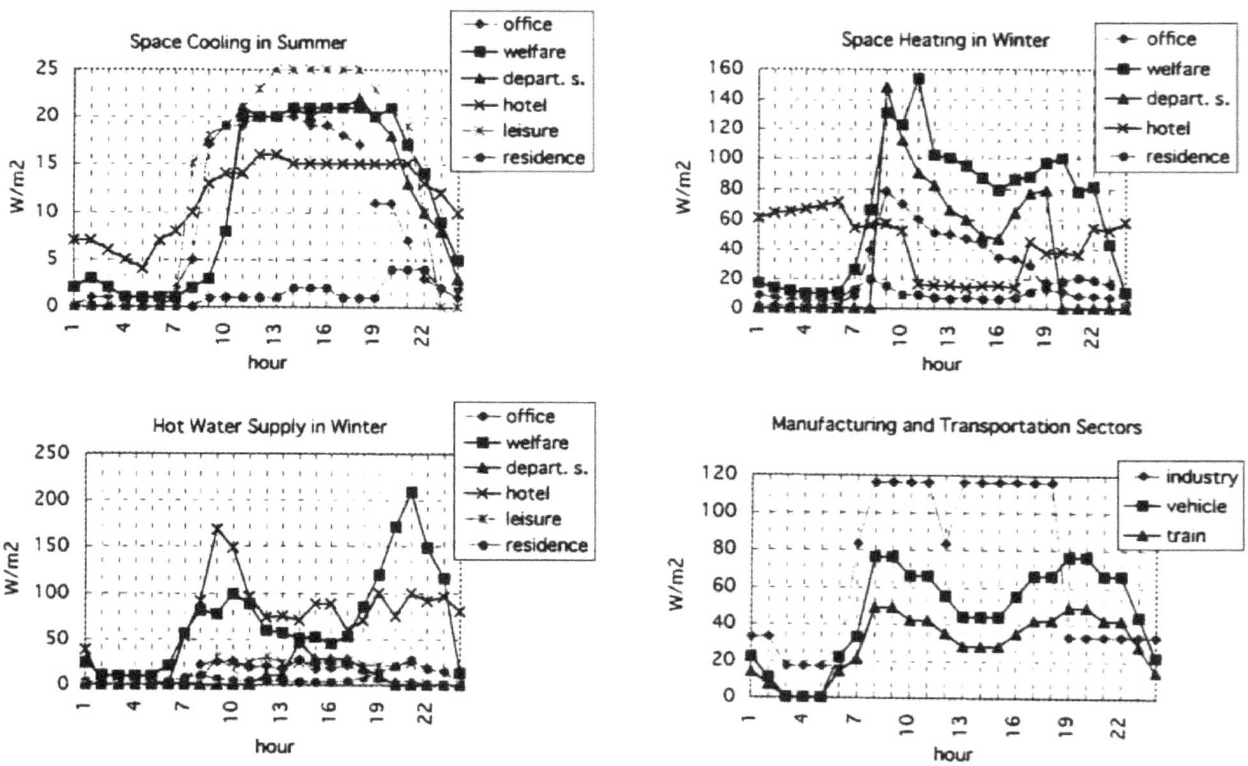

Figure 6. Diurnal variability of demand for space cooling in summer, space heating in winter and hot water supply in winter. Additionally diurnal variability of areal energy consumption in the manufacturing and transportation sectors is also shown [12].

about 30 W m^{-2} at night. Numerical simulations of urban climate in Tokyo were performed by referring to these maps. A heat island appeared evident in winter because of weakness of the sea breeze from Tokyo Bay. At 8 p.m., several peaks of high-temperature appeared, around the areas with the largest anthropogenic heat fluxes (**Figure 8**). In winter, the shortwave radiation was weak, and the influence of anthropogenic heat was relatively large.

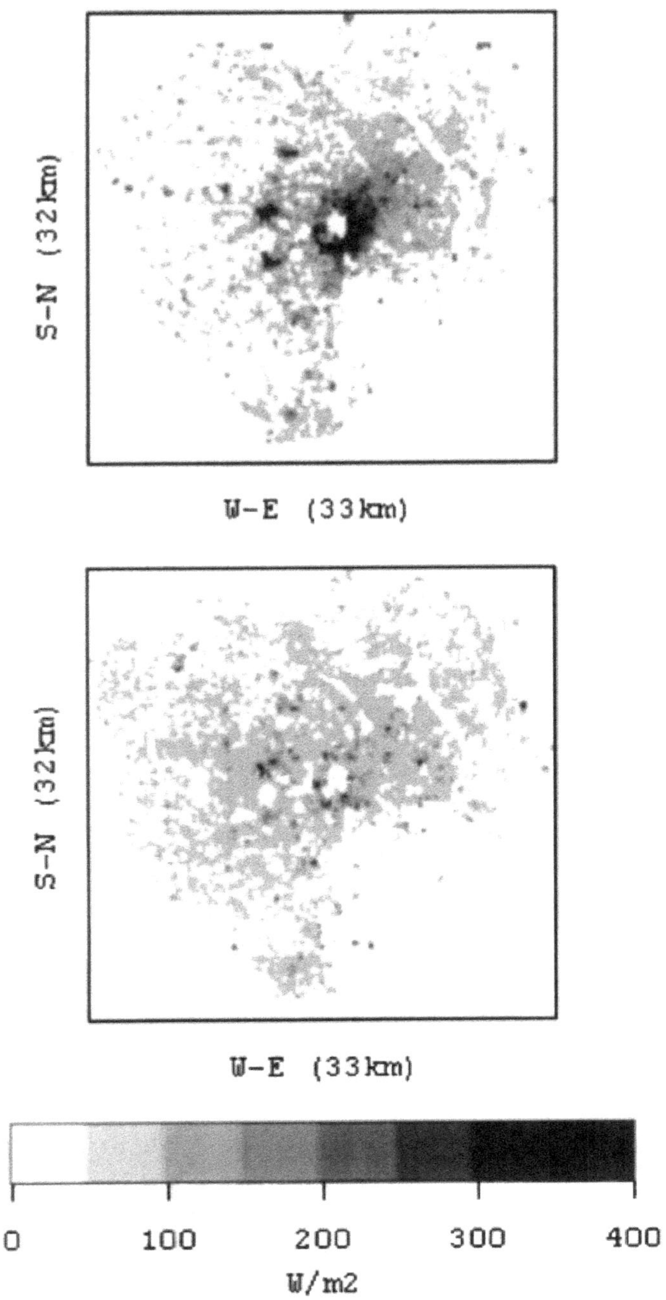

Figure 7. Distribution of anthropogenic heat in Tokyo at 2 p.m. (upper) and 9 p.m. (lower) in winter [12].

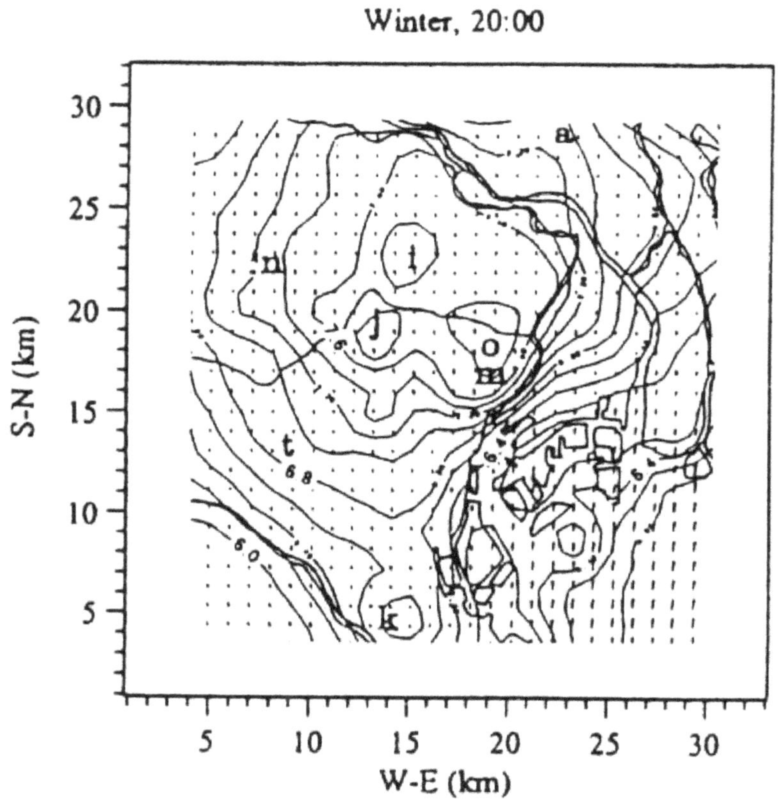

Figure 8. Horizontal distribution of air temperatures (°C) and wind system at 7.5 m above the ground in winter midnight [12]. Symbols as follows: n: Nakano, t: Setagaya, j: Shinjuku, i: Ikebukuro, k: Kamata, m: Marunouchi, o: Otemachi, a: Adachi.

The spatiotemporal data in high resolution on human activity like energy consumption in urban area enable us to evaluate the impact of anthropogenic heat to urban climate by a numerical simulation model of the local climate.

3.2. Urban green and its contribution

The following progress of RS and GIS technology brought a more advanced application of the author's approach. Hirano et al. [17] simulated local climate with surface boundary conditions based on satellite RS data. Most previous mesoscale meteorological modelling studies use land-use data instead as the surface boundary conditions. However, small patches of vegetation cover, such as garden trees and roadside trees, are excluded from the land-use data. Therefore, they made a fractional vegetation cover (FVC) data set with these small patches of vegetation cover from RS data (**Figure 9**), and then simulated the urban heat island in Tokyo with FVC data as new surface boundary conditions. The air temperature with the new boundary condition is up to 1.5°C lower than that with the old one (**Figure 10**). Furthermore, the new boundary condition led to predicted air temperatures closer to the measured temperatures than those with the old one. Therefore, it is important for urban climate simulations to include small vegetation cover.

The recent precise geographic information of urban vegetation cover derived from RS data enables us to evaluate the contribution of urban green to cool our community with higher accuracy.

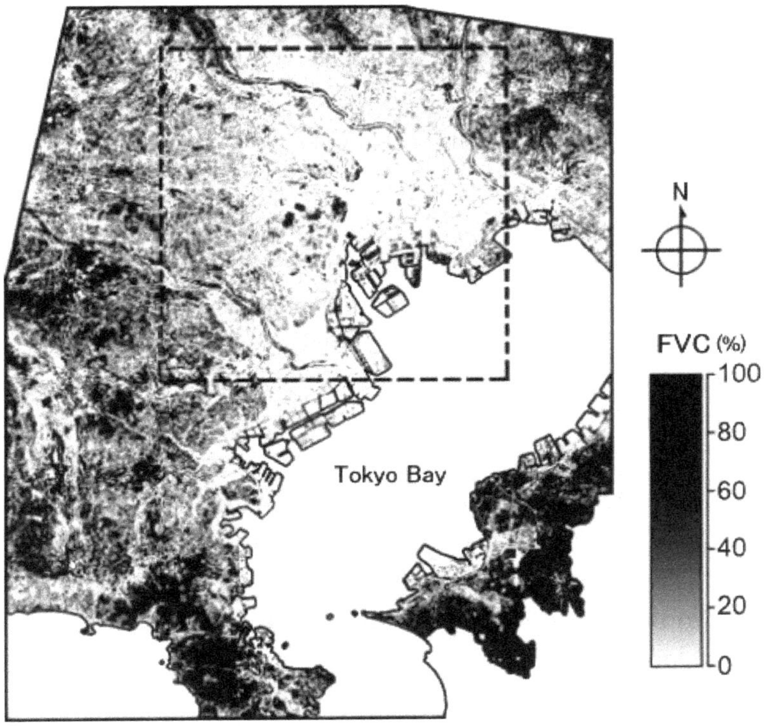

Figure 9. Estimated FVC (Fractional Vegetation Cover) around Tokyo [17]. The dashed-square indicates the target area of **Figure 10**.

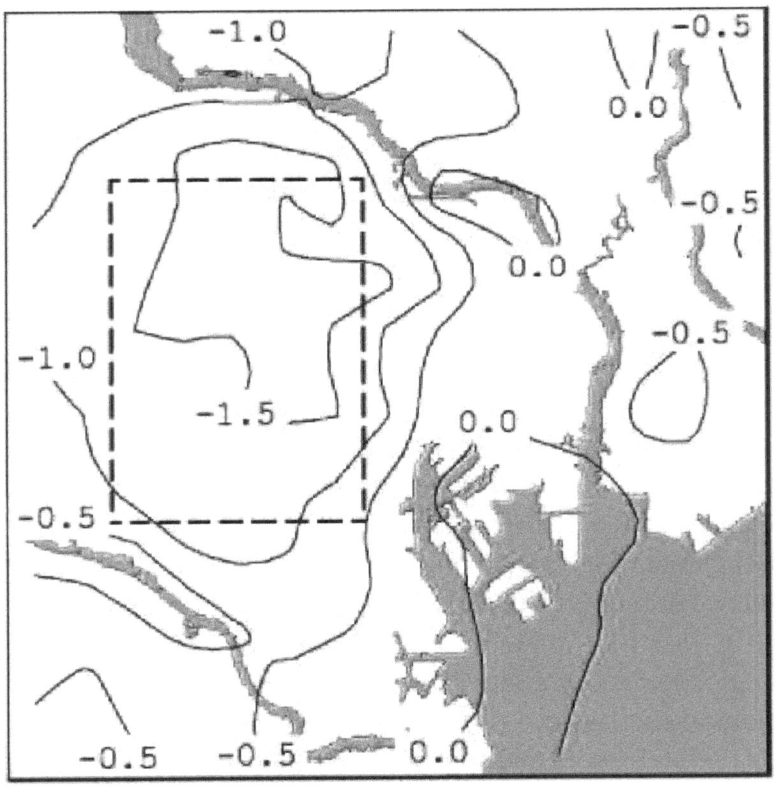

Figure 10. Temperature difference in °C between the case applying FVC data and the case not applying at 15:00 LST at 4.5 m above ground level [17]. Negative values show cooler temperatures under Case RS.

4. Estimation of past urban climate

About 120 years have passed since meteorological observations began in Asia; since then, researchers have used various methods to reconstruct the climate of earlier times. One such method has been to use notations about weather in ancient journals and documents. Since the early modern times, the Japanese have left many usable records, and, although these are mostly qualitative accounts, researchers have learned how hot in summer, or how cold in winter, it was in the past. Although the information collected from these documents contains many spatial and temporal gaps, recent advances in computerized meteorological modelling have made it possible to compute local distributions of climatologic parameters; for example, the distribution of surface air temperature is determinable if the surface boundary conditions, like the land-use distributions, are known [12, 15]. The author used numerical simulations with a mesoscale climate model referenced to digital land-use data (2 km grid) covering all of Japan to attempt to isolate the influence on surface air temperature of regional warming related to land-use change during a recent 135-year period [18]. During this period, areas of regional warming related to land use changes expanded around Osaka and Tokyo. However, the validation of the results of this study was difficult because the amount of available data on long-term climate fluctuations was limited.

To validate the results of modelling such as the author's [18], observed climatologic data (e.g., air temperature) are needed to provide initial values for the modelling. The weather stations' network and other observation facilities that have been established throughout Japan since 1876 provide long-term temperature fluctuations to the present. Data for Tokyo have often been used to associate changes in local climate in conjunction with urbanization [19]. Many instances of global warming [20] and changes of air temperature due to the regional climate change have been identified over broader areas and longer periods than those caused by local land-use changes [21]; comparing these to modeled temporal changes in air temperatures may elucidate the contribution of land-use change to climate change in Japan. Comparisons like this will also help to overcome the problem of spatial and temporal discontinuity in historical climatologic information.

Another reason such research is important in Japan is that urbanization in large Japanese cities has increased local air temperatures by about 1°C during the twentieth century [22]. Until now researchers have relied on statistical methods to distinguish between the influence on air temperature of localized warming caused by urbanization, and the influence of global warming and other broader scale changes of air temperature [23]. Of course, such methods cannot be applied to time periods or regions without any observed data. Much pioneering research has eliminated the influence of urbanization as noise when estimating temperature changes over large areas [24]. The use of a mesoscale model with past land-use distributions as surface boundary conditions to simulate past climatologic parameters (e.g., air temperature distributions) might help to distinguish the influence of localized warming due to urbanization from that of global warming and regional-scale (e.g., Far East Region, Eastern Asia, etc.) fluctuations of air temperature for time periods and regions without enough observed data [18]. Certainly, the urban-scale and mesoscale meteorological model have been widely used in other studies, such as Kusaka et al. [25] and Doan et al. [26], which created the weather research and forecasting model (WRF) to examine the phenomena of urban climate in different scales.

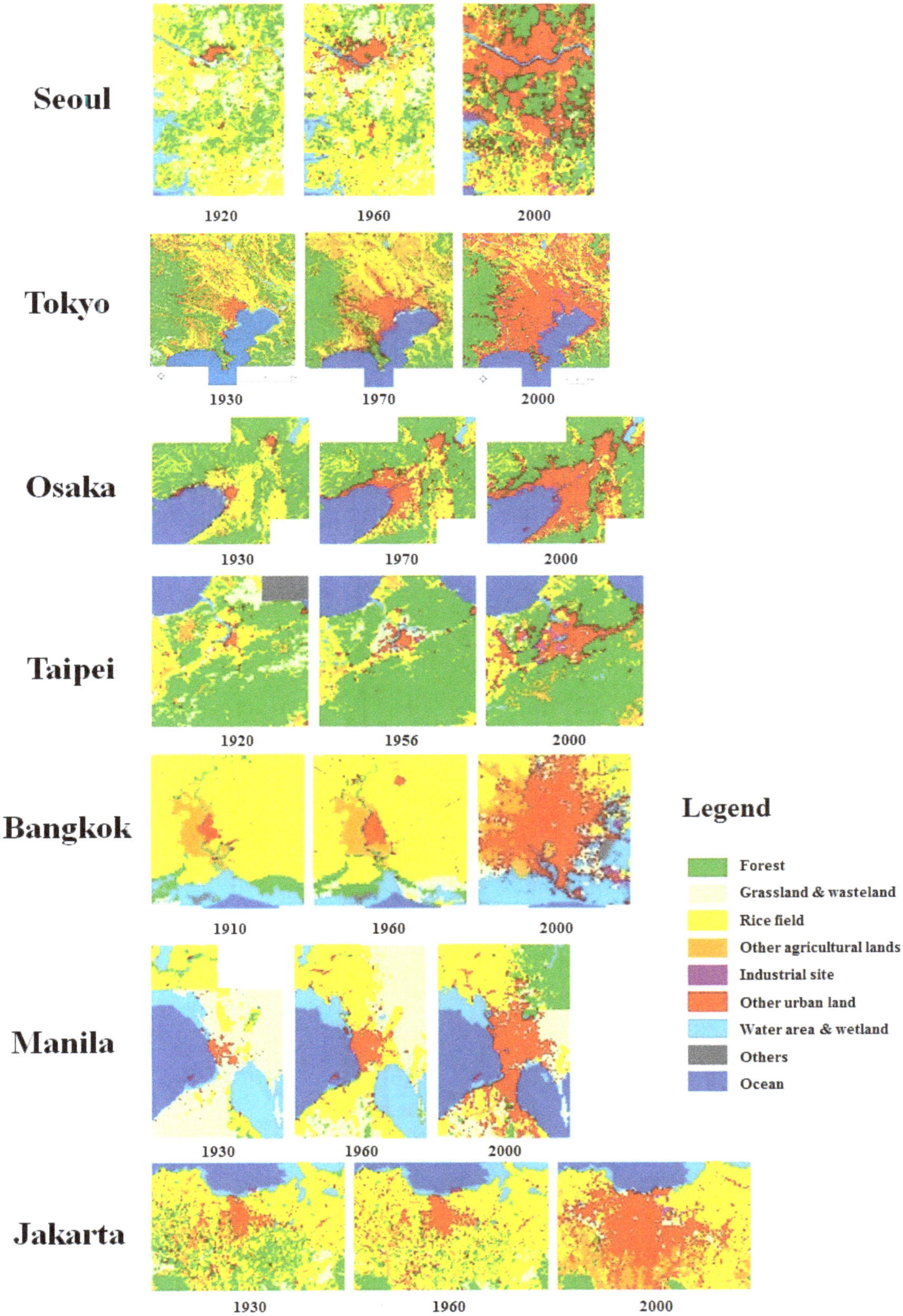

Figure 11. Urbanization around seven megacities in Asia (RIHN Research Project C-5 "Human Impacts on Urban Subsurface Environments") [28].

Case studies of the area around the upper reaches of the Rhine River as it was in 1710 [27] and of the Edo area (Tokyo was called "Edo" until the middle of the nineteenth century) during the first half of the nineteenth century [18] are examples of attempts to quantify the influence of urbanization-induced changes in the surface heat budget on surface air temperatures, as opposed to research from a historical climatology perspective. These researchers derived land-use distributions for the periods they studied from historical maps and other historical sources. Because the periods they studied predate the beginning of meteorological records, comparisons with observed meteorological data are impossible.

The author is focusing on both the past ground surface temperatures (T_{sfc}) of 100 years ago and the current one, in order to clarify the relationship between the urban development and subsurface warming in seven Asian megacities based on the numerical simulation of local climate, considering the land-use data in these megacities for three discrete years of the twentieth century [28], because subsurface temperatures are affected by surface warming (heat of the ground level is conducted downwardly). The author used the Colorado State University Mesoscale Model (CSU-MM) [29, 12, 18] and digital land-use data (2 km grid) from seven Asian megacities (Seoul, Tokyo, Osaka, Taipei, Bangkok, Manila, and Jakarta) to simulate T_{sfc} for three discrete years of the twentieth century (**Figure 11**).

Based on the vertical profile of subsurface temperature in boreholes, which is regarded to record the past T_{sfc}, Taniguchi et al. [30] identified surface warming during the twentieth

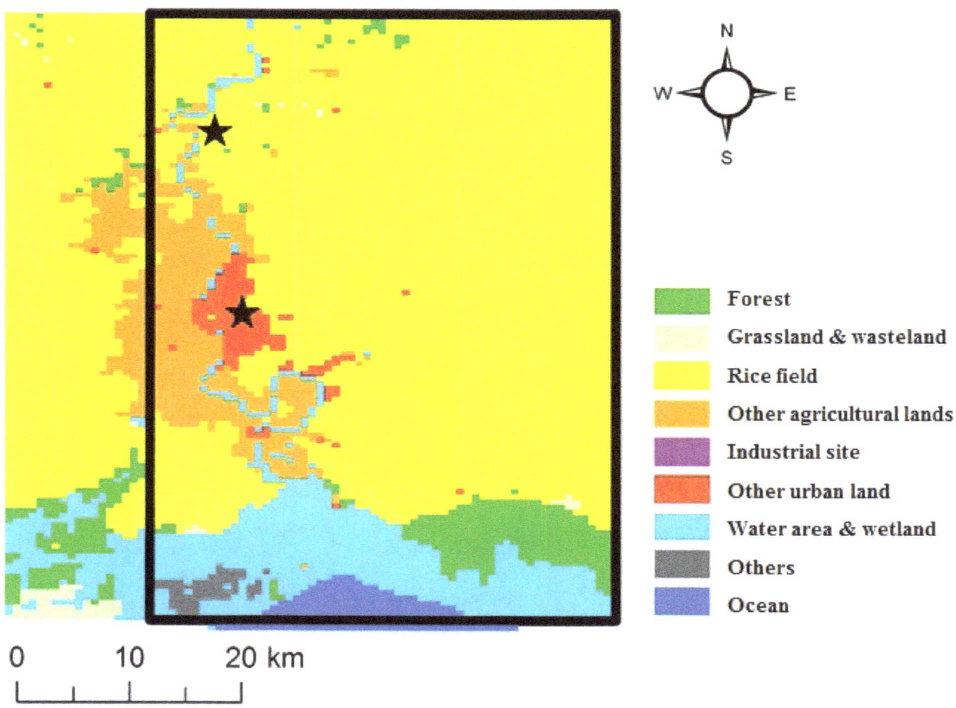

Figure 12. Land-cover/use around Bangkok in 1910 with its legend [28]. Stars show "City Center" and "Northern Suburb". Inside of solid rectangle line shows the areas displaying computed results in **Figure 13**.

century of 2.8°C in Tokyo and 1.8°C in Bangkok. The seven land-use data sets (**Figure 11**) revealed the major characteristics of land-use change during the twentieth century expansion of urbanized areas in the seven megacities considered here. Such large-scale surface changes would be expected to bring about localized climatic changes in response to the resultant changes to the surface heat budget. In the modelling on a calm and clear day of the hottest season, the central Bangkok showed T_{sfc} to be increasing by about 1.1 K 100 y^{-1} during the twentieth century [28] (**Figures 12** and **13**).

There is almost no difference between the graphs of diurnal change of T_{sfc} between 1960 and 2000 in the city center of Bangkok (**Figure 14**), whereas in the northern suburbs of Bangkok, where land use between 1960 and 2000 has changed from rice paddy to urban, there has been a rapid increase of T_{sfc}. In particular, T_{sfc} in the northern suburbs of Bangkok in 2000 was higher than in 1960, by about 15 K in the daytime and by about 5 K at dawn. Thus, it appears that subsurface temperature profiles differ according to the start time of urbanization periods.

Thus, the author's approach is available not only to evaluate the current urban environment but also to the past one. If a certain scenario of future's urban land use and urban activity, this approach will be available for the future prediction.

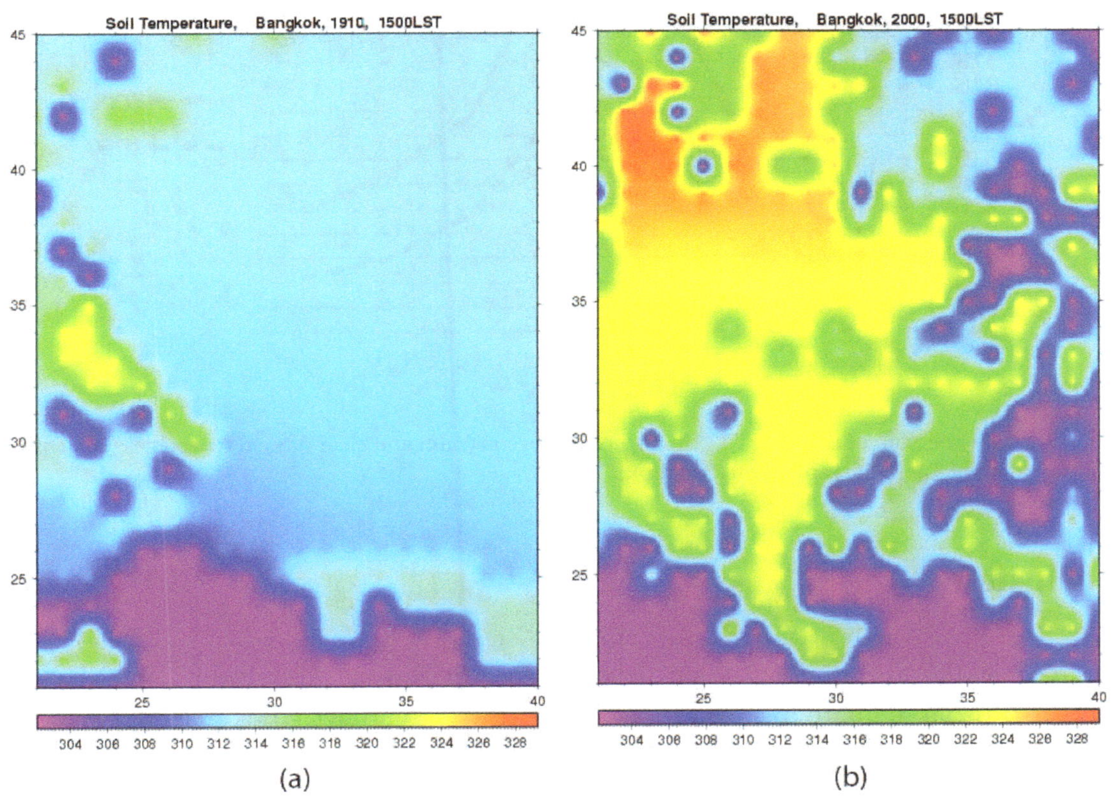

Figure 13. Computed T_{sfc} (K) around Bangkok in the hottest season (calm and clear day at the end of March) [28]. Left: 3 p.m. in 1910, right: 3 p.m. in 2000.

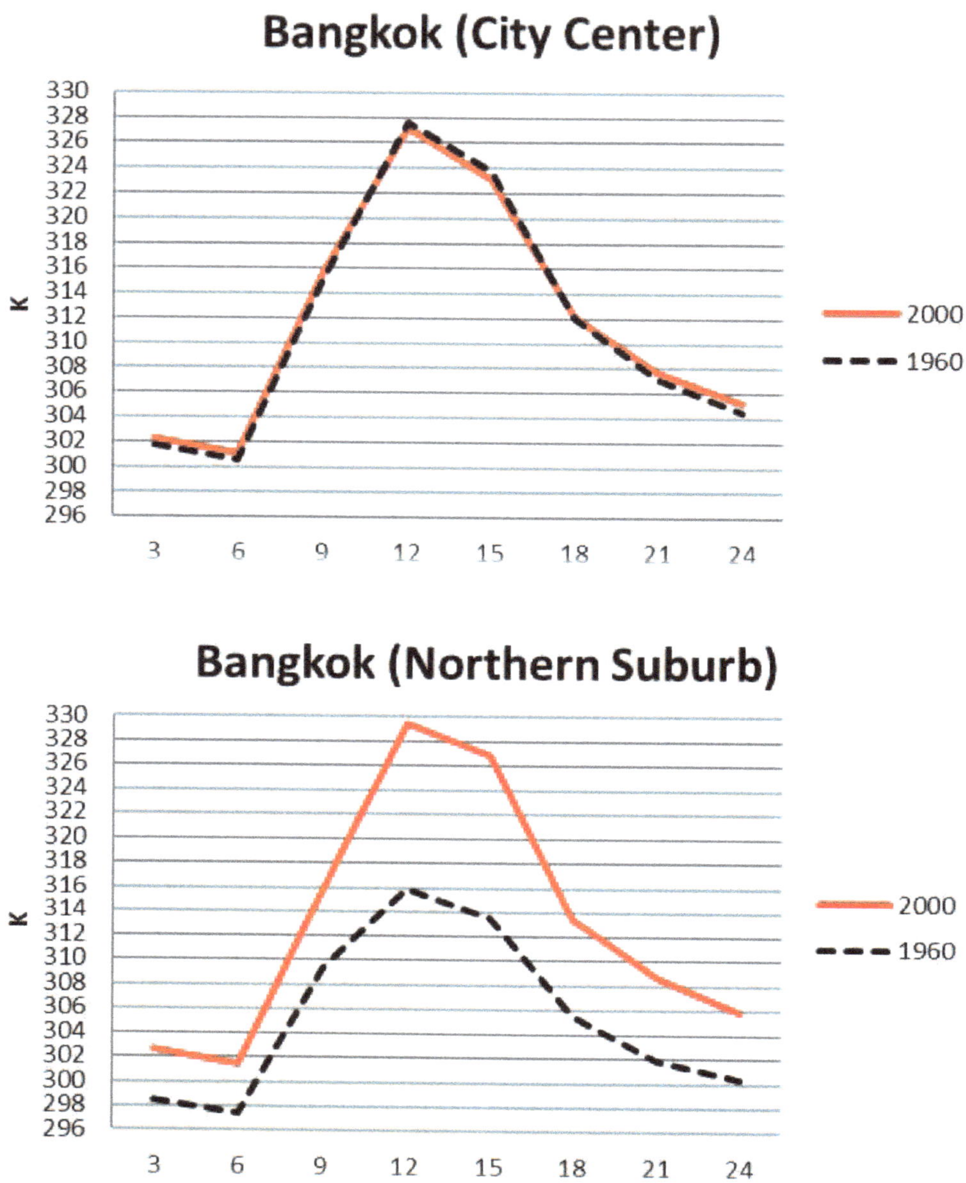

Figure 14. Diurnal variability of computed T_{sfc} in the hottest season (calm and clear day) in two stages [28]. Difference of computed warming of T_{sfc} in the same term (Bangkok).

5. Conclusion

In this chapter, the author focuses on urban area and shows case studies on urban energy planning and urban climate induced by concentrated human activities like energy use based on geographic information like land use data and numerical climate model linked to GIS. The author shows the following four outcomes: (1) In finding the candidate places to settle district heat supply facilities that make effective reuse of heat obtained from sewage, the usage of GIS gives the reasonable solution in scientific view point, (2) The spatiotemporal data in high resolution on human activity like energy consumption in urban area enable us to evaluate the impact of anthropogenic heat to urban climate by a numerical simulation model of the local climate, (3) The recent precise geographic information of urban vegetation cover derived from RS data enables us to evaluate the contribution of urban green to cool our community with higher accuracy, (4) The author's approach is available not only to evaluate the current urban environment but also to the past one. If a certain scenario of future's urban land use and urban activity, this approach will be available for the future prediction. Thus, these outputs will give some sustainable urban design in viewpoint of climate change (mitigation and adaptation).

Acknowledgements

The author sincerely thanks to Mr. Kai Liu, Research Assistant of National Institute for Environmental Studies for his great contribution to the editorial process of this manuscript. The author states that some contexts and figures of this manuscript come from his previous publications like [4, 12, 17, 28] and are permitted to use.

Conflict of interest

The author declares there is no conflict of interest on this manuscript.

Author details

Toshiaki Ichinose

Address all correspondence to: toshiaki@nies.go.jp

National Institute for Environmental Studies/Nagoya University, Tsukuba/Nagoya, Japan

References

[1] Wolman A. The metabolism of cities. Scientific American. 1965;**213**(3):179-190

[2] Kennedy C, Cuddihy J, Engel-Yan J. The changing metabolism of cities. Journal of Industrial Ecology. 2007;**11**(2):43-59

[3] Bureau of Sewerage, Tokyo Metropolitan Government. Master Plan of Sewerage System in the Second Generation; 1992 [in Japanese]

[4] Ichinose T, Kawahara H. Regional feasibility study on district sewage heat supply in Tokyo with geographic information system. Sustainable Cities and Society. 2017;**32**:235-246

[5] Connolly D, Lund H, Mathiesen BV, Leahy M. A review of computer tools for analysing the integration of renewable energy into various energy systems. Applied Energy. 2010; **87**(4):1059-1082

[6] Manfren M, Caputo P, Costa G. Paradigm shift in urban energy systems through distributed generation: Methods and models. Applied Energy. 2011;**88**(4):1032-1048

[7] Gormally AM, Whyatt JD, Timmis RJ, Pooley CG. A regional-scale assessment of local renewable energy resources in Cumbria, UK. Energy Policy. 2012;**50**:283-293

[8] Underwood CP, Ramachandran J, Giddings RD, Alwan Z. Renewable-energy clusters for remote communities. Applied Energy. 2007;**84**(6):579-598

[9] Byrne J, Zhou A, Shen B, Hughes K. Evaluating the potential of small-scale renewable energy options to meet rural livelihoods needs: A GIS- and lifecycle cost-based assessment of western China's options. Energy Policy. 2007;**35**(8):4391-4401

[10] Yue CD, Wang SS. GIS-based evaluation of multifarious local renewable energy sources: A case study of the Chigu area of southwestern Taiwan. Energy Policy. 2006;**34**(6):730-742

[11] Gils HC, Cofala J, Wagner F, Schöpp W. GIS-based assessment of the district heating potential in the USA. Energy. 2013;**58**:318-329

[12] Ichinose T, Shimodozono K, Hanaki K. Impact of anthropogenic heat on urban climate in Tokyo. Atmospheric Environment. 1999;**33**:3897-3909

[13] Bureau of Urban Planning, Tokyo Metropolitan Government. 1986. [in Japanese]

[14] Bureau of Environmental Protection, Tokyo Metropolitan Government. 1991. [in Japanese]

[15] Kimura F, Takahashi S. The effects of land-use and anthropogenic heating on the surface temperature in the Tokyo metropolitan area: A numerical experiment. Atmospheric Environment. 1991;**25B**:155-164

[16] Stanhill G, Kalma JD. Solar dimming and urban heating at Hong Kong. International Journal of Climatology. 1995;**15**:933-941

[17] Hirano Y, Yasuoka Y, Ichinose T. Urban climate simulation by incorporating satellite-derived vegetation cover distribution into a mesoscale meteorological model. Theoretical and Applied Climatology. 2004;**79**:175-184

[18] Ichinose T. Regional warming related to land use change during recent 135 years in Japan. Journal of Global Environment Engineering. 2003;**9**:19-39

[19] Yoshino M. Development of urban climatology and problems today. Energy and Buildings. 1990/1991;**15/16**:1-10

[20] Jones PD. Hemispheric surface air temperature variations: Recent trends and an update to 1987. Journal of Climate. 1988;**1**:654-660

[21] Maejima I, Nogami M, Aoyama T, Oka S, Tagami Y, Hogetsu T, Sugihara Y, Yamashita S, Nishiyama Y, Umemoto T. Recent climatic change and urban growth in Tokyo and its environs. Geographical Reports of Tokyo Metropolitan University. 1980;**14/15**:27-48

[22] Fujibe F. Temperature rising trends at Japanese cities during the last hundred years and their relationships with population, population increasing rates and daily temperature ranges. Papers in Meteorology and Geophysics. 1995;**46**:35-55

[23] Kato H. A statistical method for separating urban effect trends from observed temperature data and its application to Japanese temperature records. Journal of the Meteorological Society of Japan. 1996;**74**:639-653

[24] Kukla G, Gavin J, Karl TR. Urban warming. Journal of Climate and Applied Meteorology. 1986;**23**:1265-1270

[25] Kusaka H, Chen F, Tewari M, Dudhia J, Gill DO, Duda MG, Wang W, Miya Y. Numerical simulation of urban heat island effect by the WRF model with 4-km grid increment: An inter-comparison study between the urban canopy model and slab model. Journal of the Meteorological Society of Japan. 2012;**90B**:33-45

[26] Doan Q, Kusaka H, Ho QB. Impact of future urbanization on temperature and thermal comfort index in a developing tropical city: Ho Chi Minh City. Urban Climate. 2016;**17**:20-31

[27] Lenz JC. Energieumsetzungen an der Erdoberfläche in Gegliedertem Gelände. Zur Erlangung des akademischen Grades eines Doktors der Naturwissenschaften von der Fakultät für Physik der Universität (TH) Karlsruhe genehmigte Dissertation; 1996 [in German]

[28] Ichinose T, Liu K. Modeling the relationship between the urban development and subsurface warming in seven Asian megacities. Sustainable Cities and Society. 2018;**38**:560-570

[29] Pielke RA. A three dimensional numerical model of the sea breezes over South Florida. Monthly Weather Review. 1974;**102**:115-134

[30] Taniguchi M, Uemura T, Jago-on K. Combined effects of urbanization and global warming on subsurface temperature in four Asian cities. Vadose Zone Journal. 2007;**6**:591-596

6

Formal Urban Dynamics, Policy and Implications on Urban Planning: Perspectives on Kampala, Uganda

John J. Williams and Fred Bidandi

Abstract

Formal urban dynamics is a holistic approach that contributes towards the delivery of relevant planning solutions for cities and towns. This chapter discusses Kampala's shifting urban dynamics and their implications on planning. It argues that the current legal, political, technical, financial and administrative dynamics are problematic in nature and generally have dynamic effects on the city's planning trajectory. Uganda Bureau of Statistics (UBOS) indicates that Kampala's population has grown from 1,189,142 in 2014 to 1,583,000 in 2017, growing at 1.74% per annum. Like other cities in Africa, Kampala presents enormous challenges to urban planners, city government, local leaders and city dwellers. Watson elucidates that rapid urbanisation experienced in Africa today and Kampala in particular requires radical planning approaches in order to address the much-needed services such as water, health, waste management and sanitation. This is an empirical study with a quantitative sample of 720 households proportionally distributed according Kampala's five divisions. Qualitative data were analysed using narrative and thematic techniques, complemented by the descriptive method. The objective of this study is to investigate formal dynamics responsible for Kampala's urbanisation from 1990 to 2013 and their policy implications on planning. The findings explain Kampala's planning challenges, government modernisation agenda, legal framework, urban policy dynamics and government interventions.

Keywords: formal-urban dynamics, policy, urban planning, Kampala, Uganda

1. Problem statement

Formal dynamics such as legal, political, technical, financial and administrative are critical to investment decisions, business/commercial and residential choices made by individuals and investors [1, 2]. These dynamics have received less attention in respect of their role in influencing urban changes according to needs of urban dwellers. They are necessary to understand because they explain the manner in which cities urbanise in terms of planning. Formal dynamics also determines how city planners and administrators react to urban changes resulting from formal dynamics [3]. This reaction subsequently determines how a city urbanises systematically in terms of attracting physical investments, housing and other related programmes. This is why it is necessary to investigate so as to understand its policy implications for Kampala's urbanisation.

Apart from the existing scholarly work, Kampala's urban policies and legal instruments are enacted with intent to ensure the city urbanises in a planned manner [4–6]. Why then are those used in Kampala failing to do so? This study seeks to answer this question by analysing formal dynamics in terms of the flaws that cause the urban policies and legal instruments applied to guide the city's planning trajectory.

2. Introduction

Though conditions vary from one city to another, the increase in urban populations globally and Kampala in particular put a strain on services and resources, leaving many city dwellers on the periphery in terms of access to social-economic opportunities. Scholars attribute this to rural-to-urban circular migration and migration from small towns, all of which appear considerably the same across the globe as people assume that there is good life, better employment as well as social-economic opportunities in cities. The above assumptions provide sharp differences in the urbanisation process, particularly those who dwell in informal settlement where basic services such water, sanitation, health and roads, among others, are usually limited or difficult to come by. In this light, urban public policy lags behind urbanisation, which often means that basic services, such as sanitation, water, housing, public transport and land rights to name but four, are not delivered to many of the people living in informal settlements. People living in informal settlements have no say and are neither informed about the urban planning process. Effective urban planning in cities like Kampala therefore depends on institutions and governance. But without understanding urban policy dynamics and differences in urbanisation, the challenge ahead is the interface in service provision which has significantly contributed to unplanned urbanisation of the Kampala city [7–9].

This study discusses the nature of the dynamics experienced in Kampala, maintaining that an empirical basis is required to develop a policy that ensures that the city urbanises in an environmentally, spatially and socially satisfactory manner. Additionally, the core of the city traditionally accommodates many commercial and retail activities and surrounding residential suburbs which are encircled and connected by important national and metropolitan roads and it is geographically divided into the Central, Nakawa, Makindye, Lubaga and Kawempe divisions. The central business district is mostly designated for commerce and residence. All these five divisions are mixed between lower-middle class, working class and underclass [8].

Figure 1 indicates Kampala's five divisions.

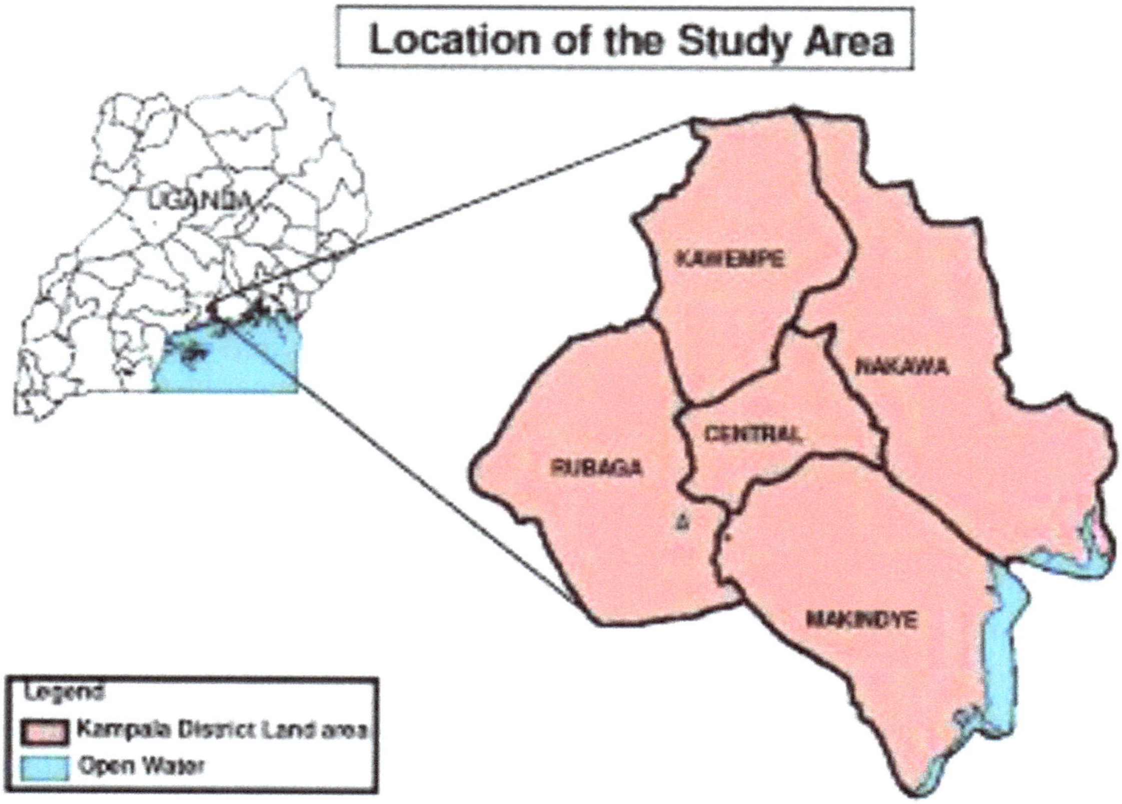

Figure 1. Location of the study area. Source: Lwasa et al. [8].

Moreover, the chapter discusses Kampala's formal dynamics as a multi-dimensional concept for urban growth between 1990 and 2013 [6].

3. Theoretical framework

For the purpose of this chapter, the modernisation theory has been utilised to underpin Kampala's urbanisation. In so doing, the theory explains systematic transformation or progressive transition from pre-modern or traditional subsistence economies to modern industrialised economies [10]. This school of thought maintains that subsistence economies develop and urbanise as they adopt more modern industrial, technological, communication and cultural practices [11, 12]. Certainly, Tettey [13] observed that urbanisation varies in line with the developmental pace of a country, and for any country to urbanise, there is a need to foster development through the adoption of technology and industrialisation. A number of scholars endorse this connection by indicating that the phenomenon and process of urbanisation is an irreversible feature of modernisation and development [14, 15].

Nonetheless, modernisation theory permits responding to the dynamics that cause transformation, including urbanisation [16]. It is transformation that should be felt by the people or for whom it occurs [17]). This argument is used in this chapter to analyse and understand Kampala's planning trajectory.

4. Formal urban dynamics contextualised

In the context of this chapter, formal urban dynamics refers to the production of urbanisation that comes through organised or planned processes, usually adopted to cover needs of urban dwellers. Formal dynamics investigated in this study include legal, political, technical, financial, administrative and policy dynamics. Literature indicates that legal dynamics are the actions undertaken according to by-laws, ordinances, acts, laws, regulations, standards and guidelines enacted to prevent, prohibit or promote urbanisation desired in terms of physical infrastructural development, settlements and service delivery [18–21, 23, 24, 26].

Administrative dynamics refer to executive actions taken to ensure that urbanisation takes place as desired by city authorities [25, 27–31, 38, 39]. Meanwhile, political dynamics refer to actions taken as a result of political decisions made either by government or by the opposition to maintain, promote or discourage urbanisation as a way of promoting political interests [32]. Turning to policy dynamics, [33] observes that the urbanisation of most African countries is explained by the apparent absence of deliberate urban policies. This absence leads to uncontrolled and unplanned settlements typified by growing slums juxtaposed with urban affluence [34–37]. Using South Africa's example, McGranahan [40] observes that formal dynamics in most cities are exclusionary in terms of spaces occupied by certain categories as we have witnessed in black communities.

The reviewed literature describes formal dynamics and their negative urban consequences, but it pays little or no attention to the policy solutions that should be adopted to deal with the consequences and ensure that the cities urbanise systematically.

5. Dynamics of urbanisation with specific reference to Kampala

Literature indicates that urban policies and budgetary allocations tend to result in changes in public investment, and these changes alter economic activities, spatial quality and environmental quality in cities and towns [36]. The changes can be positive or negative, depending on the promoted public investments and the political interests underlying the investments [36]. Annez and Buckley [42] note that governments could also pursue development in partnership with the private sector for planning for a conducive urban environment [43]. Moreover, Kampala's urbanisation is assumed to be explained by formal dynamics, which are conceptualised as all forces and processes that are officially sanctioned to cause urban changes [18]. According to Alem [44], these dynamics are important to understand because they inform urban planners on how to improve service delivery in cities. Mabogunje [33] argues that urbanisation is explained by the apparent absence of deliberate urban policies. This absence leads to unplanned settlements typified by growing slums juxtaposed with urban affluence. Besides, according to [9, 45–48], the legal dynamics creates a complicated and multiple land tenure systems which form one of the dynamics in explaining the city's unplanned urbanisation. However, constitutional recognition and protection of private land (Milo land, crown land and permanent land) ownership rights in themselves lead, for example, to unplanned urbanisation, thus leading to urban policy implications considering the city's population growth and expansion. However, literature does not provide clear policy solutions regarding legal dynamics. On an inclusive policy note, Somik [49], UN-Habitat [9] and Katembwe [50] advocate for a shift in technical planning for urbanisation in countries like Uganda to ensure systematic urbanisation.

Brown [51] and Lambright [36] observes that the planning of the National Urban Policy for Uganda (UNUP) does not pay attention to full participation of the different key stakeholders, including NGOs, CBOs, the academic community, the private sector and the different levels of government. This means that UNUP as a guiding document needs to be revisited since its implementation appears not to be pragmatic. Lambright expresses a similar view by observing that the deviation from policies leads to a complex situation given the dynamic trends of urbanisation [52–55]. This study therefore provides a comprehensive understanding of formal dynamics in relation to Kampala's urbanisation.

6. Research methodology

This is an empirical study with quantitative data comprising 720 households distributed per division as shown in **Table 1**. Quantitative data were collected using a survey method. This method was used to collect a comprehensive and consistent data from many respondents and in a relatively short period of time [57]. A self-administered, semi-structured questionnaire was used under this method. This instrument was used because Kampala's literacy rate of 91% [58] suggested that most of the respondents could read and write. The instrument was designed and administered to the selected heads of households. Focus group discussions (FGDs) were used to collect qualitative data from a group of respondents simultaneously [59]. A total of 19 respondents were interviewed, 5 of whom participated in the focus group discussions.

Divisions	Number of participant households
Central	54
Kawempe	162
Makindye	185
Nakawa	139
Lubaga	180
Total	**720**

Source of population size: Uganda Bureau of Statistics [56].

Table 1. Proportional distribution of selected household heads by divisions in Kampala city.

7. Analysis of the study findings

The selected city residents were asked to use a Likert scale of responses running from strongly disagree to disagree and not sure and agree to strongly agree to indicate whether the statements administered to them embedded formal dynamics that accounted for Kampala's urbanisation. The responses were subjected to exploratory factor analysis following Field's [60] observation by analysing the correlation between variables in a data set. The findings obtained from the analysis are summarised in **Table 2**.

7.1. Formal dynamics explaining Kampala's urbanisation (1990–2013), as reported by city residents

The findings indicate that as cities urbanise they require strong administrative systems, which ensure attention that is paid to the needs of urban residents in a satisfactory manner. However, Sachs-Jeantet [27] argues that politics appears to be a hindering factor for city planners due to bickering and misunderstandings among political office bearers. Meanwhile, Adhikari [38] points out that though investment policies pursued by governments account for how cities urbanise, implementation is usually problematic. The findings also support the observations made by World Bank [25] and John [32] that the dynamics that account for urbanisation tend to be administrative, political and legal in nature. The findings also concur with Annez and Buckley [42] and Nattrass [41], both of whom argue that the investment policy pursued by government is one of the dynamics that explains the urbanisation of cities like Kampala.

These findings suggest that the formal dynamics that account for Kampala's urbanisation are mostly administrative issues followed by politics, government investment promotion policy, and then by legal dynamics. However, the indicator that relates highest with the government

Dynamics	Indicators of dynamics	Description (N = 720)			
		Min	Max	Mean	Std.
Official administrative dynamics	It has since 1990 been easy to put up a business kiosk in Kampala city without an approved plan	1	5	4.77	.098
Official political dynamics	The incumbent government has since 1990 been releasing much more money to finance Kampala's budgets only when the city's leaders are politically affiliated to the ruling party	1	5	4.56	.113
Government investment promotion policy dynamics	Establishing a foreign-based company anywhere in Kampala city has since 1990 been easy because of government policy of attracting foreign investors	1	5	4.84	.044
Legal dynamics	My right of ownership of the land where I built my residence has never been tampered with throughout the period 1990–2013 because it is constitutionally protected	1	5	3.56	3.918

Source: Bidandi [61].

Table 2. City residents' description of the main indicators of formal dynamics explaining Kampala's urbanisation.

investment promotion policy is one which involves permitting the establishment of foreign-based companies anywhere in Kampala city as a way of attracting foreign investors without considering the implications on the city's planning process. The indicator that related highest with legal dynamics involves not tampering with city residents' right of ownership of the land where they built their residences, including informal settlements, because of being constitutionally protected. Since these are the indicators that relate highest with each of the identified formal dynamics, they are further investigated to establish how city residents perceive the manner in which they occur.

Findings in **Table 2** indicate that to city residents, the dynamics that strongly explained Kampala's urbanisation during the period 1990–2013 include administrative dynamics manifested in the form of people finding it easy to establish business kiosks without approved plans. This suggests that there is administrative laxity in terms of enforcing the construction of kiosks based on approved plans. City residents further include official political dynamics that occur in the form of the incumbent government releasing much of the money required to finance Kampala's budgets only when the city's leaders are politically affiliated to the ruling party. The policy implications of these findings are discussed together with those depicted by the findings obtained from the key informants as presented below.

Findings reflected in **Figure 2** indicate that the formal dynamics explaining Kampala's urbanisation included administrative measures, government political intervention and modernisation agenda, the legal framework and urban policy. These findings are essentially consistent with the studies of Nattrass [41], Gervase [37] and Ndengwa [20]. Each of these studies indicates that at least one of the dynamics shown in **Figure 1** explains how cities like Kampala urbanise.

The findings support the argument made by Omwenga [28] and Braun et al. [29] in which they argue that administrative dynamics involve directorial actions either to promote desired urbanisation or against urban changes deemed unnecessary, unplanned, disfiguring or leading to undesired spatial development, environmental degradation, socio-economic chaos or poor

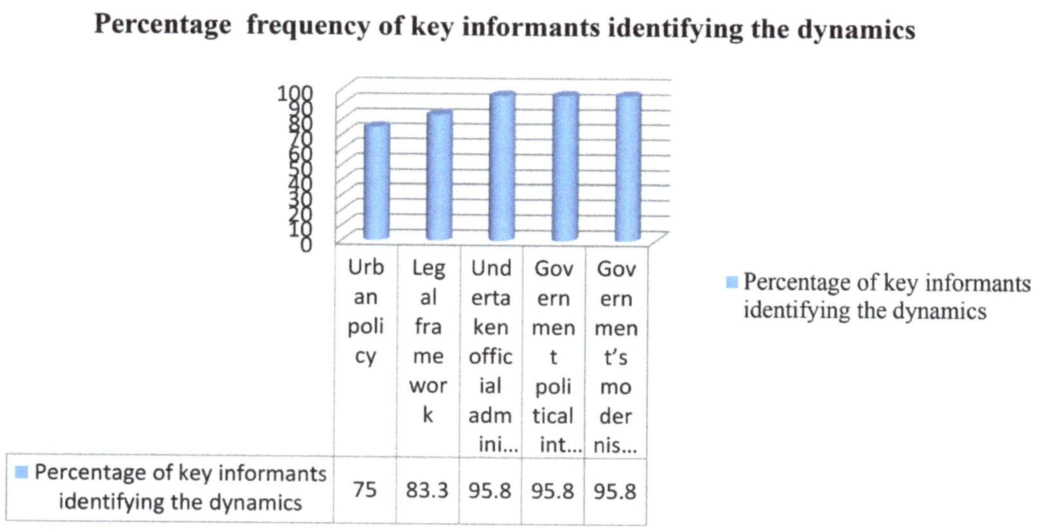

Figure 2. Frequency distribution of formal dynamics explaining Kampala's urbanisation from 1990 to 2013, as reported by key informants (policy makers & implementers). Source: Bidandi [61].

environmental health. Most of Kampala's urban dynamics are a clear manifestation of unplanned urbanisation due to the administrative weaknesses working against their effectiveness. The weaknesses include deliberate indifference, lack of strategic focus, working politically rather than professionally, staff demoralisation and vulnerability to bribery and failure to pay attention to social justice when executing many of the official administrative measures.

As the study findings indicate, this weakness needs to be minimised by ensuring that, instead of redeveloping Kampala based on only an inward-looking strategy, the city is reorganised while kampala capital city authority (KCCA) is working with local governments in the neighbouring districts. This is well explained by You-Tien [62] when discussing governance and planning of mega-cities. You-Tien [62] vividly provides that some level of satisfaction needs to be reached by city residents especially when urban changes taking place in their locations help urban physical planners, policymakers and implementers to improve locations, thereby promoting urbanisation that meets residents' expectations.

7.1.1. Government intervention

Findings from the interviews indicate that the nature of government intervention programmes, such as decentralisation and creation of more administrative units, contributes to Kampala's haphazard urbanisation. The findings confirm the observation made by Sulkin and Larsen [63] that government can halt planned urbanisation in order to derive political capital. The finding also confirms King and Wybrow's [64] observation that government intervention of this kind usually occurs after noticing that the urbanising actions taken administratively are politically costly. The review led to the findings summarised in **Figure 3**.

The fiscal years in **Figure 3** begin from 1996/1997 because, after staying in limbo since 1980, multiparty politics was rejuvenated in Uganda in 1996. The trend in **Figure 3** indicates that

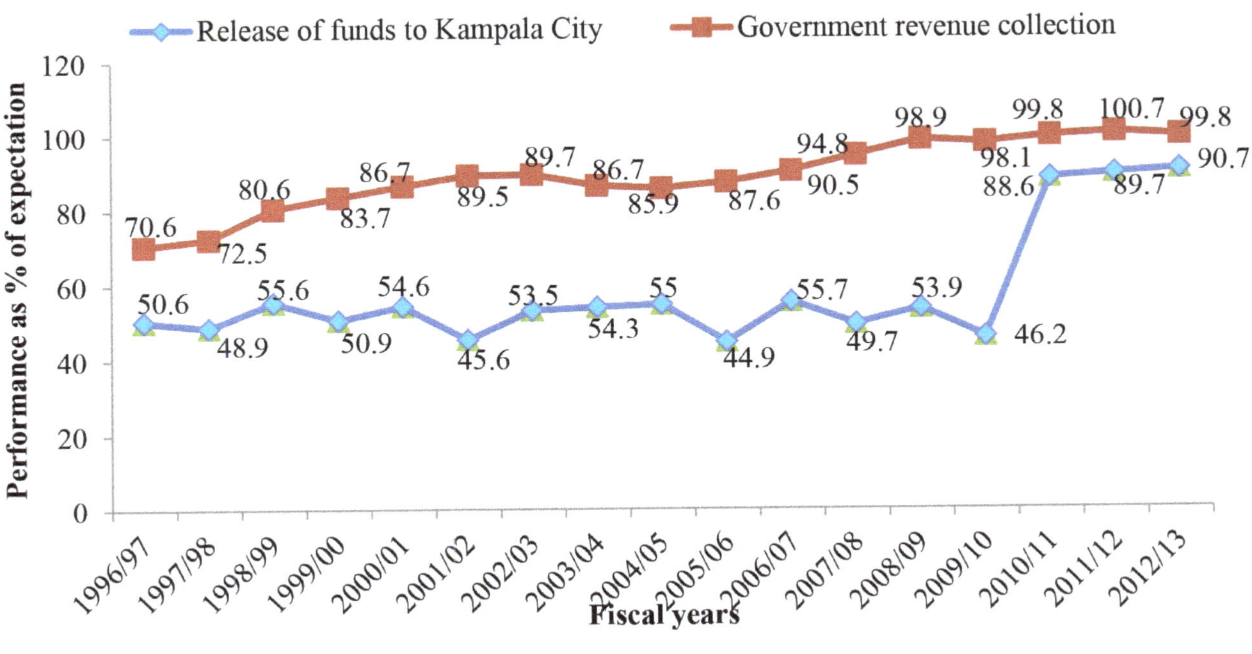

Figure 3. Performance of government revenue collection and financial releases to KCCA as percentages of expectation. Source: Background to the budget and Kampala city budgets for the shown fiscal years.

from the 1996/1997 to the 2009/2010 fiscal year, the government of Uganda released between 49.7% and 60.7% of the budgets proposed to finance Kampala's urbanisation activities. During this time, Kampala's executive leadership was in the hands of the opposition [65]. During the period 2011/2012–2013/2014, the release performance rose and started oscillating between 88.6% and 90.7%. This period coincides with the time when the ruling party under the president's office [66] directly controlled Kampala's executive leadership. Evidently, the proportion of funds that government releases to facilitate the urbanisation of Kampala depends on whether the city's executive leadership is under the ruling party or the opposition.

Since Uganda operates a cash budget [67], the performance and provision of services depend on the revenue mobilised or funds mobilised by Kampala city authorities. This argument, however, becomes untenable when released funds are compared to revenue collection. The trend in **Figure 2** indicates that the performance of expected revenue collection generally increased throughout the period 1996/1997–2012/2013. It should, however, be noted that as a matter of policy government, intervention into the urbanisation of Kampala does not need to depend on the political interests of the ruling party but on national interests.

7.1.2. Government modernisation agenda

Thematic and descriptive analysis of the views given by key informants to substantiate the modernisation agenda pursued by the Ugandan government is one of the main dynamics explaining Kampala's urbanisation that leads to the results shown in **Figure 4**.

A scrutiny of the study findings reveals that Kampala's urbanisation is being influenced by the modernisation agenda and as a result expanding the city's industrial, communications, power and trade sectors.

These findings support [38] who indicates that modernisation boosts industrialisation. They also support the observations made by [37] that modernisation alters the already existing infrastructure, communication networks and supply of social services. For Kampala, findings in **Figure 3** indicate that this alteration takes the form of improving physical infrastructure and attracting foreign investors' expanding businesses. The expansion adds to what Harvey [68, 69] refers

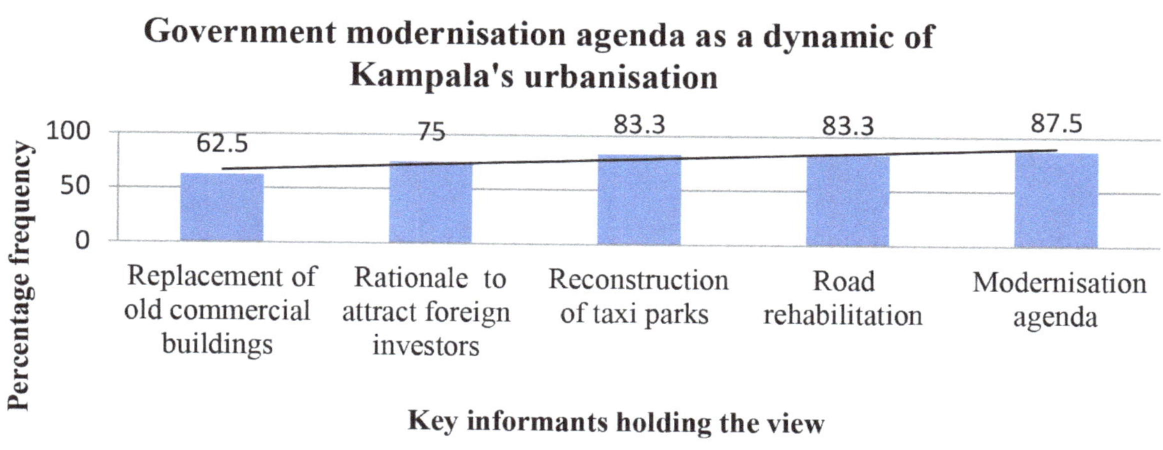

Figure 4. Frequency distribution of the views substantiating the government modernisation agenda as a dynamic of Kampala's urbanisation. Source: Bidandi [61].

to as "built environment". However, encouraging foreign investors to establish businesses in Kampala in environmentally undesirable locations implies contributing to environment degradation. Not only does degrading the environment spoil the natural beauty of a city, but it also contributes to adverse climatic change, which, according to [70], is a threat to life.

7.1.3. Legal framework

Table 3 indicates the legal framework dynamics key respondents identified that explains Kampala's urbanisation. These respondents supporte d their view by citing a number of legal instruments that had been enacted to guide what to do and how to do it in order to regulate and control all the activities and services by which urbanisation is to take place in Kampala.

The frequency distribution in **Table 3** indicates that all the key informants (100%) mentioned the 1995 Constitution of the Republic of Uganda and the 2010 Kampala Capital City Act as legal instruments that had been enacted to regulate and control the conducting of urban activities and services in Uganda, particularly in Kampala. Findings indicate that the legal framework that determines how Kampala is urbanising includes the 1995 Constitution of the Republic of Uganda, as well as the different acts, regulations and ordinances passed at different times. The findings support the observations made by [18, 21, 22, 25, 26]. Each of these authors cites at least one of the legal instruments shown in **Table 3** observing that they are the instruments that regulate and control the legal actions by which a city urbanises.

Legal instrument	Percentage of key informants identifying the instrument (N = 24)
Electricity Act, 1999	33.3
Market Act, Cap 94	50.0
National Physical Planning Standards and Guidelines, 2011	62.5
Water Act 1997, Cap 152	50.0
The Kampala Capital City (Taxi Management) Ordinance, 2013	50.0
The Uganda National Roads Authority Act, 2006	87.5
Local Government Act, 1997	87.5
Solid Waste Management Ordinance, 2005	95.8
National Environment Act, Cap 153	79.2
The Condominium Property Act, 2001	50.0
The Physical Planning Act, 2010	79.2
Land Act, 1998	62.5
Kampala Capital City Act, 2010	100.0
The Constitution of the Republic of Uganda, 1995	100.0

Source: Bidandi [61].

Table 3. Frequency distribution of legal instruments explaining Kampala's urbanisation as revealed by key informants.

A review of the Kampala Capital City Act, 2010, reveals that it mandates KCCA to urbanise Kampala in a planned manner. However, the powers to enforce and monitor this urbanisation are not clearly demarcated among the top-most recognised officials who include the minister for the presidency, the executive director and the Lord mayor.

What this minister, mayor and director should respectively do with respect to the approval, supervision, monitoring and evaluation of the implementation of Kampala city's urban development policies is not clearly demarcated. This constrains the implementation of the Act, thereby rendering the realisation of its intended purpose (the proper urban development of the city) difficult. This is exacerbated by the fact that the Act takes the implementation and control of this function in an omnibus way, which separates who should implement and who should control, thus making it difficult to effectively address Kampala's urban dynamics [4, 71]. One key informant substantiated the above situation as follows:

> *Proper urbanisation of Kampala City is difficult to achieve when Uganda's constitution recognises and protects privately owned land anywhere in Uganda, even within the capital city. While it is in order to protect people's rights in land, it works against planned development and urbanisation when ownership is not conditioned to land development. The Constitution of Uganda recognises Mailo (individually owned), customary, freehold and leasehold as the four systems of land tenure, but most of the land in Kampala is Mailo or customarily owned by Buganda Kingdom. Private land owners determine what to do with their land. They choose which activities to carry out on their land. Some people even put up buildings without approved plans, and the law of the land does not allow demolition of any building erected on privately owned land. Some of the Mailo land is owned ancestrally and current occupants are too poor to develop it. Even when KCCA makes efforts to compensate these owners on a negotiated basis, they set quite high prices. Some land is owned by the Buganda Cultural Institution, and the process of compensating this institution faces a lot of cultural resistance. In fact, most of the areas in Kampala City are informally developed or left undeveloped because of the constitutional protection of private ownership of land. officials in the Physical Planning Department on March (Interview held in February with KCCA officials in the Physical Planning Department on March 5, 2014).*

The preceding findings indicate that the constitutional protection of private land ownership is one of the dynamics that explained the uneven and informal urbanisation of Kampala between 1990 and 2103. The findings therefore concur with the observations made by [9, 46–48] in which, for example, [46, 48] indicate that the Constitution of the Republic of Uganda and all land acts recognise and protect private land ownership as opposed to land development. UN-Habitat [9] argues that these legal dynamics create complicated and multiple land tenure systems which form one of the dynamics explaining the city's unplanned urbanisation. The multiplicity of the tenure systems is also highlighted by [47] as a major factor underlying urbanisation. However, these authors analyse these dynamics as a basis for developing a comprehensive strategy and action plan for only slum upgrading (UN-Habitat), dealing with informal settlements (Lwasa) and improving access to housing (Mukiibi).

7.1.4. Urban policy dynamics

The study findings indicate that urban policy is in the form of administrative, political, modernisation and legal and policy dynamics. The administrative dynamics occur mainly in the form of directorial measures that are undertaken to deal with activities that are deemed illegal or contribute to the urbanisation process. The effectiveness of these dynamics is, however, compromised by a number of dynamics which include Kampala city authority officials' deliberate indifference to developments that are being established under the unofficial influence of high-ranking government officials. Other weaknesses included the fact that some Kampala city officials are working politically rather than professionally. There is also low staff motivation which has encouraged vulnerability to bribery and laxity that job makers exploited to erect business kiosks in Kampala without approved plans. Since these dynamics compromise the effectiveness of the official administrative dynamics that are meant to promote planned urbanisation, they need to be curtailed through adopting policy measures that can empower not only KCCA to operate independently of politicians but also its employees to become vigilant and invulnerable to bribery. This view was expressed by one respondent as follows:

> *The policy used to guide Kampala's urbanisation exists in fragmented pieces. We have to refer to particular policies. For instance, when the issue is about approving physical constructions, we use the physical planning policy. When the issue is about land, we apply the land policy. When it is about water supply, we appeal to the water supply and sanitation policy; when it is about energy, the energy policy is called into force, and so on. We hear that the MLHUD is developing a comprehensive urban policy. This implies that at the moment, we do not have an integrated urban policy that can be applied to guide the urbanisation of Kampala city all is waste of resources. (Interview held with a KCCA official in the Executive Director's office on January 14, 2014).*

The findings earlier support Mabogunje's [33] observation in which he points out that the urbanisation of most regional cities such as Kampala and Nairobi is explained by the apparent absence of deliberate urban policies. This absence leads to uncontrolled and unplanned settlements typified by growing slums juxtaposed with urban affluence.

To ensure that this policy is comprehensive, it is necessary to develop it based on the understanding of not only formal dynamics but also residents' satisfaction with the services rendered in the city.

8. Discussion of the research results

Findings indicate that the formal dynamics explaining Kampala's urbanisation between 1990 and 2013 include administrative dynamics, government political intervention dynamics, government modernisation agenda, the legal framework and urban policy dynamics. By revealing these forms of dynamics, the findings concur with those of [20, 37, 41]. Each of these scholars indicates that at least one of these dynamics explains how cities urbanise. For instance, [41] clarifies that cities develop as a result of social dynamics officially sanctioned in the form of observed culture, racism and social classes created by the official system of education, income distribution, property ownership and access to jobs.

The formal administrative dynamics in this study have both positive and negative features. There are those activities that are taken to ensure that Kampala urbanises in a planned manner. These include the cancellation of contracts of companies that manage the city's public transport services and market places poorly, demolition and abolition of the informal economic activities and structures, decisive dismissal and replacement of corrupt KCCA officials and the closure of arcades constructed without following the approved plans. These dynamics are intended to restore planned urbanisation in Kampala. They have, however, achieved little success because of the negative features which include administrative weaknesses. These weaknesses comprise deliberate indifference, lack of strategic focus by senior KCCA officials working politically rather than professionally, that is, staff demoralisation and vulnerability to bribery and failure to pay attention to social justice when executing many of the official administrative measures. According to Turok [72], leaving unplanned developments untouched is itself a sign of failure to enforce strategic urbanisation. Besides, staff demoralisation and vulnerability to bribery are also linked since, according to Mills [73] and Cooper et al. [74], poor motivation increases susceptibility to accepting bribes.

Government political intervention has made a twofold contribution to Kampala's urbanisation, that is, intervention to protect administrative actions that ensure Kampala is urbanised in a planned manner and developing an inclusive urban policy. It is, however, compromised by another form of government intervention, which involves halting administrative decisions that are intended to promote planned urbanisation but in a politically costly manner. The halting is tactical as it intends to derive political capital instead of serving to promote planned urbanisation in the national interest. It therefore confirms [63] who argue that the government can halt planned urbanisation in order to derive political capital. While the government's intention is to derive political capital from these latter two forms of intervention, they limit the realisation of planned urbanisation. The intentions are thus not good for Kampala's planned urbanisation, their political intentions notwithstanding. They need to be replaced by other interventions from which the government can derive political capital while promoting planned urbanisation.

The dynamics of the modernisation agenda explaining Kampala's urbanisation involves attracting investors and allowing them to establish factories and business companies at any locations of the investors' choice. These dynamics also involve reconstruction of taxi parks and roads and the replacing of old commercial buildings with new ones, including supermarkets, malls and arcades. The manner in which these dynamics occur supports [38], who indicates that modernisation boosts industrialisation. It also supports [37] who argues that modernisation alters the already existing infrastructure. Indeed, the attracted investors contribute to the expansion of Kampala's built environment as [68, 69] prefer to call it. The way the expansion is taking place is, however, environmentally unfriendly because it involves constructing the developments by the infilling of swamps, destroying greenbelts and blocking natural drainage. Not only does this degrade Kampala's natural beauty but it also contributes to adverse climatic change, which, according to [70], is a threat to life.

The legal framework dynamics occur in the form of different legal instruments that are enacted to guide Kampala's urbanisation. These instruments include acts, by-laws, ordinances and regulations.

9. Conclusion

The formal dynamics which explain Kampala's urbanisation during the period 1990–2013 include official administrative dynamics, government political intervention, a modernisation agenda implemented through a government investment promotion programme, a legal framework and urban policy dynamics. Many of these dynamics are associated with weaknesses that cause Kampala to urbanise in an unplanned manner. Consequently, the weaknesses need to be rectified. The effectiveness of formal dynamics is, however, compromised by administrative weaknesses, politics and land tenure system.

Author details

John J. Williams and Fred Bidandi*

*Address all correspondence to: 2257858@myuwc.ac.za

University of the Western Cape, Cape Town, South Africa

References

[1] Uganda Bureau of Statistics. Household Census Survey, 2017: Population Report. Kampala: UBOS; 2017

[2] Watson V. The planned city sweeps the poor away…: Urban planning and 21st century urbanisation. Progress in Planning. 2009;**72**(3):151-193

[3] Aulia DN, Ismail AM. Residential satisfaction of middle income population: Medan city. Procedia-Social and Behavioral Sciences. 2013;**105**:674-683

[4] Ministry of Lands, Housing and Urban Development. Issues Paper for the Development of the National Urban Policy. Kampala: CME Consult Group Ltd.; 2013

[5] Kapala Capital City Authority. Solid Waste Management Policy and Strategy. Kampala: Kampala Capital City Authority; 2010b

[6] Kiechel W. The Lords of Strategy. New York: Harvard Business Press; 2010

[7] Ministry of Lands, Housing, and Urban Development. Uganda National Urban Policy (Draft 1). Kampala: Ministry of Lands and Urban Development; 2013

[8] Lwasa S, Nyakaana JB, Senyendo H. Population, Urban Development and the Environment in Uganda: The Case of Kampala City and its Environs. Kampala Paris, Unesco: Makerere university; 2007

[9] UN-Habitat. What are slums and why do they exist? Documento presentado en la Vigésima primera Sesión del Consejo de Administración; 2007. pp. 16-20

[10] Sheri B. Modernization in historical perspective: The case of imperial Germany. World Politics. 2001;**53**(3):431-462

[11] Schwab K, Xavier S. The Global Competitiveness Report. Geneva: World Economic Forum; 2014

[12] Milkovich GT. Organization and management in the midst of societal transformation: The People's Republic of China. Organization Science. 2004;**15**(2):133-144

[13] Tettey C. Urbanisation in Africa in relation social-economic development: A multifaceted quantitative analysis [PhD thesis]. University of Akron; 2005

[14] Usman B. Confronting the haphazard urbanization challenges in Anambra State of Nigeria through urban structure planning/re-planning and implementation. Settlements Discussion Paper Series No.4; 2011

[15] Sano JC. Urban infrastructure in Kigali city, Rwanda: Challenges and opportunities for modernized decentralized sanitation system in poor neighbourhoods [MSc thesis]. Wageningen University; 2007

[16] Harding R. Urban-rural differences in support for incumbents across Africa. Afrobarometer Working Paper No. 120; 2010

[17] Fox SR. The political economy of urbanisation and development in sub-Saharan Africa [PhD thesis]. Department of International Development of the London School of Economics; 2013

[18] Sullivan E, Richter CA. The Legislature Makes Significant Structural Changes to the Oregon Land Use Program, New York: Garvey Schubert Barer; 2013

[19] Carter T. Panhandling in Winnipeg legislation Vs support services, Volume 2: Literature and legislation review. A Study for the Public Interest Law Centre; 2007

[20] Ndengwa S. Decentralisation in Africa: A Stocktaking Survey. Washington DC: World Bank; 2002

[21] Kombe JW. Land use dynamics in Peri-urban areas and their implications on the urban growth and form: The case of Dar es Salaam. Dar es Salaam: University of Dar es Salaam, Department of Urban and Regional Planning; 2005

[22] Connolly J. Institutional Change in Urban Environmentalism: A Case Study Analysis of State-Level Land Use Legislation in California and New York. New York: Columbia University press; 2012

[23] UN-Habitat. State of the World's Cities 2010/2011—Cities for All: Bridging the Urban Divide; 2010

[24] Houston NR. Legislative Summary: Emergency Management. Carolina: University of North Carolina, School of Government; 2014

[25] World Bank. Priorities for City Leaders. Urbanisation Review Flagship Report. Washington, DC: World Bank; 2012

[26] Kingo'ori ZI. Nairobi Urban Transformation Challenges: Learning from Japan. Nairobi: Roads Department; 2007

[27] Sachs-Jeantet C. Managing social transformations in cities: A challenge to social sciences. Paris: UNESCO. Discussion Paper No. 2; 2006

[28] Omwenga M. Nairobi-emerging metropolitan region: Development planning and management opportunities and challenges. University of Nairobi: Department of Urban and Regional Planning; 2010

[29] Braun E, van den Berg L, van der Meer J, editors. National Policy Responses to Urban Challenges in Europe. Ashgate Publishing, Ltd; 2007

[30] Olima WHA. The dynamics and implications of sustaining urban spatial segregation in Kenya: Experiences from Nairobi Metropolis. Paper Presented at the International Seminar on Segregation in the City, Held at Lincoln Institute Land Policy in Cambridge; July 25-28; 2001

[31] Williams JJ. South Africa: Urban transformation. Cities. 2000;**17**(3):167-183

[32] John P. Explaining policy change: The impact of the media, public opinion and political violence on urban budgets in England. Journal of European Public Policy. 2006;**13**(7):1053-1068

[33] Mabogunje AL. Positioning Nigerian cities for economic development. Keynote Address to the Nigerian Urban Summit Held at Abuja on 1-3 September; 2008

[34] Oucho J. Devolved governance and futuristic urbanisation in Kenya. A paper presented at the conference on urbanisation and rural-urban migration in sub-Saharan Africa held 26-27th November at Nairobi, Kenya; 2012

[35] Peng X, Chen X, Cheng Y. Urbanisation and its Consequences. Fundan University, School of Development and Public Policy; 2000

[36] Lambright GMS. Opposition politics and urban service delivery in Kampala, Uganda. Development Policy Review on Urban Governance and Service Delivery in sub-Saharan Africa. 2014;**32**(1):39-60

[37] Gervase C. Housing the Poor: A Case Study of Kisumu Town. Liverpool: Liverpool University Press; 2010

[38] Klopp J. University/City Partnerships: Creating Policy Network for Urban Transformation in Nairobi. Columbia: Centre for Sustainable Development; 2002

[39] Adhikari J. Urbanisation, government policies and growing social and environmental problems in Pokhara, Nepal. Studies in Nepali History and Society. 2011;**5**(1):43-87

[40] McGranahan G. Let's See what the BRICS Teach Us about Urbanisation and Economic Growth. London: International Institute for Environment and Development; 2014

[41] Nattrass J. The dynamics of urbanisation in South Africa. Working Paper No. 2. University of Natal, Durban: Development Studies Unit, Centre for Applied Social Sciences

[42] Annez PC, Buckley RM. Urbanization and growth: Setting the context. Urbanization and Growth. 2009;**1**:1-45

[43] Lindeboom W, Muzzini E. The urban transition in Tanzania. Building the Empirical Base for Policy Dialogue. 2008. pp. 1-166

[44] Alem Y. Life-Satisfaction in Urban Ethiopia: The Role of Relative Poverty and Unobserved Heterogeneity (no. dp-14-04-efd); 2014

[45] Hipp J. What Is the 'Neighborhood' in Neighborhood Satisfaction? Comparing the effects of structural characteristics measured at the micro-neighborhood and tract levels; 2014

[46] Lwasa S. Informal Land Markets and Residential Housing Development in Kampala: Processes and Implications to Planning. Bonn: IHDP Urbanisation and Sustainability; 2002

[47] Mukiibi S. The Effect of Urbanisation on the Housing Conditions of the Urban Poor in Kampala, Uganda. Noida, India: The second International conference on advances in engineering & technology; 20th-21st December. 2011

[48] Muinde DK. Assessing the effects of land tenure on urban developments in Kampala. University of Twente Faculty of Geo-Information and Earth Observation (ITC); 2013

[49] Somik L. Planning for Uganda's urbanisation. Inclusive Growth Policy Note 4; 2012

[50] Katembwe C. Planning Ahead: A Look at the New Physical Planning Law in Uganda; 2011

[51] Brown AM. Uganda's National Urban Policy: The Emerging Response to Poverty, Food Security and Gender in Urban Uganda. Centre for International Governance Innovation; 2013

[52] Uganda Investment Authority. The Big Push Strategy. Kampala; 2004

[53] Yusuf S, Evenett S, Wei J, editors. Facets of globalization. International and local dimensions of development. Discussion Paper No. 415. World Bank; 2001

[54] Kahangirwe P. Linking environmental assessment and rapid urbanization in Kampala City. Impact Assessment and Project Appraisal. 2012;**30**(2):111-115

[55] Katusiimeh MW, Mol APJ, Burger K. The operations and effectiveness of public and private provision of solid waste collection services in Kampala. Habitat International. 2012; **36**:247-252

[56] Uganda Bureau of Statistics. Uganda national household survey 2009/2010: Socio-Economic Module: A Bridged Report. Kampala: UBOS; 2010

[57] Amin ME. Social Science Research: Conception, Methodology and Analysis. Kampala: Makerere University Printery; 2005

[58] Uganda Bureau of Statistics. Statistical Abstract. Kampala: UBOS; 2013

[59] Morgan DL. Focus Group as a Qualitative Research. London: Sage Publications; 2006

[60] Field AP. Reliability Analysis, Discovering Statistics using SPSS. 2nd ed. London: Sage; 2005

[61] Bidandi F. The Dynamics (Pembroke, Ont.) of Urbanisation in Kampala, Uganda: Towards a Possible Alternative Urban Policy. Cape Town: University of the Western Cape; 2015

[62] You-Tien H. The Great Urban Transformation: Politics of Land and Property in China. Oxford: Oxford University Press; 2010

[63] Sulkin JBT, Larsen H. Policy punctuations in American political institutions. American Political Science Review. 2003;**97**:151-169

[64] King A, Wybrow R. British Political Opinion 1937-2000. London: Politicos; 2001

[65] Goodfellow T. The institutionalisation of 'noise' and 'silence' in urban politics: Riots and compliance in Uganda and Rwanda. Oxford Development Studies. 2013;**41**(4):436-454

[66] Conroy-Krutz J, Logan C. Museveni and the 2011 Ugandan election: Did the money matter? The Journal of Modern African Studies. 2012;**50**(4):625-655

[67] MFPED (Ministry of Finance, Planning and Economic Development). Background to the Budget. Kampala: Ministry of Finance, Planning and Economic Development; 2012

[68] Harvey D. The Urbanization of Capital: Studies in the History and Theory of Capitalist Urbanization. Oxford: Blackwell; 1985

[69] Harvey D. The right to the city. New Left Review. 2008;**53**:23-40

[70] Lwasa S. Managing African urbanisation in the context of environmental change. Interdisciplina. 2014;**2**(2):263-280

[71] Ugandans at Heart. Implications, contradictions and implementation failures of Kampala Capital City Act, 2010; 2013

[72] Turok I. Urbanisation and development in South Africa: Economic imperatives, spatial distortions and strategic responses. Urbanization and Emerging Population Issues Working Paper No. 8; 2012

[73] Mills A. Causes of Corruption in Public Sector Institutions and its Impact on Development. 2012. [Electronic Resource].–Mode of Access: http://unpan1.un.org/intradoc/groups/Public/documents/undpadm/unpan049589.pdf

[74] Cooper D, Krieckhaus J, Lusztig M. Corruption, democracy and economic growth. International Political Science Review. 2006;**27**(2):122-143

7

Political Economy and the Work of Kenneth Arrow

Norman Schofield

Abstract

This chapter contrasts two domains of human activity: LOGOS, the principle of enlightenment rationality and MYTHOS, the search for meaning through religion. LOGOS has created our modern world, and we have many successes as a result: the general equilibrium result for economics, general relativity, and quantum mechanics (but as yet no combined theory of the two). Even Newtonian mechanics has led to the notion of chaos. The Hilbert program to show the consistency and completeness of Mathematics has been invalidated by Godel's Theorem, while the attempt to extend the economics general equilibrium theorem fails because of Arrow's Impossibility Theorem. Darwinian evolutionary theory is another success of Logos, but even here, there are many difficulties, particularly how genes work Logos has allowed us to create our industrial society, but has also led us to climate change, without indicating how we can avoid the collapse of civilization. Here we suggest that we may be able to use Mythos, our collective beliefs in what we should do, to help us make a wise choice about the future. The greatest failure of Logos is that we have no understanding of the nature of consciousness. If we can develop such a theory, then perhaps we can construct a theoretical political economy. Without this, it appears likely that climate change could induce a Malthusian trap for us unless we pay heed to Pope Francis's call for us to "Care for Our Common Home." Since this presents us with a common goal, it is possible that we can make a wise choice over our future.

1. Introduction

The foundations of western science in the most general sense of the term were perhaps handed down by Thomas Hobbes (1588–1679) in his Leviathan of 1651 [78] and by Isaac Newton (1642–1727) in his Philosophiae naturalis principia mathematica of 1687. Newton's work, particularly the Optiks, as well as his underlying philosophy of science, was transmitted throughout Europe by Voltaire's (1694–1778) book on the Elements of Newton's Philosophy (published in 1738).

The human sciences, and especially political economy and moral philosophy, were developed further in France by Condillac's (1715–1780) Essay on the Origin of Human Knowledge (1746) and Turgot's (1727–1781) Reflections on the Formation and Distribution of Wealth (1766), and in Scotland by David Hume's (1711–1776) Essays Moral and Political (1742) and Adam Smith's (1723–1790) Wealth of Nations of 1776. At roughly the same time, Condorcet (1743–1794) published his Essay on the Application of Analysis to the Probability of Decisions (1785) and the Esquisse d'un tableau historique des progres de l'esprit humain (1794). The former essay had little widespread impact at the time (although Thomas Jefferson, a friend of Condorcet, while in Paris almost certainly read it). The latter essay was used by Thomas Malthus (1766–1834) as the point of departure for his pessimistic book, the Essay on the Principle of Population [106], where he argued against what he saw as Condorcet's excessively optimistic, "Smithian," viewpoint.

Since then, of course, political economy developed apace in the work of Ricardo, Pareto, Walras and Marshall, culminating in the mathematical existence theorems for a competitive equilibrium (von Nuemann, 1935 (1946); Wald, 1935; [7]; McKenzie, 1954).

In contrast to the theoretical efforts on the economic side of political economy, almost no work on formalizing Condorcet's insights, in his Essay of 1785 on the political side of political economy, was attempted until the late 1940s, when Duncan Black and Kenneth Arrow published seminal papers on this topic.

In 1948, Duncan Black published his paper "On the Rationale of Group Decision Making," [20] specifically addressed to the question of existence of a voting equilibrium. He followed this in 1958 with his monograph on The Theory of Committees and Elections. The monograph emphasized the importance of Condorcet's work in voting theory but paid much less attention to the so-called Condorcet Jury Theorem. In contrast, recent research has suggested that this latter theorem gives a justification for majority rule as a "truth seeking" device.

Arrow's paper on "A Difficulty in the Concept of Social Welfare," [6] derives, I believe, from quite a different tradition of formal political economy, namely the work in welfare economics of Bergson (1938), Hicks (1939) and Lange (1942). (It should perhaps be emphasized that both welfare economics and political economy, viewed in the larger sense, became the arena for sometimes vigorous arguments in the 1930s and 1940s in the work of Schumpeter, Hayek, Popper and von Mises, etc.).

Arrow's famous paper of 1950 shows essentially that any social welfare function (that maps families of weak individual orderings to a weak social order) is either imposed or dictatorial. To obtain what Arrow termed this "possibility theorem," he assumed that the social welfare function had universal domain and satisfied a property of positive association of preferences. Reading this paper and a related one by Arrow on "Welfare Economics" [6], I infer that Arrow's realization of the applicability of the social choice paradox to welfare economics came about from his deep understanding of Scitovsky's "Note on Welfare Properties in Economics" (1942). A typical assumption in economics is that a move from a restricted trade situation, x, say, to a "free-trade" situation y is "welfare preferred." Even though not all individuals may prefer y to x, nonetheless winners in y may compensate losers in y, so that the post-compensation outcome y*, say, is unanimously preferred to x. If this holds, then y is said to be "welfare preferred" to x, even when compensation is not implemented. Scitovsky observed that "welfare preferred" can be badly behaved, since y may be preferred to x, and x to y.

As Arrow commented in his paper, the negative result of the "possibility theorem" was "strongly reminiscent of the intransitivity of the concept of domination in the theory of multiperson games" as presented in von Neumann and Morgenstern (1947). Since intransitivity of domination occurs most obviously in constant sum voting games, I also infer that Arrow means that the "possibility theorem" was derived from the fact that all welfare judgments are implicitly based on transfers of wealth. In his 1950 paper, (on Social Welfare), Arrow also emphasizes that he views the theorem as relevant to a situation where individuals make value judgments rather than to the more typical economic context where agents make choices based on their tastes. Since all political choices are based, to some degree or other, on the aggregation of values, I further infer that the "possibility theorem" addresses not just the traditional questions of welfare economics, but the larger issue of the interaction between the political and economic realms. In other words, the relevance of the theorem is not simply to do with the Condorcetian question of voting cycles, or intransitivies, but concerns the larger questions of political economy that were discussed earlier by Schumpeter, Hayek, Popper, and so on in the period before World War II.

The formal exercise of proof of existence of an economic equilibrium (obtained between 1935 and 1954) leaves unanswered many questions. For example, can the existence proof be extended from the domain of private commodities to include public goods? More particularly, can democratic procedures be devised to ensure that preference information be aggregated in an "efficient" fashion so that social choice is welfare maximizing. Arrow's possibility theorem suggests that democracy itself may be flawed: indeed it suggests that democratic institutions may (as Madison foresaw in Federalist X) be perverted or turbulent. Thus, difficult questions of institutional design need to be addressed. Third, since Arrow's theorem required an ordering of social states, it implicitly brings into question the nature of the stability of price equilibria (even when they exist). The example of Scarf (1954) indicates that the dynamical system defined by tatonnement may be structurally unstable. This raises the possibility that both economic and political systems may be chaotic.

Since 1950, these issues have been discussed at varying levels of intensity. All of them come back in one sense or another to an interpretation of Arrow's Theorem. In the rest of this essay, I shall attempt to outline my sense of the current state of the debate, and the relationship with Arrow's Theorem, along the following lines:

1. Extension of equilibrium theories in economics to the larger realm of political economy (neo-Smithian theories);
2. Democratic institutions and the compatibility of economics and politics (neo-Condorcetian theories);

3. "Chaos" in political economies, and recent notions of complex behavior in dynamical systems (from Newton to Laplace to Poincare);
4. Large-scale prediction or "post diction," and macroeconomic history (neo-Malthusian theories).
5. The effect of climate change on these theories.
6. The possibility that complex societies such as ours can reach a tipping point and collapse (as the Roman empire did).
7. Extension of equilibrium theories and much of economic theory emphasise the importance of economic growth. But after 200 years of growth, we now face the likelihood that climate change will make growth in the future impossible. Recent work by Stern [160, 161] suggests how we can estimate the costs of ameliorating the ravages of climate change. Ultimately, we need a new theory of economic externalities. Stern [162] goes on to suggest that we need a new theory of social welfare to indicate how we can structure a rational political economy. However Arrow's Impossibility theorem indicates that a social welfare function may not exist. Devising a social welfare function may necessitate incorporating the following themes:
8. Aggregation procedures: We can first mention the work of Don Saari [138, 139] who has shown how almost all aggregation procedures can give rise to counter intuitive or "chaotic" outcomes. In particular, the work of Sonnenschein [158], Mantel [108] and Debreu [46] made it clear that devising a method to translate rational consumer behavior into a price vector was anything but obvious.
9. Chaos: Philosophers such as Joseph Schumpeter, Friedrich von Hayek, and Karl Popper in the period around World War II were all involved in the debate about whether a social welfare function could be constructed. Arrow's Theorem gave a negative answer to this debate. In order to understand Arrow's theorem, it is necessary to set out the fundamental problem of political economy, namely, the nature and evolution of the social relationship between human beings. Since I understand this debate to focus on the possibility of equilibrium in contrast to disequilibrium (or disorder), I shall also mention what I judge to be a significant anti-equilibrium discovery of this century: "chaos." To do this, I think it appropriate to briefly comment on my perception of the main themes of this debate, going back to the time of the time of Hobbes. For Hobbes, society could fall into disorder in the absence of a Leviathan, able to maintain the peace. I understand this debate to focus on the possibility of equilibrium in contrast to disequilibrium (or disorder), I shall also mention what I judge to be a significant anti-equilibrium discovery of this century: "chaos." The idea of "deterministic" chaos only developed in the last 50 years and came about because of a better understanding of the solar system. "To illustrate, astronomers since the time of Pierre-Simon Laplace (1799) have believed that the solar system is structurally stable: In other words, small perturbations in each planetary orbit (induced by other planets) cannot dramatically change the nature of the orbit. Although Isaac Newton was aware of the problem of perturbations (Newton 1687), even Henri Poincare in his treatise of 1890 could not solve the differential equations." However, Poincare's work led to the beginning of differential topology and the work of Marston Morse, John Milnor, and Stephen Smale in this century. If the solar system were structurally

unstable, or indeed chaotic, then it would be impossible to predict its evolution. In fact, it is not chaotic, although subsystems (such as asteroids) are[1]. As a result of popular books [60], we can conceive of natural phenomena (hurricanes) or even large-scale dynamic systems (such as climate) as potentially chaotic (Lorenz, 1993)[2]. Although still a young science, human evolutionary theory suggests that chaotic transformations in weather may have had a profound effect on the human diaspora "out of Africa" (Boaz 1997; Calvin 1990; Stanley 1996). "Equilibrium"-focused evolutionary theory may also need revision (Eldridge and Gould 1972; Gould 1996). **Figure 1** shows the chaotic variation in temperature over the last 100,000 years.

10. Malthus and evolution: Malthus published his book on Population in 1798, arguing that population will tend to grow exponentially while resources are bounded arithmetically. This insight provided Darwin [44] with the logic of natural selection. Although Darwin favored the notion of slow evolutionary change, it has been pointed out that evolution has in the past been driven by cataclysmic extinction events when nearly all species go extinct (Eldridge and Gould 1972; Gould 1996). This notion of punctuated equilibrium has been applied at the level of the entire biosphere to suggest that the evolution of the climatic system can undergo dramatic transformations. Indeed Greer [59] suggests that the current episode of anthromorphic climate change will force on us such a transformation sufficient to destroy our civilization. Our civilization has depended on the availability of cheap energy, and as we use up this energy, we will face a neo-Malthusian catastrophe. Indeed as Diamond [47] pointed out many societies have in the past faced such a Malthusian collapse.

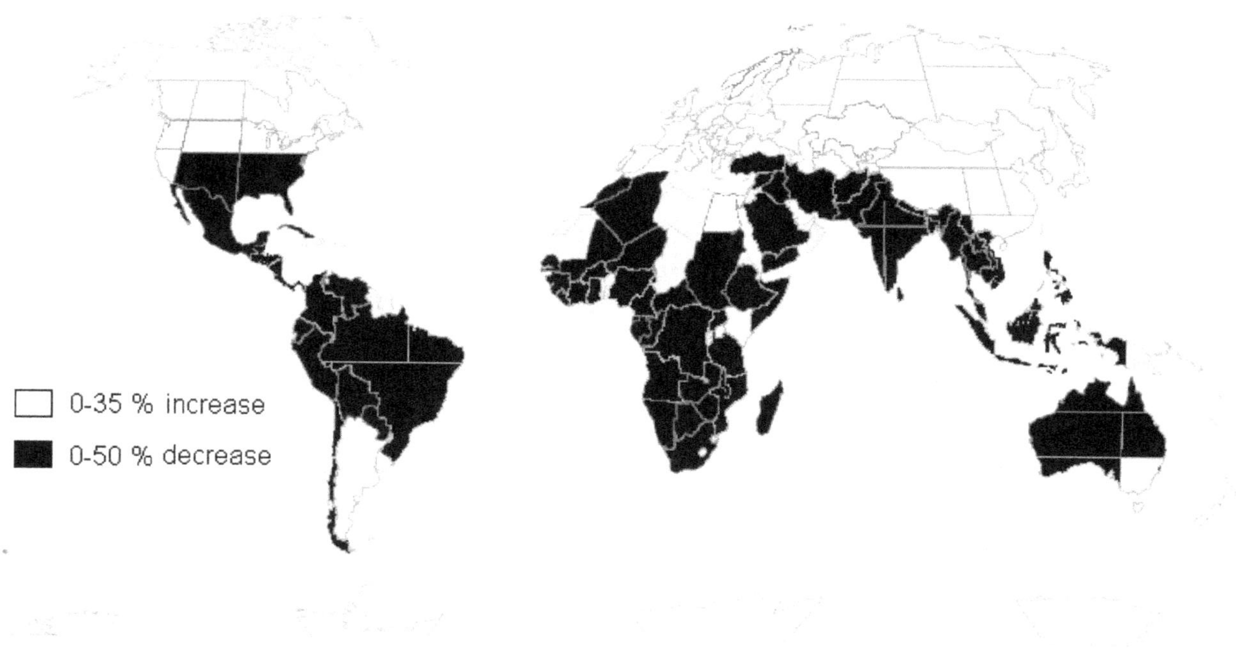

Figure 1. The effect of climate change on agricultural production [36].

[1] Chaotic phenomena can, of course, have profound consequences. A chaotic event, an asteroid collision, may have led to the extinction of the dinosaurs.

[2] See the discussion on theories of the solar system in Peterson (1993).

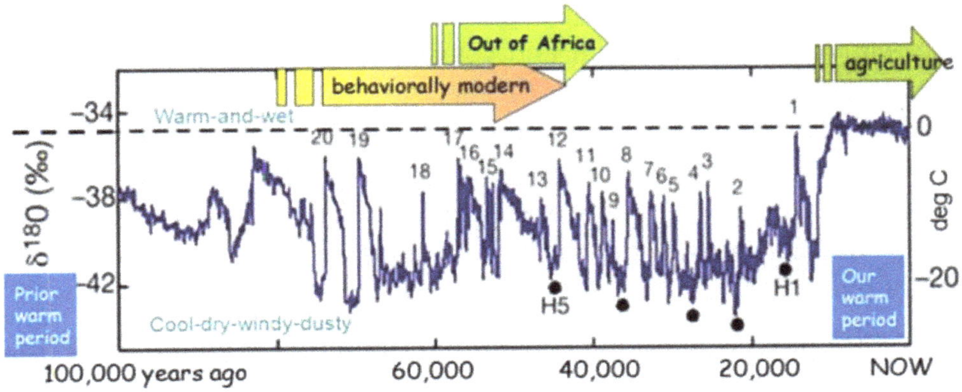

Figure 2. Climate 100KYBP to now: chaos from 90 to 10KYBP (source: Global-Fever.org).

Zhang [186] provided us with some estimates of the Malthusian effects of climate change. Indeed Tainter and Renfrew [166] argued that all complex societies, including our own, can or indeed will hit a tipping point after which they go into decline[3]. Ahmed [5] ties this tipping point to the combined effect of peak energy (the increasing cost of obtaining energy), food disruption (due to climate change) and social dislocation. To illustrate, **Figure 2** shows the estimated losses of agricultural production due to climate change.

11. Collapsing political systems In recent years, we have become used to political disorder in the Middle East, Africa, the Ukraine, and we can expect more of the same (Pakistan?). The countries that have experienced such disorder have also been subject to climatic transformations and food disruptions, as can be seen in **Figure 2**.
12. Limits to growth in 1972, Meadows et al. [111] published a book arguing that population growth would eventually give us a world where pollution and the boundedness of the world's resource would have to be faced. Now, 16 years later, the science of climate change has shown the validity of this argument. Hawken [74] covered some of the ways in which we can limit climate change. However, climate change induces a prisoner's dilemma in the behavior of nation states, since no state is willing to adopt costly behavior which will limit the effects of climate change.
13. Markets as prisoners' dilemmas Cassidy [27] argues that markets involve prisoner's dilemmas, what he calls rational irrationality. Since a prisoner's dilemma is fundamentally chaotic, it is not surprising that we can have chaotic market behavior, like the recent turmoil in financial markets. The economic growth that we have experienced since 1945 can be related to the willingness of the United States to act as a hegemon in the creation of the Bretton Woods system of institutions facilitating trade, the protection of property rights and the rule of law.
14. Complexity theory: The notion of chaos has also given rise to the idea of complexity, the possibility that dynamical systems can generate complex phenomena perhaps through evolutionary transformations [89]. A reason to believe this is because of the Godel [61]

[3] Schofield [147] offers the example of the decline and collapse of the Roman Empire.

incompleteness/inconsistency theorem that mathematics, able to include arithmetic cannot prove that it is both consistent and complete. In addition, Turing's Theorem [175] showed that Mathematics is a subject to a halting problem unable to determine whether an algorithm has or has not been completed, Penrose [132] uses these two results to argue that the mind is not algorithmic, and indeed is creative. We still have no general theory of consciousness, but without doubt we will probably have such a theory in the future that will allow us to construct a better theory of rationality and thus to construct a theory of social choice that will extend Arrow's Theorem [141].

15. MYTHOS As we have suggested, Logos provides humanity with a set of useful clues as to how to understand and thus manipulate the natural world. Knowledge, in the sense of the enlightenment, allowed for the innovations of the scientific revolution. For policy makers, the upshot of these innovations was an often-unstated belief that technology and science would explain and redress the crises that would come to confront humankind. However, thus far, the opposite is true. Even if there was general agreement regarding markets, political methods, and so on, our models are insufficient to capture the scope of these phenomena. Indeed, all our scientific riches put into relief just how defective our model of choosing is.

Therefore, it may be the case that our model of decision-making does not consider the influence of Mythos on human decision-making. Perhaps humans need something beyond rationality. It could be, therefore, useful to look to thinkers like Pope Francis who take Mythos as the starting point for their thought but have deep respect for scientific inquiry. As he writes in his encyclical Laudato [100], pp. 119:

> *Doomsday [regarding climate] predictions can no longer be met with irony or disdain. We may well be leaving to coming generations debris, desolation, and filth. The pace of consumption, waste, and environmental change has so stretched the planet's capacity that our contemporary lifestyle, unsustainable as it is, can only precipitate catastrophes such as those which even now periodically occur in different areas of the world. The effects of the present imbalance can only be reduced by our decisive action, here and now. We need to reflect on our accountability before those who will have to endure the dire consequences.*
>
> *Succinctly this captures the quandary policy makers find themselves in. Our climate models indicate disaster on the horizon. On the other hand, our models of decision-making indicate that there is no way to rank policy outcomes in an orderly fashion without influence of a hegemon [Arrow]. However, obviously Francis's thoughts are from an ethical vantage point—the Christian is to be a noble steward of her habitat. Thus, perhaps a religious ideology sympathetic to science, like Francis's theology, can point the way to a better model of choosing.*

Author details

Norman Schofield

Address all correspondence to: schofield.norman@gmail.com

Department of Political Science, Washington University, Saint Louis, USA

References

[1] Acemoglu D, Robinson J. Persistence of power, elites, and institutions. American Economic Review. 2008;**98**:267-293

[2] Acemoglu D, Johnson S, Robinson J, Yared P. Reevaluating the modernization hypothesis. Journal of Monetary Economics. 2009;**56**:1043-1058

[3] Acemoglu D, Ozdaglar A, Tahbaz-Salehi A. Cascades in networks and aggregate volatility. NEER working paper # 16516; 2010

[4] Ackerman B. Decline and Fall of the American Republic. Boston, MA: Harvard University Press; 2011

[5] Ahmed NM. Falling States, Collapsing Sytems. Berlin: Springer; 2017

[6] Arrow. A difficulty in the concept of social welfare. Journal of Political Economy. 1950;**58**:326-346

[7] Arrow K, Debreu G. Existence of an equilibrium for a competitive economy. Econometrica. 1954;**22**:265-290

[8] Arrow K. Social Choice and Individual Values. New York: Wiley; 1951

[9] Arrow, Kenneth, Hahn F. General Competitive Analysis. San Francisco: Holden Day; 1971

[10] Armstrong K. The Eattle for God. New York: Random House; 2011

[11] Austen-Smith D, Banks J. Information aggregation, rationality, and the Condorcet jury theorem. American Political Science Review. 1996;**90**:34-45

[12] Axelrod R. The emergence of cooperation among egoists. American Political Science Review. 1981;**75**:306-318

[13] Axelrod R. The Evolution of Cooperation. New York: Basic; 1984

[14] Axelrod R, Hamilton WD. The evolution of cooperation. Science. 1981;**211**:1390-1396

[15] Barbera R. The Cost of Capitalism: Understanding Market Mayhem. New York: McGraw Hill; 2009

[16] Belich J. Replenishing the Earth. Oxford: Oxford University Press; 2009

[17] Bellah. Religion in Human Evolution. Cambridge, MA: Belknap Press; 2011

[18] Belich J. Replenishing the Earth: The Settler Revolution 1783-1939. Oxford: Oxford University Press; 2009

[19] Bikhchandani S, Hirschleifer D, Welsh I. A theory of fads, fashion, custom, and cultural change as information cascades. Journal of Political Economy. 1992;**100**:992-1026

[20] Black D. On the rationale of group decision making. Journal of Political Economy. 1948;**56**:23-34

[21] Bowles S et al. The Co-Evolution of Individual Behaviors and Social Institutions. Journal of Theoretical Eiology. 2003;**223**:135-147

[22] Boyd J, Richerson PJ. The Origin and Evolution of Culture. Oxford: Oxford University Press; 2005

[23] Broodbank C. The Making of the Middle Sea. Oxford: Oxford University Press; 2013

[24] Burkhart JM, Hrdy SB, van Schaik CP. Cooperative breeding and human cognitive evolution. Evolutionary Anthropology. 2009;**18**:175-186

[25] Calvin WH. The ascent of mind. New York: Bantam; 2003

[26] Carothers T. The end of the transition paradigm. Journal of Democracy. 2002;**13**:5-21

[27] Cassidy J. How Markets Fail: The Logic of Economic Calamities. New York: Farrar, Strauss and Giroux; 2009

[28] Cavallli-Sforza L, Feldman M. Cultural Transmission and Evolution. Princeton, NJ: Princeton University Press; 1981

[29] Chichilnisky G. The topology of fear. Journal of Mathematical Economics. 2009;**45**:807-816

[30] Chichilnisky G. Avoiding Extinction: Equal Treatment of the Present and the Future, Working Paper. Columbia University; 2009

[31] Chichilinsky G. The foundations of statistics with black swans. Mathematical Social Science. 2010;**59**:184-192

[32] Chichilinsky G. The topology of change: foundations of statistics with black swans. Forthcoming in Topics in Mathematical Economics: Essays in Honor of J. Marsden, Fields Institute Communication Volume. Providence, Rhode Island: American Math Society; 2014

[33] Chichilinsky G, Eisenberger P. Asteroids: Assessing catastrophic risks. Journal of Probability and Statistics. 2010. Article ID 954750

[34] Clark G. What made Brittania Great? How much of the Rise of Britain to World Dominance by 1850 does the Industrial Revolution explain? In: Rourke KO', Taylor A, editors. Comparative Economic History: Essays in Honor of Jeffrey Williamson. Cambridge, MA: MIT Press; 2007

[35] Clark G. A Farewell to Alms. Princeton, NJ: Princeton University Press; 2007

[36] Cline W. Global Warming and Agriculture Impact Estimates by Country. Peterson Institute: Washington, DC; 2007

[37] Collier P et al. Climate Change and Africa. Working Paper. Oxford University; 2008

[38] Collier P. Wars, Guns and Votes. New York: Harper; 2009

[39] Collier P. The Plundered Planet. Oxford: Oxford University Press; 2010

[40] Condorcet N. Essai sur l'application de l'analyse a la probabilite des deci- sions rendues a la pluralite des voix. Paris: Imprimerie Royale; 1994 [1785] [Translated in part

In: McLean I, Hewitt F, editors. Condorcet: Foundations of Social Choice and Political Theory. Aldershot, UK: Edward Elgar Publishing

[41] Condorcet N. Esquisse d'un tableau historique des progres de l'esprit humaine. Paris: Yves Gravier; 1798

[42] Cooper. Money, Elood and Revolution. Petersfield, UK: Harriman; 2014

[43] Corcos et al. Imitation and contrarian behavior: Hyperbolic bubbles, crashes and chaos. Quantitative Finance. 2002;**2**:264-281

[44] Darwin C. The Origin of Species by Means of Natural Selection. London: John Murray; 1859

[45] Darwin C. The Descent of Man. London: John Murray; 1871

[46] Debreu G. Excess demand functions. Journal of Mathematical Economics. 1974;**1**:15-21

[47] Diamond J. Collapse. London: Penguin; 2011

[48] Deutscher G. The Unfolding of Language. New York: Holt; 2006

[49] Easley D, Kleinberg J. Networks, Crowds and Markets. Cambridge: Cambridge University Press; 2010

[50] Ferguson N. Introduction. In: Ferguson N, editor. Virtual History. London: Picador; 1997

[51] Ferguson N. Empire: The Rise and Demise of the Eritish World Order. London: Penguin Books; 2002

[52] Ferguson N. Civilization. London: Penguin; 2011

[53] Ferguson N. The Great Degeneration. London: Penguin; 2012

[54] Fox J. The Myth of the Rational Market. New York: Harper; 2009

[55] Fukuyama F. The Origins of Political Order. New York: Ferrar, Strauss and Giroux; 2011

[56] Fukuyama F. Political Order and Political Decay. New York: Ferrar, Strauss and Giroux; 2014

[57] Gazzaniga MS. The Ethical Brain. New York: Harper; 2006

[58] Gintis H. Strong reciprocity and human sociality. Journal of Theoretical Eiology. 2000;**206**:169-179

[59] Greer JM. Dark Age America: Climate Change, Cultural Collapse, and the Hard Future Ahead. Vancouver, Canada: New Society; 2015

[60] Gleick J. Chaos: Making a New Science. New York: Viking; 1987

[61] Kurt G. Uber formal unentscheidbare Satze der principia Mathematica und verwandter Systeme. Monatschefte fur Mathematik und Physik. 1931;**38**:173-98 [Translated as On formally undecidable propositions of principia mathematica and related systems. In:

van Heijenoort J, editor. Frege and Godel: Two Fundamental Texts in Mathematical Logic. Cambridge, MA: Harvard University Press]

[62] Golub B, Jackson M. Naive learning in social networks and the wisdom of crowds. American Economic Journal: Microeconomics. 2010;**2**:112-149

[63] Gore A. The Assault on Reason. London: Bloomsbury; 2007

[64] Gray J. Enlightenment's Wake. London, Routledge; 1995

[65] Gray J. Endgames. London: Blackwell; 1997

[66] Gray J. False Dawn. London: New Press; 2000

[67] Gribbin J. Alone in the Universe. NewYork: Wiley; 2011

[68] Hamilton W. The genetical evolution of social behavior I and II. Journal of Theoretical Eiology. 1964;**7**:1-52

[69] Hamilton W. Selfish and spiteful behavior in an evolutionary model. Nature. 1970;**228**: 1218-1220

[70] Hardin G. The tragedy of the commons. In: Daly HE, editor. Towards a Steady State Economy. San Francisco: Freeman; 1968 [1973]

[71] Hardin R. Collective action as an agreeable prisons' dilemma'. Behavioral Science. 1971; **16**:472-481

[72] Hardin R. Collective Action. Baltimore, MD: Johns Hopkins University Press; 1982

[73] Harper K. The environmental fall of the Roman Empire. Daedalus. Spring; 2016;101-111

[74] Hawken. Drawdown, New York, Penguin; 2017

[75] Hays JD, Imbrie J, Shackleton NJ. Variations in the Earth's orbit: Pacemaker of the ice ages. Science. 1976;**194**(4270):1121-1132

[76] Henrich J et al. Foundations of Human Sociality. Oxford: Oxford University Press; 2004

[77] Henrich J et al. 'Economic man' in cross-cultural perspective: Behavioral experiments in 15 small-scale societies'. Behavioral and Brain Sciences. 2005;**28**:795-855

[78] Hobbes T. Leviathan; or the Matter, Forme, and Power of a Common-wealth, Ecclesiastical and Civil. In: Gaskin, editor. Oxford: Oxford University Press; 2009 [1651]

[79] Hrdy SB. Mothers and Others: The Evolutionary Origins of Mutual Understanding. Cambridge, MA: Harvard University Press; 2011

[80] Hsiang. Quantifying the influence of climate on human conflict. Science Express. 2013. 10.1126

[81] Israel J. Radical Enlightenment. Oxford: Oxford University Press; 2002

[82] Israel J. Enlightenment Contested. Oxford: Oxford University Press; 2006

[83] Israel J. Revolution of the Mind. Princeton, NJ: Princeton University Press; 2010

[84] Israel J. Revolutionary Ideas. Princeton, NJ: Princeton University Press; 2014

[85] Jacques M. When China Rules the World. London: Penguin; 2009

[86] Jamieson D. Reason in a Dark Time. Oxford: Oxford University Press; 2014

[87] Kahneman D. Thinking Fast and Slow. New York: Ferrar Strauss and Giroux; 2011

[88] Karklins R, Petersen R. Decision calculus of protestors and regime change: Eastern Europe 1989. Journal of Politics. 1993;**55**:588-614

[89] Kauffman S. Humanity in a Creative Universe. Oxford: Oxford University Press; 2016

[90] Kennedy P. The Rise and Fall of the Great Powers. New York: Random House; 1987

[91] Keohane R, Nye R. Power and Interdependence. New York: Little Brown; 1977

[92] Keynes JM. Treatise on Probability. London: Macmillan; 1921

[93] Keynes JM. The General Theory of Employment, Interest and Money. London: Macmillan; 1936

[94] Kindleberger C. The World in Depression 1929-1939. Berkeley, CA: University of California Press; 1973

[95] Kolbert E. The Sixth Extinction. New York: Holt; 2014

[96] Kreps DM et al. Rational cooperation in the finitely repeated prisoners' dilemma. Journal of Economic Theory. 1982;**27**:245-252

[97] Ladha K. Condorcet's jury theorem, free speech and correlated votes. American Journal of Political Science. 1992;**36**:617-674

[98] Ladha K, Miller G. 'Political discourse, factions and the general will: Correlated voting and Condorcet's jury theorem. In: Schofield N, editor. Collective Decision Making. Boston: Kluwer; 1996

[99] Landemore H. Democratic Reason. Princeton, NJ: Princeton University Press; 2012

[100] Laudato S. Encyclical Letter of the Holy Father Francis "on Care for our Common Home". Rome: The Holy See; 2015

[101] Leakey R, Lewin R. The Sixth Extinction. New York: Anchor; 1995

[102] Leakey R. The Origin of Humankind. New York: Basic; 1994

[103] Lohmann S. The dynamics of information cascades. World Politics. 1994;**47**:42-101

[104] Maddison A. Contours of the World Economy 1-2030 AD: Essays in Macro-Economic History. Oxford: Oxford University Press; 2007

[105] Madison J. Federalist X. In: Rakove J, editor. Madison: Writings. New York: Library Classics; 1999 [1787]

[106] Malthus T. An Essay on the Principle of Population. London; 1798

[107] Johnson J, Mandelbrot B, Hudson R. The (Mis)Behavior of Markets. New York: Perseus; 2004

[108] Mantel R. On the characterization of aggregate excess demand. Journal of Economic Theory. 1974;**7**:348-353

[109] Margolis H. Selfishness, Altruism and Rationality. Cambridge: Cambridge University Press; 1982

[110] McWhorter J. The Power of Eabel. New York: Holt; 2001

[111] Meadows D, Meadows DL, Randers J, Behrens W. Limits to Growth. New York: Signet; 1972

[112] Miller G, Schofield N. Activists and partisan realignment in the U.S. American Political Science Review. 2003;**97**:245-260

[113] Miller G, Schofield N. The transformation of the republican and democratic party coalitions in the United States. Perspectives on Politics. 2008;**6**:433-450

[114] Minsky H. John Maynard Keynes. New York: Columbia University Press; 1975

[115] Minsky H. Stabilizing an Unstable Economy. New Haven: Yale University Press; 1986

[116] Mokyr J. The intellectual origins of modern economic growth. Journal of Economic History. 2005;**65**:285-351

[117] Mokyr J. The Enlightened Economy: An Economic History of Eritain 1700-1850. New Haven, CT: Yale University Press; 2010

[118] Mokyr J, Nye VC. Distributional coalitions, the industrial revolution, and the origins of economic growth in Britain. Southern Economic Journal. 2007;**74**:50-70

[119] Morris I. Why the West Rules. New York: Ferrar, Strauss and Giroux; 2010

[120] Nagel T. Mind and Cosmos. Oxford: Oxford University Press; 2012

[121] Nordhaus. Climate Casino. New Haven: Yale University Press; 2013

[122] North DC. Institutions, Institutional Change and Economic Performance. Cambridge: Cambridge University Press; 1990

[123] North DC, Weingast BR. Constitutions and commitment: The evolution of institutions governing public choice in seventeenth century England. Journal of Economic History. 1989;**49**:803-832

[124] North DC, Wallis B, Weingast BR. Violence and Social Orders: A Conceptual Framework for Interpreting Recorded Human History. Cambridge: Cambridge University Press; 2009

[125] Nowak M. Supercooperators. New York: Free Press; 2011

[126] Ormerod P. Butterfly Economics. New York: Basic; 2001

[127] Orestes N, Conway E. The Collapse of Western Civilization. New York: Columbia University Press; 2014

[128] Pagden A. The Enlightenment. New York: Random; 2013

[129] Parfit D. On What Matters. Oxford: Oxford University Press; 2011

[130] Parker G. Global Crisis. New Haven, CT: Yale University Press; 2013

[131] Penn E. A model of far-sighted voting. American Journal of Political Science. 2009;**53**:36-54

[132] Penrose R. The Emperor s New Mind. London: Oxford University Press; 1989

[133] Piketty T. Capital. Cambridge, MA: Harvard University Press; 2014

[134] Putnam RD, Campbell DE. American Grace: How Religion Divides and Unites Us. New York: Simon and Schuster; 2010

[135] Rae D. Decision rules and individual values in constitutional choice. American Political Science Review. 1969;**63**:40-56

[136] Rees M. Our Cosmic Habitat. Princeton University Press: Princeton; 2001

[137] Reynolds. Energy Civilization. Fairbanks, Alaska: Alaska Chena Publisher; 2011

[138] Saari D. The aggregated excess demand function and other aggregation procedures. Economic Theory. 1992;**2**:359-388

[139] Saari D. Chaotic exploration of aggregation procedures. SIAM Review. 1995;**37**(1):37-52

[140] Sagan C, Drake F. The search for extra terrestial intelligence. Scientific American. 1975;**232**:80-89

[141] Searl J. Minds brains and programs. Behavioral and Brain Sciences. 1980;**3**:417-457

[142] Schofield N. Is majority rule special? In: Niemi RG, Weisberg HF, editors. Probability Models of Collective Decision-Making. Columbus, OH: Charles E. Merrill Publishing Co.; 1972

[143] Schofield N. Ethical decision rules for uncertain voters. British Journal of Political Science. 1972;**2**:193-207

[144] Schofield N. Evolution of the constitution. British Journal of Political Science. 2002;**32**:1-20

[145] Schofield N. Madison and the founding of the two party system in the US. In: Kernel S, editor. James Madison: The Theory and Practise of Republican Government. Stanford, CA: Stanford University Press; 2002. pp. 302-327

[146] Schofield N. Is the political economy stable or chaotic? Czech Economic Review. 2011;**5**:76-93

[147] Schofield N, Malthus. Climate change and collapse. Working Paper. Saint Louis: Washington University; 2018

[148] Schofield N, Gallego M. Leadership or Chaos. Berlin: Springer; 2011

[149] Scharf C. The Copernicus Complex. New York: Ferrar, Strauss and Giroux; 2014

[150] Schweitzer F et al. Economic networks: The new challenges. Science. 2009;**325**:422-425

[151] Sen A. Poverty and Famines. Oxford: Clarendon Press; 1982

[152] Shiller R. The New Financial Order. Princeton, NJ: Princeton University Press; 2003

[153] Shiller R. Irrational Exuberance. Princeton, NJ: Princeton University Press; 2005

[154] Smale S. Structurally stable systems are not dense. American Journal of Mathematics. 1966;**88**:491-496

[155] Smith A. The Theory of Moral Sentiments. Indianapolis, IN: Liberty Fund; 1984 [1759]

[156] Sperber J. Karl Marx: A Nineteenth Century Life. New York: Liveright; 2011

[157] Smolin L. The Trouble with Physics. New York: Houghton Mifflin; 2007

[158] Sonnenschein H. Do Walras' identity and continuity chracterize the class of excess demand functions. Journal of Economic Theory. 1973;**6**:345-354

[159] Stringer C. Lone Survivors. New York: St. Martins; 2013

[160] Stern N. The Economics of Climate Change. Cambridge: Cambridge University Press; 2007

[161] Stern N. The Global Deal. New York: Public Affairs; 2009

[162] Stern N. Why Are We Waiting: The Logic Urgency and Promise of Tackling Climate Change. Cambridge: MIT Press; 2015

[163] Sunstein CR. A Constitution of Many Minds. Princeton, NJ: Princeton University Press; 2009

[164] Sunstein CR. Going to Extremes. Oxford: Oxford University Press; 2011

[165] Surowiecki J. The Wisdom of Crowds. New York: Anchor; 2005

[166] Tainter JA, Renfrew C. The Collapse of Complex Societies. Cambridge: Cambridge University Press; 1988

[167] Taleb NN. The Elack Swan. New York: Random; 2007

[168] Taleb NN, Blyth M. The black swan of Cairo. Foreign Affairs. 2011;**90**(3):33-39

[169] Taylor M. Anarchy and Cooperation. London: Wiley; 1976

[170] Taylor M. Community, Anarchy and Liberty. Cambridge: Cambridge University Press; 1982

[171] Tegmark M. The mathematical universe. Foundations of Physics. 2008;**38**:101-150

[172] Tegmark M. Our Mathematical Universe. New York: Random House; 2014

[173] Trivers R. The evolution of reciprocal altruism. Quarterly Review of Etiology. 1971;**46**:35-56

[174] Trivers R. Social Evolution. Menlo Park, CA: Cummings; 1985

[175] Turing A. On computable numbers with an application to the Entscheidungs problem. Proceedings of the London Mathematical Society. 1937;**42**:230-265 [Reprinted in Copeland J. The Essential Turing. Oxford: The Clarendon Press]

[176] Vogel K. Big Money. New York: Public Affairs; 2014

[177] Wallace AR. Natural Selection. New York: Classics US; 1898

[178] Waltham D. Lucky Planet. New York: Basic; 2014

[179] Weitzman M. Additive damages, fat-tailed climate dynamics, and uncertain discounting. Economics. 2009;**3**:1-22

[180] Wells S. Pandora's Seed. New York: Random House; 2011

[181] Wilson EO. The Meaning of Human Existence. New York: Norton; 2014

[182] White T et al. Ardipithecus ramidus and the paleobiology of early hominids. Science. 2009;**326**:65-86

[183] Wolf M. The Shifts and the Shocks. New York: Penguin; 2014

[184] Yanofsky NS. The Outer Limits of Reason. Cambridge, MA: MIT Press; 2013

[185] Zeeman EC. Catastrophe Theory: Selected Papers, 1972-77. New York: Addison Wesley; 1977

[186] Zhang. Global climate change, war, and population decline in recent human history. Proceedings of the National Academy of Science. 2007;**104**(49):19214-19219

[187] Keohane R. After Hegemony. Princeton, NJ: Princeton University Press; 1984

8

Risk Analysis and Land Use Planning

Valentina Svalova

Abstract

Natural hazards are potentially damaging physical events and phenomena, which may cause the loss of life; injury or human life disruption; property damage; social, economic, and political disruption; or environmental degradation. Systematic approach to the natural hazard research on the base of risk concept is a very fruitful and progressive method. Areas of possible disaster events could be the places of the highest risk at the natural risk maps of the territories. It is necessary to use big databases and data banks and GIS technologies for such map constructions. Sometimes people have to live in such dangerous places. It is necessary for people living under natural risk to understand and estimate this risk and to know how to overcome it and how to act in case of crises events. Risk management concept is a good instrument for systematic approach to the problems of the rational land use.

Keywords: risk, risk management, risk analysis, risk assessment, mapping, land use planning

1. Introduction

Natural hazards are potentially damaging physical events and phenomena, which may cause the loss of life; injury or human life disruption; property damage; social, economic, and political disruption; or environmental degradation.

Earthquakes, volcano eruptions, tsunamis, karst, suffusion, coast erosion, and landslides belong to geological hazards [1–3].

About 20% of the world population, approximately 1.2 billion people, live in earthquake dangerous area of about 10 million km^2 that is near 7.5% of the total area of the planet.

93 million population of such countries as Iceland, Japan, the Philippines, Indonesia, the United States, Mexico, Central America, Colombia, Ecuador, and Chile occupy about 0.4 million km^2 of land where volcanic activity is concentrated.

3.7 million km^2 of land is susceptible to sliding, while the population exposed is in the order of 300 million. Areas of high risk of landslides are inhabited by 66 million of inhabitants, occupying a land surface of 820,000 km^2.

14 million people are exposed to tsunamis. The major potentially affected areas are located along the coasts of countries facing the oceans and seas (UNISDR 2009).

Millions of human lives are lost due to earthquakes and volcano eruptions, and property damage has exceeded hundreds of billions USD. It is not possible to make reliable earthquake forecast now, but there exist a few success examples.

"Earthquake early warning systems" alert people of the hazardous ground shaking. Developing earthquake scenarios, as what would happen if an earthquake repeats, where it had occurred in the past, is also very effective in developing earthquake-resilient societies.

It is important to develop earthquake-resilient societies. It could be useful to investigate earthquake scenarios as it had occurred in the past.

Volcanic eruptions have great societal impacts connected with damages, disruptions, health problems, ash fall, lava flows, gases, hot ash clouds, lahars, and hazard to aviation.

Predicting a volcanic eruption is possible on the base of interdisciplinary approach that includes continuous observation of different parameters such as earthquakes and changes in ground and water conditions. Joint efforts of scientists and cooperation of volcano seismology, geodesy, magnetic studies, and hydrology lead to eruption forecasting.

The effects of tsunamis are widely distributed; the consequences can be global. So it was in 2004 during Sumatra earthquake-induced tsunami. Many countries around the Indian Ocean were affected.

A global tsunami warning system was set up to tackle with the challenging problems of tsunami disasters. Also local and regional warning systems generate scientific-based information. Scientific modelling and tsunami forecasting are still to be improved so that the time available between warning and action can be used in the best possible way.

Hydrometeorological and climatological hazards include heavy rains, storms, hurricanes, droughts, tropical cyclones, rainstorm floods, heat waves, low-temperature disasters, lightning, tornadoes, dust storms, hail, frost, fog, haze, and others. They are the most frequent causes of the disaster events among all natural hazards.

Local authorities must be ready for constant monitoring and technical-engineering works in such areas. Good examples of monitoring organization and engineering works are demonstrated and suggested in some different areas. But sometimes people do not pay enough attention to the problems. Sometimes it is necessary to evaluate whether to reconstruct the object after disaster event or to change the place for another similar construction or living.

Sure the best way is to forecast disaster events and provide protective measures in advance. Life and work in areas of high natural risk demand knowledge, resources, equipment and willing to be ready for prognosis, forecast, people education, and information. In case of disaster events, it is necessary to be ready for the consequence liquidation and the territories and object reparation. The most important thing is to provide help to people. Sometimes people have to live in such dangerous places. It is necessary for people living under natural risk to understand and estimate this risk and to know how to overcome it and how to act in case of crises events. It is necessary to elect and appoint responsible people with good knowledge and special education for managerial posts. Risk management concept is a good instrument for systematic approach to the problems of the rational land use decision.

1.1. World Conference on Disaster Risk Reduction

The World Conference on Disaster Risk Reduction is a series of United Nations conferences focusing on disaster and climate risk management in the context of sustainable development.

There were three conferences: in Yokohama in 1994, in Kobe in 2005, and in Sendai in 2015. As requested by the UN General Assembly, the United Nations Office for Disaster Risk Reduction [United Nations International Strategy for Disaster Reduction (UNISDR)] served as the coordinating body for the Second and Third UN World Conference on Disaster Reduction in 2005 and 2015.

The second conference accepted the Hyogo Framework for Action 2005–2015: Building the Resilience of Nations and Communities to Disasters in 2005 and the Yokohama Strategy and Plan of Action for a Safer World in 1994.

The Third UN World conference accepted the Sendai Framework for Disaster Risk Reduction 2015–2030.

2. Natural risk

Natural risk is a relatively new and not fully explored concept. There are many definitions of natural risk. And often a scientific study or a scientific approach to the problem begins with a presentation of the author's position and the choice of the definition of natural risk for the problem [4–13].This individualistic approach is difficult to avoid. Spores are carried out so far, for example, if there is a risk without material damage to people or not.

If one of the main systematic approaches to hazard research is their classification, so now also the concept of risk **management** can be considered as new step of science development and new basement for systematic hazards investigations.

Development of the **risk** concept demands the promotion of the methods for **risk assessment** and calculation. It makes the theory of **risk** the scientific discipline with good mathematical background. It is necessary to elaborate common approaches to the risk calculation for different types of natural hazards. The methods of seismic risk assessment as the most promoted ones must be spread to landslides, karst, suffusion, flooding, pollution, and other types of natural hazards and risks and also to complex and multi-risk.

Arising from everyday life, gambling, finance, business, and building the **risk** concept became the subject for scientific research and basement for systematic investigations of natural and man-made hazards and disasters.

Risk management is an important way to risk reduction. The main aspects of natural risk management could be considered as risk assessment and mapping, monitoring, and engineering methods for rational land use.

Geological risk management includes:

1. Hazard identification
2. Vulnerability evaluation
3. Risk analysis
4. Concept of acceptable risk
5. Risk assessment
6. Risk mapping
7. Measures for risk reduction:
   - Legislative
   - Organizational and administrative
   - Economic, including insurance
   - Engineering and technical
   - Modelling
   - Monitoring
   - Information

Vulnerability to natural hazards and disasters depends on location, frequency of dangerous events, type of human activity in the area, and other factors.

Systematic approach to the natural hazard research on the base of risk concept is a very fruitful and progressive method.

According to the most common definition, the risk is the probability of the natural hazard event multiplied by the possible damage:

$$R = P \times D, \tag{1}$$

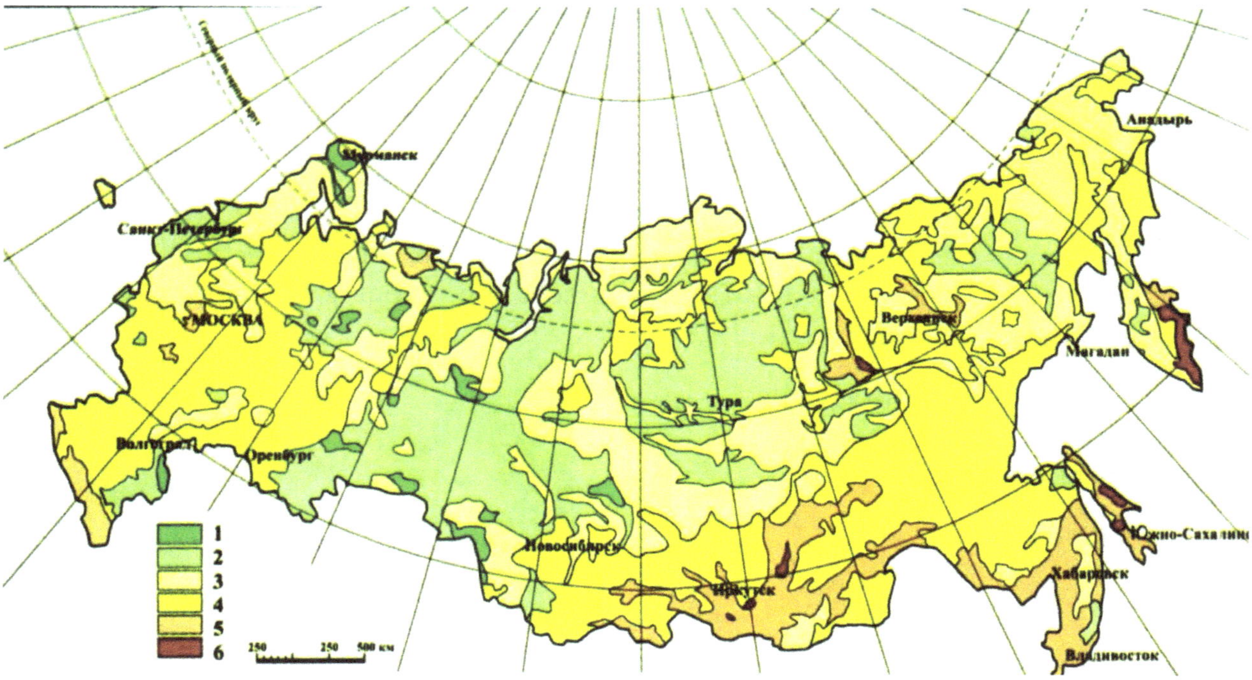

Figure 1. Map of the natural economic risk of construction development and land use of the territory of the Russian Federation [5]. One point corresponds to the average annual damage of 1 million rubles per year (in 1990 prices) on an area of 20 thousand square km): 1, very small (<2). 2, small (10–2). 3, medium (20–10). 4, significant (80–20). 5, large (200–80). 6, huge (>200).

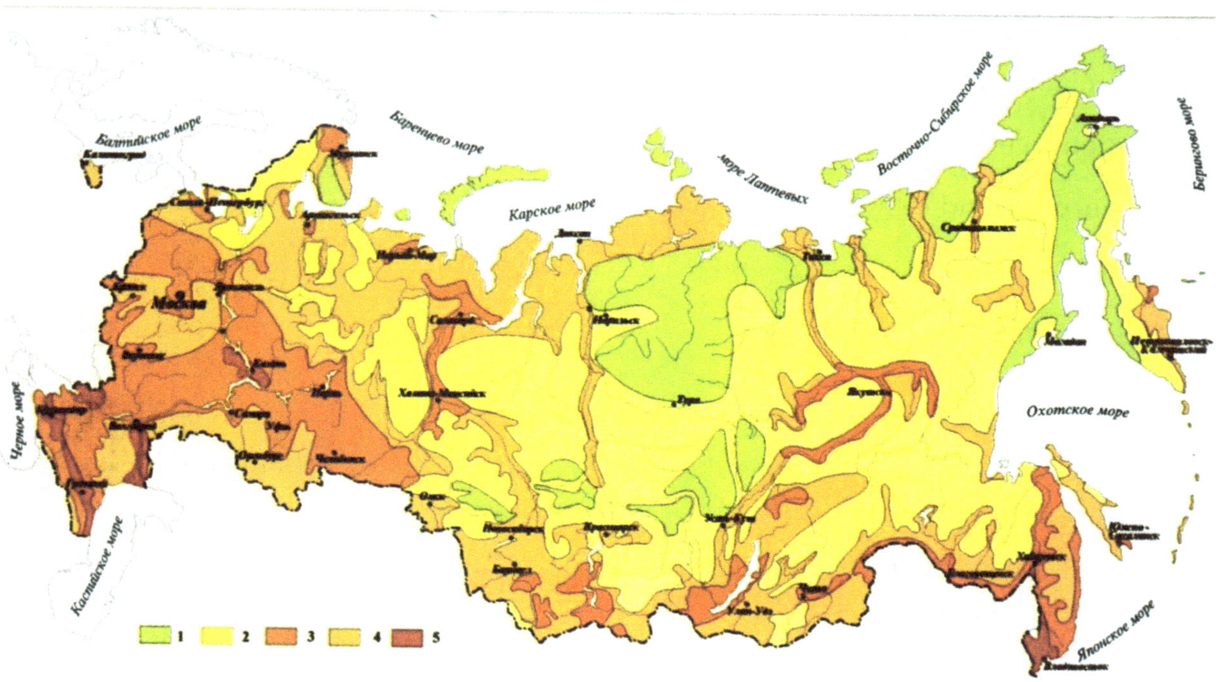

Figure 2. Map of the natural disasters on the territory of Russia, caused by earthquakes, floods, cyclones, squalls, tornados, heavy rains, snowfalls, snowstorms, hail, snow avalanches, and landslides (A. L. Shnyparkov). Frequency of occurrence (cases/year): 1, $<10^{-5}$; 2, 10^{-5}–10^{-4}; 3, 10^{-4}–10^{-3}; 4, 10^{-3}–10^{-2}; 5, $>10^{-2}$.

Figure 3. Map of individual seismic risk for Russian Federation, 2005. IEG RAS (www.geoenv.ru).

where R is risk, P probability, and D damage.

For multi-risk assessment it is possible to use sum of risks of different hazards:

$$R = \sum R_i \tag{2}$$

For risk map construction, it is necessary to use the natural hazards maps and maps of possible damage. These maps can be of local, regional, federal (sub global), and global levels. It is necessary to use big databases and data banks and GIS technologies for such map constructions.

Areas of possible disaster events could be the places of the highest risk at the natural risk maps of the territories.

On the base of this approach, different risk maps and natural hazard maps can be constructed (**Figures 1–3**).

3. Monitoring systems for natural hazards

3.1. Topsides Induced Acceleration Monitoring System (TIAMS) for oil and gas offshore platforms

Early warning system can be elaborated on the base of analysis of seismological phone changes.

The "system monitoring acceleration induced on the upper part of the offshore oil and gas platforms" was developed for deposits Lunskoe-A (LUN-A) and Piltun-Astokhskoye (PA-B) for Sakhalin-2 project.

The system was developed by Sergeev Institute of Environmental Geoscience RAS (IEG RAS) and was intended to ensure the safety of the operation of these platforms [14–17].

This system can be used for safety operation of environmentally hazardous facilities like pipelines, nuclear power plants, chemical industry, etc.

According to the Sakhalin II Project, Sakhalin Energy Investment Company was building offshore oil and gas platforms PA-B and LUN-A at the Sakhalin Island shelf within the seismically dangerous area where destructive earthquakes are likely to occur (**Figure 4**).

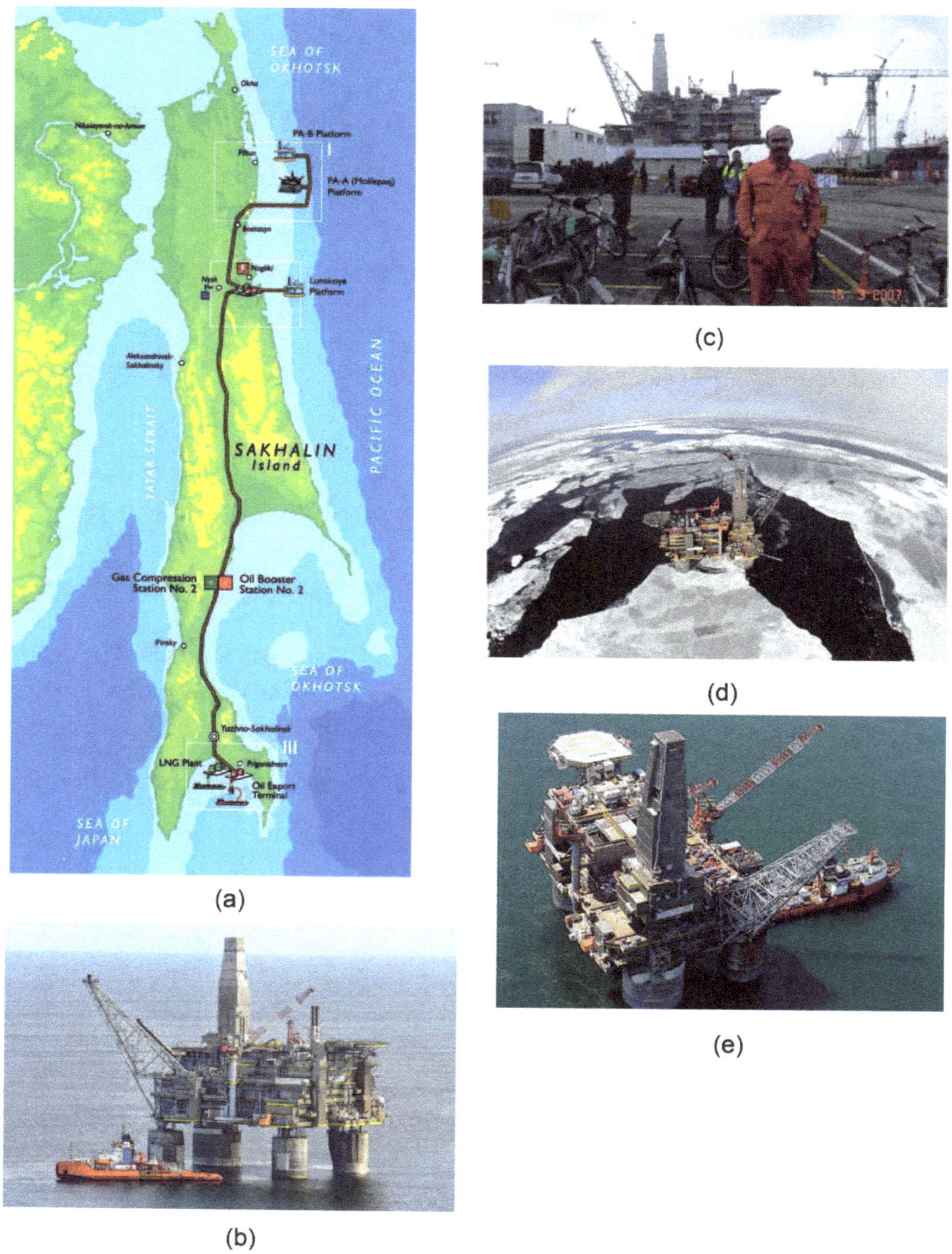

Figure 4. (a) Sakhalin Island and oil-gas platforms. (b) Platform for PA-B deposit. (c) Commissioning works in South Korea. (d, e) Platform for Lunskoe-A (LUN-A) deposit.

To reduce the risk of environmental accidents that can appear during oil and gas production as a result of destructive earthquake, Client took a decision to provide platforms with Topsides Induced Acceleration Monitoring System (further referred as TIAMS).

Function of the TIAMS is to distinguish dangerous earthquakes from other impacts induced to the platform. Ice impacts, ship impacts, wave impacts, drill snatch, etc. also can cause accelerations at the topsides of the platforms. The TIAMS will initiate the emergency shutdown signal (ESD) in case the destructive earthquake has been detected. The signal will be done if acceleration level will exceed the threshold of 0.5 g in any key point of the platform.

3.2. Landslide monitoring system for coastal slope of the river Yenissei

Experience of the system creation was used for real-time early warning landslide monitoring system construction. This system was successfully used for landslide monitoring of coastal slope of the river Yenissei.

Geohazard monitoring system designed to monitor landslide at the coastal slope of the river Yenissei in real time [15–17]. The system provides rapid collection of measurement data on the state of the observed landslides, processing, and analysis of monitoring results (**Figure 5a**).

The monitoring system provides collecting, processing, and distributing data. It includes eight mass displacements of ground points, two points of monitoring changes in the level of groundwater, and automatic workplace of a geologist (**Figure 5b–d**).

Equipment set deep frame is designed to measure linear displacement by its transformation into a digital code (**Figure 5b**). Complete registration of groundwater level is designed for continuous automated measurement level, water temperature, and atmospheric pressure well, and transfer of the measurement results in digital form (**Figure 5b–d**).

The monitoring system has two operating modes: normal and abnormal. If the ground speed displacement mass or velocity of groundwater level changes less than threshold, the information is recorded, analyzed, and compared with data obtained previously. When the speed of the displacement of soil mass or rate of change of groundwater level exceeds a predetermined threshold, the equipment sends an alarm. Alarm is the basis for decisions on a more detailed examination of the coastal slope and, if necessary, the evacuation of people from the building and further strengthening of the coastal slope (**Figure 5e,f**).

3.3. Landslide monitoring system for objects of the 2014 Olympics in Sochi

It is necessary to elaborate specific monitoring system for every type of landslide. One of the case studies was mountain area of the 2014 Winter Olympics in Sochi, Caucuses. The main geologic hazards along the road from Adler to Krasnaya Polyana are landslides. Monitoring systems are installed at a number of Olympic structures (**Figure 6**).

Also monitoring systems were used during construction of the roads.

It was determined that the most widespread type of landslides within the study area is a debris slide. Several sites, especially ones at the beginning of the route, exhibited block-type landslides of compression extrusion.

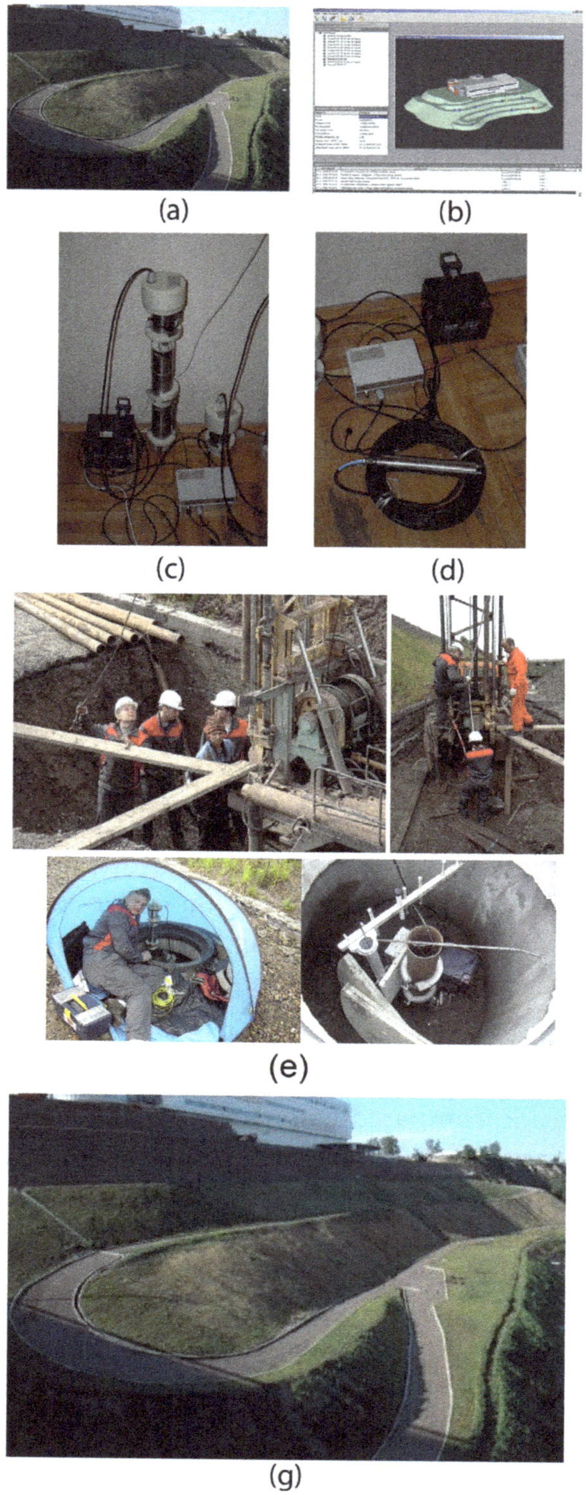

Figure 5. (a) Trade and amusement complex "JUNE," located on the monitored coastal slope of the river Yenissei. (b) The main window of the work program. (c) Equipment set deep frame. (d) Complete registration of groundwater levels. (e) Installation and commissioning of geohazard monitoring system (Ginzburg). (f) After installation and commissioning of geohazard monitoring system (Ginzburg).

Two general methods of observations were accepted in the automatic monitoring system along the combined highway and railway: (1) extensometer and (2) inclinometric measurement in

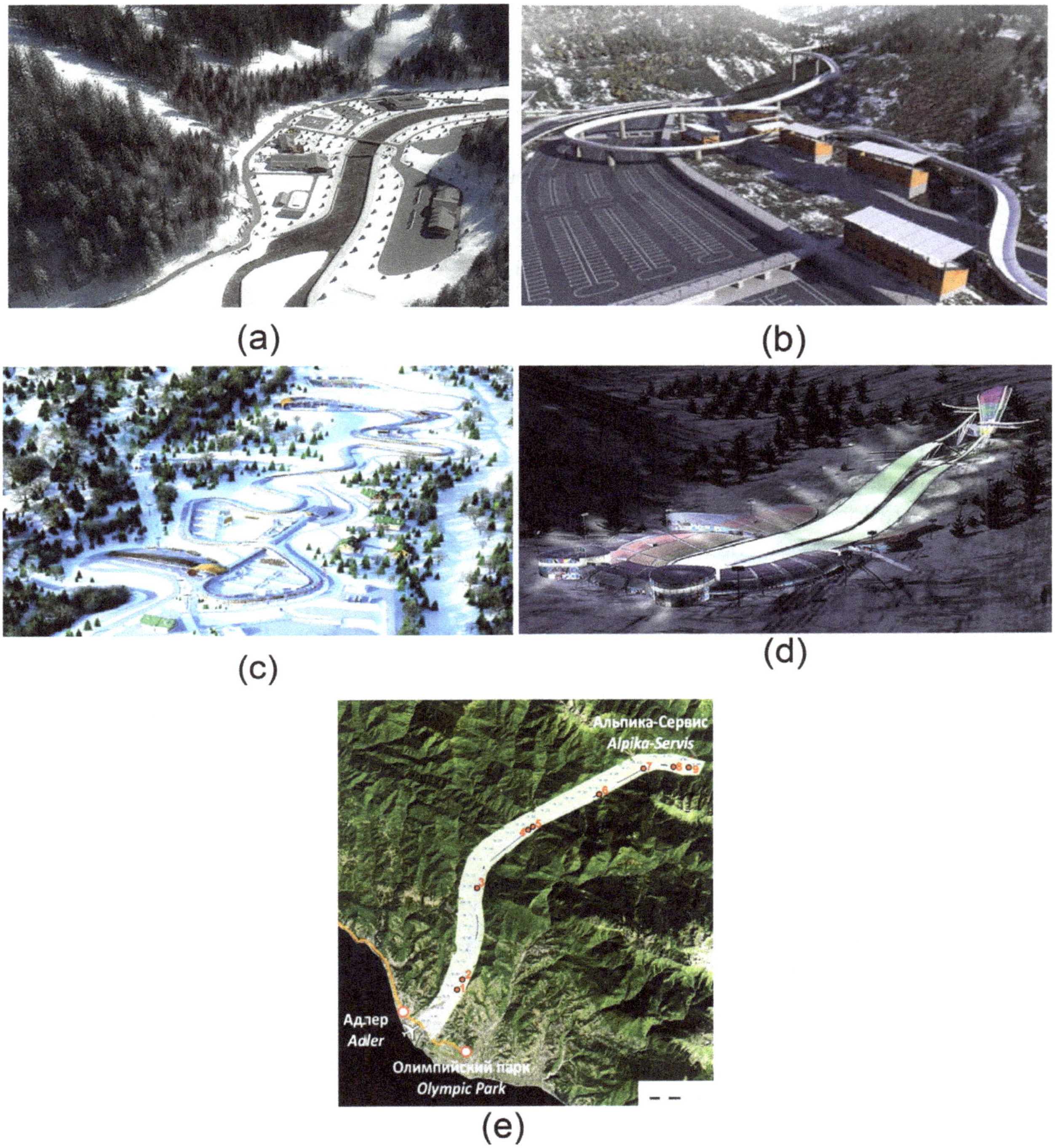

Figure 6. (a) Ski slope. (b) Bobsleigh track. (c) Bobsleigh track. (d) Ski jump. (e) The location of landslide sites along the combined road from Adler to Alpika service (Krasnaya Polyana).

drills. The results of research have shown that the most useful parameters related to the characterization of an active landslide state and sliding dynamics are landslide displacement velocity, depths of slip surfaces, and propagation of active displacements within the territory [17].

Landslide hazard criteria were proposed for the constructions of the road based on the monitoring data of an active landslide along the railway from Adler to Krasnaya Polyana. Several monitoring methods as related to the landslide hazard were recommended along the

Adler-Krasnaya Polyana railway: automatic observations of displacements over the slope surface using extensometers and inclinometers (during site visits and in partly automatic mode).

3.4. Russia-Turkey gas pipeline "Blue Stream" monitoring system

The construction of the Russia-Turkey gas pipeline "Blue Stream" was accomplished in 2002 (**Figure 7**).

The pipeline route has a total length of 1226 km. It crosses the Black Sea. The pipeline crosses the northwestern slopes of the Big Caucasus Ridge. Thirty-five landslides are registered there; seven of them are very hazardous. The online operating automatic control system of landslide processes was developed for these sites. The next registering devices were installed at each of the seven sites: the seismic acoustic control unit, the inclinometer control unit, and the groundwater-level control unit.

The measurement complex included the gauge of seismic acoustic emission and two units of data registration and collection. Three-point extensometer was applied for rock mass displacement measurement. The measured data are transferred to the monitoring center, where they are processed using the special software.

Also the remote sensing control based on the high-resolution space and aerial survey is used. The remote survey data are also processed using the special software. The developed monitoring system permits to control the conditions of the landslide-prone slopes and thus ensure the safety of pipeline operation at the site of high geological risk.

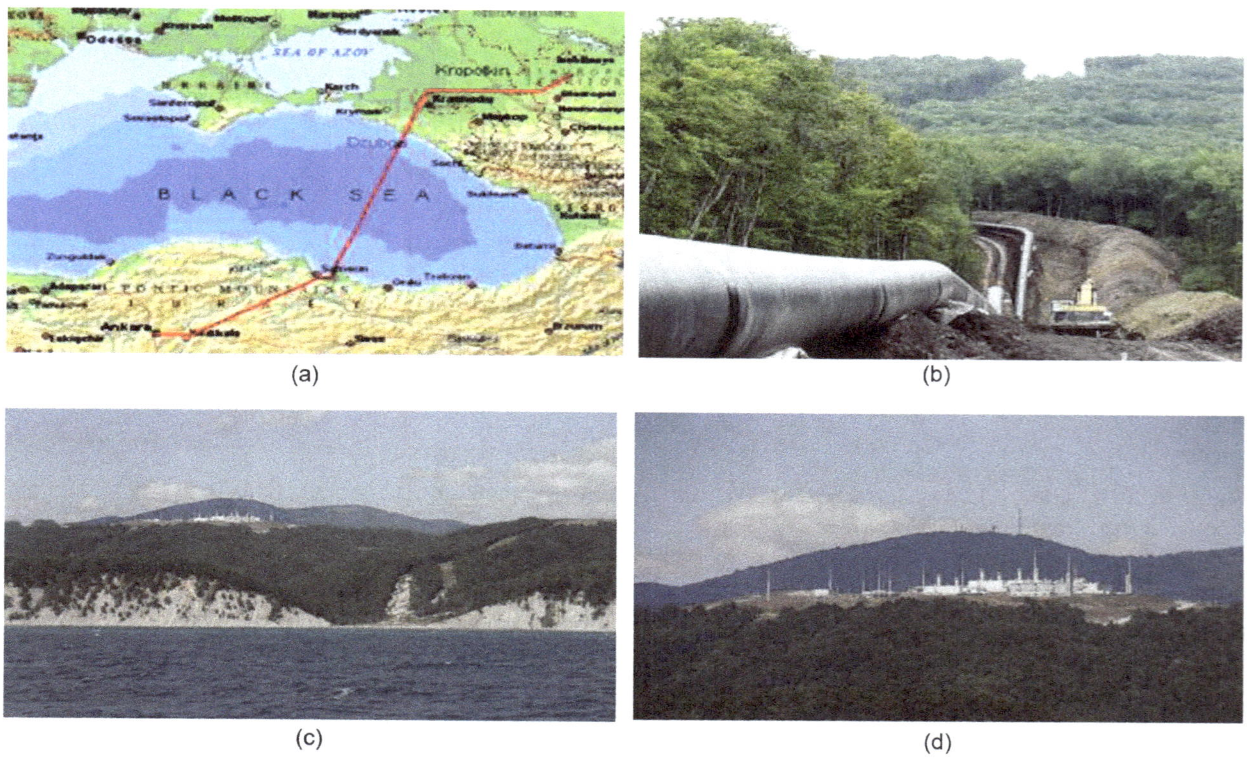

(a) (b) (c) (d)

Figure 7. (a) "Blue Stream" scheme. (b) "Blue Stream" during construction. (c) The compressor station "Beregovaya," view from the sea. (d) The compressor station "Beregovaya," visible glade in which the pipeline is buried.

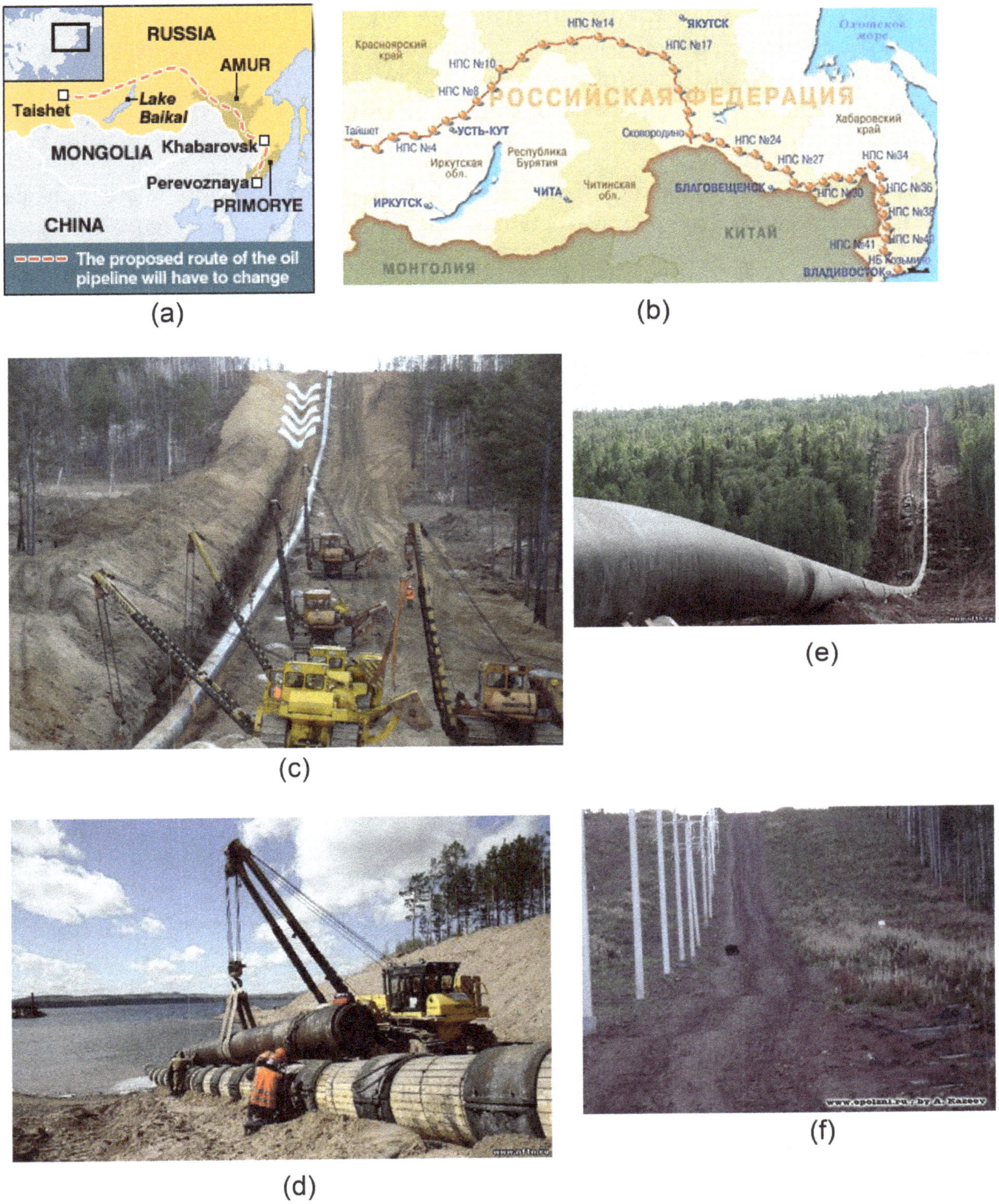

Figure 8. (a, b) Eastern Siberia-Pacific ocean oil pipeline (c, d, e, f). ESPOOP construction.

3.5. Eastern Siberia-Pacific ocean oil pipeline

Similar monitoring system was elaborated and constructed for ESPOOP. The Eastern Siberia-Pacific Ocean oil pipeline (ESPO pipeline or ESPOOP) is a pipeline system for exporting

Russian crude oil to the Asia-Pacific markets (Japan, China, and Korea). The pipeline is built and operated by Russian pipeline company Transneft. The 4857-kilometer pipeline is being laid by the route of Taishet-Kazachinskoye-Skovorodino-Kozmino. Because of protests of environmental organizations, the initial pipeline route was moved 40 kilometers north of Lake Baikal (**Figure 8**).

During the construction of pipelines, the necessity of laying tracks on sloping areas or near them (at the intersection of rivers, construction along the coast, etc.) arises (**Figure 8**).

IEG RAS provided geological research before and during ESPOOP construction and elaborated ESPOOP monitoring system.

4. Conclusions

Risk management concept is a good instrument for systematic approach to the problems of the rational land use. Measures for risk reduction could be legislative; organizational and administrative; economic, including insurance; engineering and technical; modelling; monitoring; and informative. Monitoring system organization and construction are two of the most important methods for natural hazard forecasting, prognosis, and early warning.

Conflict of interest

There is no conflict of interests in the chapter.

Author details

Valentina Svalova

Address all correspondence to: inter@geoenv.ru

Sergeev Institute of Environmental Geoscience, Russian Academy of Sciences, Moscow, Russia

References

[1] Kutepov VM, Sheko AI, Anisimova NG, Burova VN, Victorov AS, et al. Natural Hazards in Russia. Exogenous Geological Hazards. Moscow: KRUK; 2002. 345 p

[2] Osipov VI, Shojgu SK, Vladimirov VA, YuL V, Avdod'in VP, et al. Natural Hazards in Russia. Natural Hazards and Society. Moscow: KRUK; 2002. 245 p

[3] Wirtz A, Kron W, Löw P, Steuer M. The need for data: Natural disasters and the challenges of database management. Natural Hazards. 2014;**70**:135-157

[4] Corominas J, van Westen C, Frattini P, Cascini L, Mallet J-P, et al. Recommendations for the quantitative analysis of landslide risk. Bulletin of Engineering Geology and the Environment. 2014;**73**(2):209-263

[5] Ragozin A, editor. Natural Hazards of Russia. Evaluation and Management of Natural Risk. Moscow: KRUK; 2003; 316 p

[6] Svalova VB. Landslide Risk: Assessment, Management and Reduction. New York: Nova Science Publishers; 2017; 253 p

[7] Svalova VB. Modeling and monitoring for landslide processes. In: Linwood K, editor. Chapter in Book: Natural Disasters—Typhoons and Landslides—Risk Prediction, Crisis Management and Environmental Impacts. NY USA: Nova Science Publishers; 2014. pp. 177-198

[8] Svalova VB. Monitoring and reducing the risk of landslides in Taiwan. Monitoring. Science and Technology. 2016;**3**:13-25

[9] Svalova VB. Landslides modeling, monitoring, risk management and reduction. East European Scientific Journal, Poland. 2016;**7**(11):43-52

[10] Svalova VB. Risk analysis, evaluation and management for landslide processes. Sciences of Europe (Praha, Czech Republic). 2016;**4**(6):15-25

[11] Svalova VB. Landslide risk analysis, management and reduction for urbanized territories. In: Proceedings of WLF4 (World Landslide Forum 4); Ljubljana, Slovenia:Springer; 2017. pp. 439-445

[12] Svalova VB, editor. Risk Assessment. Rijeka: In-Tech; 2018. 380 p

[13] Vranken L, Vantilt G, Van Den Elckhaut M, Vandekerckhove L, Poesen J. Landslide risk assessment in densely populated hilly area. Landslides. 2015;**12**(4):787-798

[14] Svalova VB. Monitoring and modeling of landslide processes. Monitoring Science and Technology. 2011;**2**(7):19-27

[15] Svalova V. TXT-tool 3.007-1.1: Mechanical-mathematical modeling and monitoring for landslide processes. In: Sassa K, Tiwari B, Liu KF, McSaveney M, Strom A, Setiawan H, editors. Landslide Dynamics: ISDR-ICL Landslide Interactive Teaching Tools. Springer; 2018. pp. 315-319

[16] Ginzburg A, Nikolaev A, Svalova V, Manukin A, Savosin V. TXT-tool 2.007-1.1: Monitoring alarm system of landslide and seismic safety for potentially hazardous objects. In: Sassa K et al., editors. Landslide Dynamics: ISDR-ICL Landslide Interactive Teaching Tools. Springer; 2018. pp. 309-325

[17] Ginzburg A, Nikolaev A, Svalova V, Postoev G, Kazeev A. TXT-tool 2.007-1.2 landslide and seismic monitoring system on the base of unified automatic equipment. In: Sassa K et al., editors. Landslide Dynamics: ISDR-ICL Landslide Interactive Teaching Tools. Springer; 2018. pp. 327-340

9

Evaluating the Effectiveness of CCTV in Baltimore, Maryland

Brian Ways and Brooks C. Pearson

Abstract

This report reviews the results of an investigation into the effectiveness of public surveillance systems in Cherry Hill, Baltimore City, Maryland. Its chief objective is to discern whether closed circuit television (CCTV) cameras are a successful crime control measure and whether their implementation displaced crime or generated a diffusion of benefits to the areas outside a camera's viewshed. Previous research on the efficacy of CCTV leaves much to be desired, and the debate is still inconclusive. Thus, this study attempts to move toward new ways of analyzing crime data gathered pre- and post-implementation of cameras in a given target area. To conduct this research, the investigation utilizes a combination of existing and new geographical information systems (GIS) spatial techniques to visualize and measure crime distribution and uses light detection and ranging (LiDAR) data to produce a precise mapping of a camera's line of sight. Findings from each analytical method are compared, contrasted, and combined to provide rounded results. These findings suggest that the implementation of GIS techniques in crime mapping have huge potential and could provide innovative uses for CCTV within law enforcement crime control programs.

Keywords: GIS, CCTV, crime displacement, crime, LiDAR, spatial, crime distributions, surveillance, WDQ analysis, spatial analysis, hot spot analysis, viewshed analysis, crime mapping

1. Introduction

Crime can be defined as a type of action that is considered wrong and punishable by society [1], although some have argued that the term "crime" does not have any basic or widely accepted definition [2]. Crimes of various types happen every day in the world, and each event occurs at an exact geographical location. Whether it is the victim's physical location, a cybercrime IP address, or a computer where data are stored, all crime has a connection to a place. Thus, with the study of crime, emerges the field of crime mapping. Crime mapping has applications to many fields, including law enforcement and geographic information systems. The technology used to map crime can also be employed to monitor, reduce, and/or displace criminal activities. One such example is the use of closed circuit television (CCTV). There are many different arguments and opinions as to whether CCTV cameras actually help to reduce crime or whether they merely displace crime around the corner.

CCTV cameras are currently used in the United States and in other parts of the world [3] with the intention of reducing criminal activity in large urban centers. The city of Baltimore has a history of high crime rates. Of late, this has been directly addressed by one of the most extensive camera systems in the United States, in the hope of improving the lives of citizens and the economic status of the city [4]. However, systematic studies to assess the validity of this approach have not yet been published [5]. For this reason, Baltimore City's extensive use of CCTVs renders it a suitable candidate to determine whether there is a correlation between the implementation of CCTVs and a decrease in criminal activity. More specifically, this study employs impact analysis, as well as geostatistical methods and theory, to characterize any positive/negative associations of crime pattern alterations in CCTV camera viewshed areas in Baltimore. This investigation has the potential to greatly improve our general understanding of the role that CCTV devices play in crime control. It also provides informed recommendations to justify the investment and systematic use of these devices to improve safety in large urban centers. Moreover, this study will potentially increase community awareness of the reliability of crime cameras and thereby influence other law enforcement agencies to follow in Baltimore's footsteps.

Baltimore has had a well-documented history of widespread criminal activity. In 1978, the Baltimore Police Department (BPD) reported 197 homicides, a number that then almost tripled by 1993 to 353 homicides. As recently as 2013, a total of 235 murders were documented in the city. Baltimore's rate of violent crime is above the nation's average, as national violent crime in 2009 summed to 1,318,398 incidents, an equivalent rate of 0.4 incidents per 100 people [6]. In 2003, Baltimore was ranked seventh in total violent crime in the United States, experiencing a total of 11,183 violent crimes and 48,653 total crimes [7].

In an attempt to control and reduce criminal activity, in September of 1995, the Baltimore Police Department was awarded the $75,000 federal Byrne Memorial Grant to start a video control project [5]. The overall expectations for the project were to decrease violence in the downtown business district and to increase business in downtown centers which had become unpopular due to high criminal activity [5]. Sixteen fixed position cameras were installed for $47,000 to view sidewalks, public parks, and streets, with the ultimate goal of monitoring crime activities [5]. Subsequently, in 2004, the mayor of Baltimore visited London and realized that public surveillance technology may help in crime solution and prevention [4]. In CCTV, he hoped for a new and effective combatant to the city's high crime rates.

Currently, the City of Baltimore has one of the most extensive camera systems in the United States. The City has spent over $16 million in acquisition, installation, and maintenance of the

CCTV system. Most of the funding came from federal homeland security grants. The cameras are composed of both fixed- and hard-wired systems and portable units, or Police Observation Devices (PODS). PODS are camera units that can be easily moved around and set up, but footage must be downloaded directly from each digital recording system for post-incident review and analysis [4]. The fixed hardwired systems, on the other hand, are mounted at certain locations and monitored by police officers from a centralized watch center. Officers at the watch center can zoom, pan, and follow a moving object. One Baltimore citizen stated that "From time to time, thanks to surveillance camera, crooks are caught in the act. It helps put down some of the crime that's going on in the area" [8]. The general consensus among Baltimore citizens who live near one of the camera locations is that the system provides a certain degree of increased comfort [5]. Thus, CCTV cameras, including PODS, are useful tools that allow us to characterize and deter criminal activity, serving the community directly.

2. Materials and methods

The methods used to determine whether CCTV cameras displace crime, decrease its frequency, or leave it unchanged are detailed in the following sections. This methodology involves both descriptive and inferential geographical statistics to evaluate criminal instances in Baltimore city, by using the procedures of data collection, study area identification, spatial analysis tools, map creation, and data analysis. Also examined are descriptive statistics for overall crime, significant changes in crime rates, and if these changes are due to the crime cameras.

3. Data collection

This study uses data collected by the Baltimore City Police Department (BPD). In order to obtain the required data for this study, a request letter was sent to the commissioner of BPD, Mr. Anthony W. Batts. Once the commissioner approved the release of requested data, a subsequent formal request to export the data was sent to the BPD Information Technology Department. The BPD provided camera locations with installation dates for all cameras installed throughout the city from 2005 to 2014. Additionally, BPD provided victim-based incident crime data for the period of January 2009 through January 2015 for the entire city of Baltimore. The crime camera data were given in a spatial events file used in GIS software (shapefile format), which contains the field attributes of camera number, location, project type, date installed, latitude, and longitude. BPD was not able to supply victim-based incident crime data from the year January 2003 to January 2009 due to a fire in the police department. Alternatively, the victim-based incident crime data from January 2003 to January 2009 were obtained from John Hopkins University GIS department's library database. The attributes that came with this data were nearly identical.

The data were collected by BPD using protocol GPS handheld devices to record the exact location of each incident. Additionally, BPD uses satellite imagery to record the latitude and longitude coordinates of the precise location of the crime for the crime camera data. The data are projected to Maryland's NAD83 state plane coordinate system. The crime events are recorded by BPD using CitiStat, which is an extension of the CompStat software used by the entire city of Baltimore. CitiStat was originally a way for the city to handle complaints but has evolved into a program capable of handling many issues within the city, such as policing, homelessness, public works, and waste management. The data in CompStat are collected by the minute and analyzed, then presented in a visual form by BPD [9]. BPD has been working to incorporate the crime cameras into CompStat to make it easier to identify suspects whose criminal activity was recorded by the cameras [4].

Light detection and ranging (LiDAR) data were collected from the US Geological Survey (USGS) website and was captured by USGS for Baltimore City in 2008. LiDAR is a type of remote sensing technology that can measure distance by lighting up a target area with a laser, then computing the reflected light [10]. LiDAR systems collect elevation (z) and positional (x, y) data for the reflected objects scanned by the laser under an aircraft, which can create three-dimensional generations of the landscape with multipoints [11]. This file was downloaded from USGS's website and then converted into multipoint data for this study using Environmental Systems Research Institute (ESRI) ArcMap.

4. Study areas

Following the work of Bowers and Johnson [12], we measured the frequency of crimes that occurred 5 years prior to and 5 years after the installation of cameras in given areas. The cameras used in this study were installed in Baltimore during the month of February in 2008. Camera cluster analysis (CCA) provides additional evaluation of appropriate areas for this study, where cameras have overlapping buffers [13]. It is important that these camera clusters are adequately distanced from individual cameras, to avoid a buffer zone that could conflict with the analysis [13]. A buffer is an area around a map feature that is measured in units of distance or time (ESRI GIS Dictionary, n.d).

Theoretically, there is no reason to examine displacement of crime if crime in the target area has not changed over time [13] because the results would not show any significance. As illustrated in Map 1, downtown Baltimore City is made up of 50 city blocks, which range from Center Street to Camden Street and from Green Street to Cathedral Street with a camera in nearly every intersection [13]. The neighborhood of Cherry Hill was chosen for this study as crime data was available from BPD for 5 years prior to installation in 2008 and 5 years post-installation. Cherry Hill as seen in **Figure 6** is located on the southwest border of Baltimore City and is mostly positioned around an outer edge of the Chesapeake Bay estuary. Cherry Hill's population consisted of 96% African American in 2000 and 94.6% in 2010, compared to 64% in the city. There was a total population of approximately 7772 in 2000 and 8367 in 2010. Cherry Hill's housing consists of low-income public housing, including the largest public housing development in the city. It is a dense urban residential neighborhood. The median rent is $442 in 2000, which is comparatively lower than the city as seen in **Table 1**. Minimal gentrification has taken place during this study's time period, based on the census data. The number of nonfatal shootings per 10,000 residents in Cherry Hill is at a rate of 96.3 compared

Figure 1. WDQ equation.

to Baltimore City at 46 between the years 2005 to 2009 [14]. Baltimore City's level of violent crime is much higher than the national average: violent crimes were reported at a rate of 1,318,398 or 0.4 incidents per 100 people in 2009 (**Figure 1**) [7].

5. WDQ and DiD analysis

This study assesses the hypothesis that the presence of the CCTV crime camera decreases the rate of crime within the visibility of the cameras. An impact analysis test is used to examine the validity of the crime cameras in Baltimore City for crime distribution. In order to perform the weighted displacement quotient (WDQ) and difference-in-differences (DiD) analyses, the following steps were implemented: (1) buffer zones were assigned around the individual selected crime cameras in the study area, (2) displacement was determined using concentric buffer zones, thus allowing for a distinction between a decrease in displacement and the influence of diffusion [12] buffer zones to be implemented to capture the presence of displacement or any change or potential displacement for the crime distribution.

The WDQ was computed using the GIS software ESRI ArcMap (Standard Desktop version 10.2.2) to produce three concentric buffer zones that are placed around each individual CCTV crime camera using the "buffer" tool in the geoprocessing toolbar. The size of each class of buffer is determined by the BPD's internal requirements for crime analysis of CCTV effectiveness for 28 days before and after the camera installation period. Buffer zone "A" represents the project area set at a 0.05 mile radius, buffer zone "B" represents displacement area with a 0.1 mile radius, and buffer zone "C" represents the control area with a 0.2 mile radius. Buffer zone "B" accounts for a proportion of crime committed in buffer zone "C." Displacement of crime is shown by movement of crime sites; area "B" should increase, while crime in buffer zone "A" should decrease. If the results of the buffer WDQ analysis are positive, then there is sufficient evidence for diffusion of benefits; if the results are negative, displacement of crime may be more likely. This analysis is a type of interpretation of the WDQ. Thus, WDQ is used as an indicator of potential diffusion of benefits and displacement of crime prior to and post-camera installation within the target area of the crime camera buffers. WDQ values can be inferior, equal to or greater than 0 and can be used to indicate whether diffusion of benefits of spatial displacement has occurred.

The crime data is separated into six categories: grand theft auto, larceny, robbery, assault, rape, and homicides. Data is then spatially joined to the buffer zones using GIS before being exported into individual feature classes in the project geodatabase. A report generated by ArcMap arranges the crimes within each buffer zone in a tabular format to further visualize the data for pattern analysis. The crime codes are generalized for each crime into one category (e.g., all types of robbery are placed into one category). This is to maintain consistency and to reduce error. The data were linked in ArcGIS using a spatial join to count the number of different crimes for each buffer. A report with charts was subsequently developed to easily visualize these findings. Additionally, a percentage of change was calculated from before and after the comparison. An additional chart was made to visualize the percentage change. Finally, a Difference-in-Differences (DiD) is calculated to quantitatively measure the amount of displacement and diffusion of benefits and to determine how much impact the crime cameras have in the distribution of crime beyond the visibility of the camera and into surrounding areas. In this study, DiD was used to compute the average percentage increase or decrease for displacement within the Baltimore City CCTV crime camera's visibility zone, taking the values for buffer zone "A" and subtracting them from the values from buffer zone "C".

6. Spatial analysis and map creation

Tools used to visually display the patterns of CCTV crime camera distribution in Baltimore City consist of spatial analyses and map generations. Environmental Systems Research Institute's (ESRI) ArcGIS ArcMap Desktop Standard 10.2.2 is used in this study to perform these two functions. Spatial data included in the project are: crime camera locations, crime events, streets where crimes occurred, neighborhood boundaries, digital elevation model, Bing aerial base map, traffic lights, neighborhood boundaries, buildings, county boundaries, and municipality boundaries. GIS software is employed in this study to create maps and to implement spatial analysis tools to evaluate whether the presence of the CCTV crime camera decreases the rate of crime within the visibility of the cameras. All geographic files for the entire study are assembled into an ArcCatalog Geodatabase. To properly map the crime data from the cameras in Baltimore City using crime mapping, hot spots must be identified. The following type of spatial analysis tools and scripts were used to identify hot spots in ESRI ArcMap: fishnet, spatial join, incremental spatial auto-correlation, and optimized hot spot analysis. In order to characterize the occurrence of crime diffusion of benefits and displacement, hot spot analysis maps are created.

The process to create these maps required the determination of a grid size for the aggregated crime data up to 5 years before and 5 years after the installation. The tool in ArcMap to create a grid is Fishnet. In order to determine the appropriate grid size for the crime data, the formula $\left(\frac{A}{N}\right)^{\frac{1}{2}}$ [15] is used, where A represents the study area, the Cherry Hill neighborhood, and N is the number of crime incidents.

Subsequently, the number of incidents within each grid cell is calculated in ArcMap using the spatial join tool. The incremental spatial auto-correlation tool is then used in ArcMap to determine the fixed distance band, which uses a critical distance to decide what neighbors to use in the grid cells. This ensures that the scale of the analysis will not change and will be consistent across the study area. Every feature within a fixed distance is included, and every feature outside that distance is excluded. The fixed distance band is ideal for point data, as it ensures that all points have a neighbor. The tool performs the Global Moran's I statistic for a series of increasing distances to measure the amount of clustering for each distance. The output of the incremental spatial auto-correlation tool will produce a report and a graph of "*Z-scores*" at each distance. The peak value of the z-score distance is used for the fixed band distance value. Then the optimized hot spot tool will be used to produce the "*Z-scores*" and *p-values* of the crime data. A small *p-value* and high "*Z-score*" will indicate a substantial hotspot. Alternatively, a low "*Z-score*" and small *p-value* will show a significant cold spot. A "*Z-score*" near zero means minimal or no spatial clustering. The optimized hot spot analysis tool uses the hot spot analysis (Getis-Ord Gi*) tool using parameters from the count of the crime incidents in each grid cell from the fishnet tool with the spatial join. The output of this tool creates a "*Z-score*", *p-value* and GI_Bin result for each feature. The GI_Bin identifies hotspots corrected for multiple testing and spatial dependences suing the false discovery rate (FDR) method.

This study also uses the spatial analysis method for calculating the viewshed of the crime cameras, using the viewshed analysis tool in ArcMap. The LiDAR data in the study area is converted into a multipoint using ArcMap. The multipoint is then converted into a raster using ArcMap point to raster tool, which will create a raster elevation model. The parameter used in the analysis for the viewshed of the crime cameras is the crime camera height of 25 feet. This is called the offset, the vertical distance in surface units to be added to the *z-value* of the location of the surface. In this case, it is called OFFSETA, which adds the vertical distance to the *z-value* of the observation point. A viewshed analysis requires two input data sets, usually a point layer of viewpoints and an input of a DEM or TIN, which represents the land surface [16]. In this case, the inputs for the tool are the raster elevation model derived from the LiDAR data and the Cherry Hill crime camera points. The result of the viewshed analysis is a raster that categorizes cells into the visible and nonvisible classes [16].

7. WDQ and DiD analysis results

The WDQ Chart as seen in **Figure 2** displays the amount of each crime incident before the installment of the crime camera and after. These reported crime incident totals for each buffer area are variables to produce the WDQ results.

It is worth noting that in 2008, the "Before" column displays a low crime incidence value, due to crime taking place in the month prior to installation in February. The "After" column

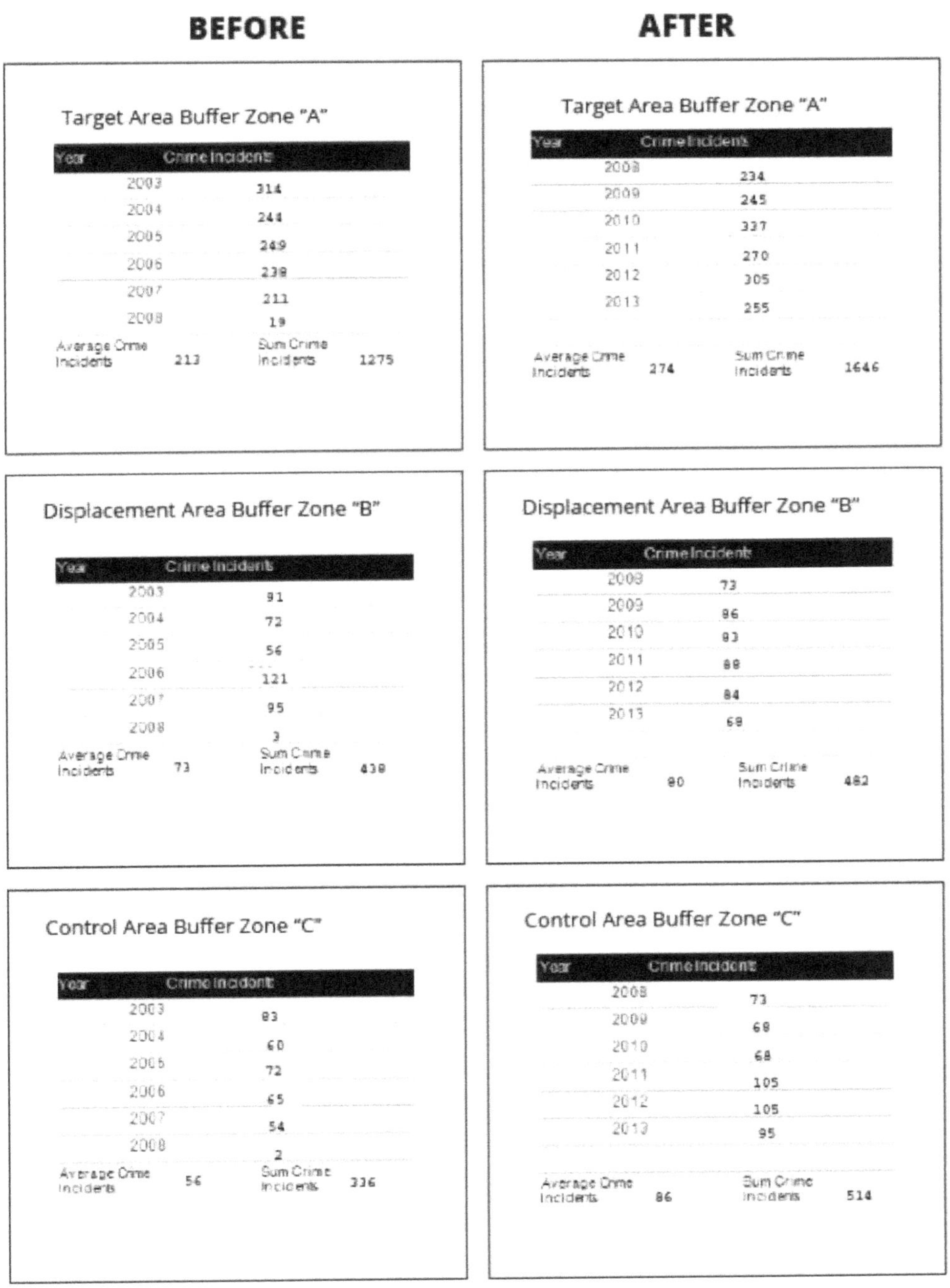

BEFORE

Target Area Buffer Zone "A"

Year	Crime Incidents
2003	314
2004	244
2005	249
2006	238
2007	211
2008	19
Average Crime Incidents 213	Sum Crime Incidents 1275

Displacement Area Buffer Zone "B"

Year	Crime Incidents
2003	91
2004	72
2005	56
2006	121
2007	95
2008	3
Average Crime Incidents 73	Sum Crime Incidents 438

Control Area Buffer Zone "C"

Year	Crime Incidents
2003	83
2004	60
2005	72
2006	65
2007	54
2008	2
Average Crime Incidents 56	Sum Crime Incidents 336

AFTER

Target Area Buffer Zone "A"

Year	Crime Incidents
2008	234
2009	245
2010	337
2011	270
2012	305
2013	255
Average Crime Incidents 274	Sum Crime Incidents 1646

Displacement Area Buffer Zone "B"

Year	Crime Incidents
2008	73
2009	86
2010	83
2011	88
2012	84
2013	68
Average Crime Incidents 80	Sum Crime Incidents 482

Control Area Buffer Zone "C"

Year	Crime Incidents
2008	73
2009	68
2010	68
2011	105
2012	105
2013	95
Average Crime Incidents 86	Sum Crime Incidents 514

Figure 2. WDQ count chart.

indicates crimes committed after installation in 2008. This chart was used to produce a trend graph as seen in **Figure 3**. The intervention line as seen in **Figure 3** displays when the cameras were installed in Cherry Hill at the end of February 2008. This graph shows the crime trend from 2003 to 2014. Overall, the average number of crimes has increased slightly, but not appreciably.

Results from the WDQ seen in **Figure 4** show that post camera installation, the reported crime incidents in the target buffer area (A) increased by 29%. The reported crime incidents in the buffer area (B) increased by 10%. In the control buffer area (C) the reported crime incidents increased by 53%. The success measure for the analysis was at −0.586. This means that the reported crime incidents in the target area were increased, but it performed better than the control area. The buffer displacement measure came in at −0.366 and the weighted displacement quotient (WDQ) at 0.624. Significantly, the buffer displacement and success measure indicate an effective scheme with a diffusion of benefits. It must be remembered that diffusion of benefits only occurs when there is a decrease of crime outside the camera's surveillance area. This means that displacement buffer area (B) suffered proportionally less crime in comparison to the control buffer area (C) after CCTV implementation.

Overall, the WDQ analysis determined that crime in the target area was higher and performed better than the control area. However, total crime in the combined buffer area and target area increased by 25%. The total net effect (TNE) for the Cherry Hill area was the prevention of the possibility of 490 crimes. Conclusively, it can be said that there is a positive benefit to the operation of installing crime cameras in the target area. It should be noted

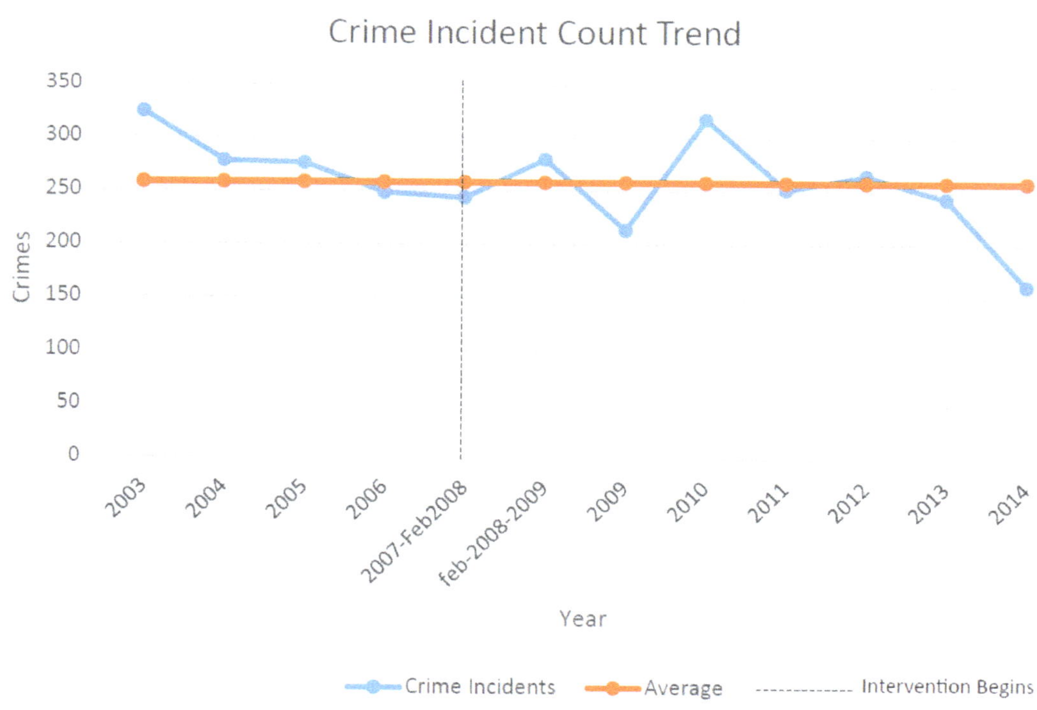

Figure 3. Crime trend in Cherry Hill, Baltimore, MD, 2003–2014.

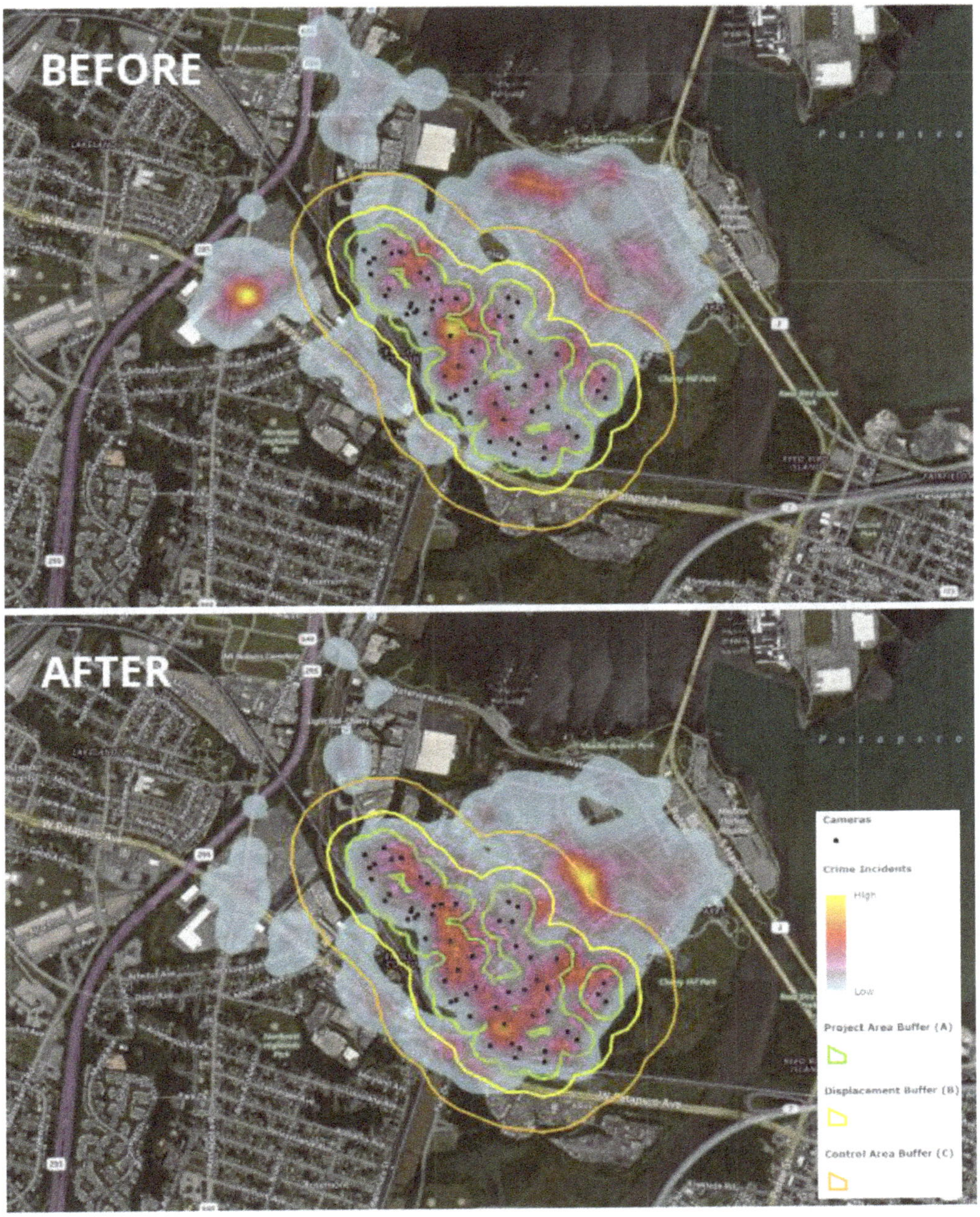

Figure 4. (a) WDQ analysis: before and after; (b) green: buffer A; yellow: buffer B; orange: buffer C.

that although crime incidents after the intervention are higher, this could be due to cameras recording more criminal activity. So, in reality, crime may have been occurring prior to CCTV interpolation but not officially recorded as an incident by police, due to lack of evidence and reportage.

8. Viewshed and hot spot analysis

The viewshed analysis provides a great visual understanding of the visible landscape from the Cherry Hill BPD crime cameras. Seen in **Figure 5** are the results of the viewshed analysis in the study area Cherry Hill, the transparent purple layer representing the area visible by the

Figure 5. Viewshed analysis.

crime cameras, and a transparent red for the non-visible areas. The identifications of buildings are placed on top of the visible or non-visible viewshed layer to demonstrate what is actually seen by the crime cameras.

The viewshed analysis result layer was then used as a parameter to calculate the Getis-Ord Gi* optimized hot spot analysis. The results of the Getis-Ord Gi* optimized hot spot analysis tool created a new feature class of the crime incident data, for pre- and post-installation of CCTV cameras. It was signified by whether it is part of a statistically significant hot spot, a statistically significant cold spot, or if it is not part of any statistically significant cluster as seen in **Figure 6(a)**. The red areas are hot spots, or areas where a high number of crime incidents have taken place. The cold spots would be the blue areas, but there are none on this map for the data. The beige areas represent a statistically insignificant cluster of crime incidents.

The magnitude results from the hot spot analysis can be visualized as a − 0.50774 change rate in average magnitude (in comparing before and after CCTV implementation), outside of the camera's viewshed as seen in **Table 1**. In the viewshed of the camera, the average rate of crime increased with a value of 0.039368. The degree of crime decreased after intervention outside CCTV view, but the level increased after intervention within the cameras' line of vision. The decrease in magnitude of crime incidences outside the camera's viewshed after the installation is analogous to diffusion of benefits. The increase of magnitude inside the camera's viewshed could be comparable to the WDQ results. In the qualitative thematic map produced as seen in **Figure 4**, the change from densest (darker) crime zones from pre- to post-installation offers a strong visual depiction of the magnitude of the measure of crime. Some of the movement

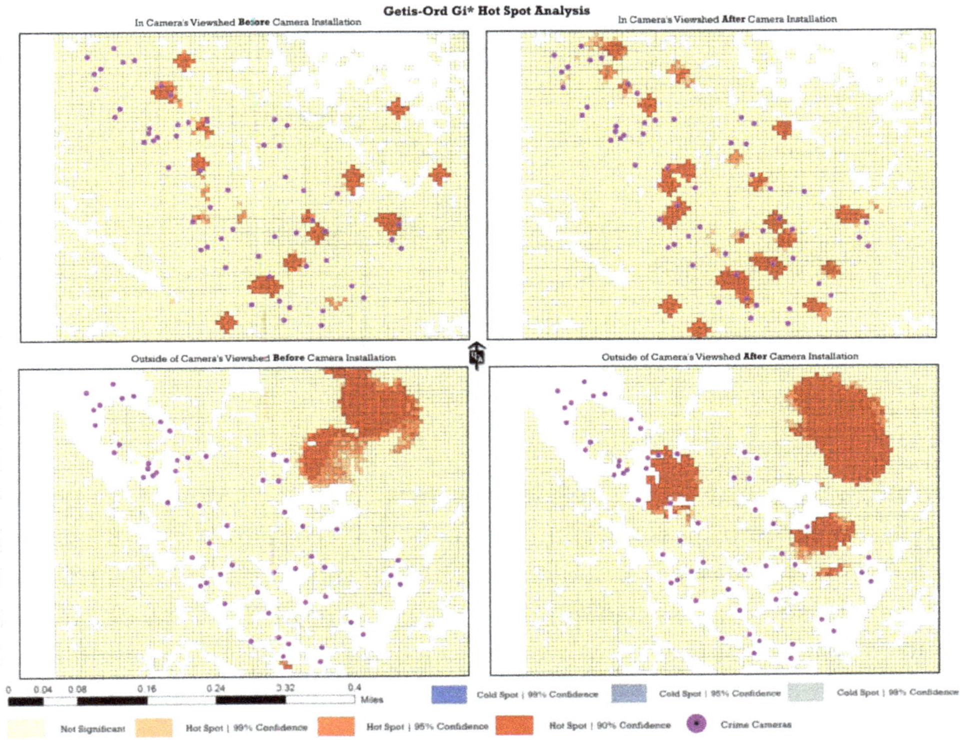

a)

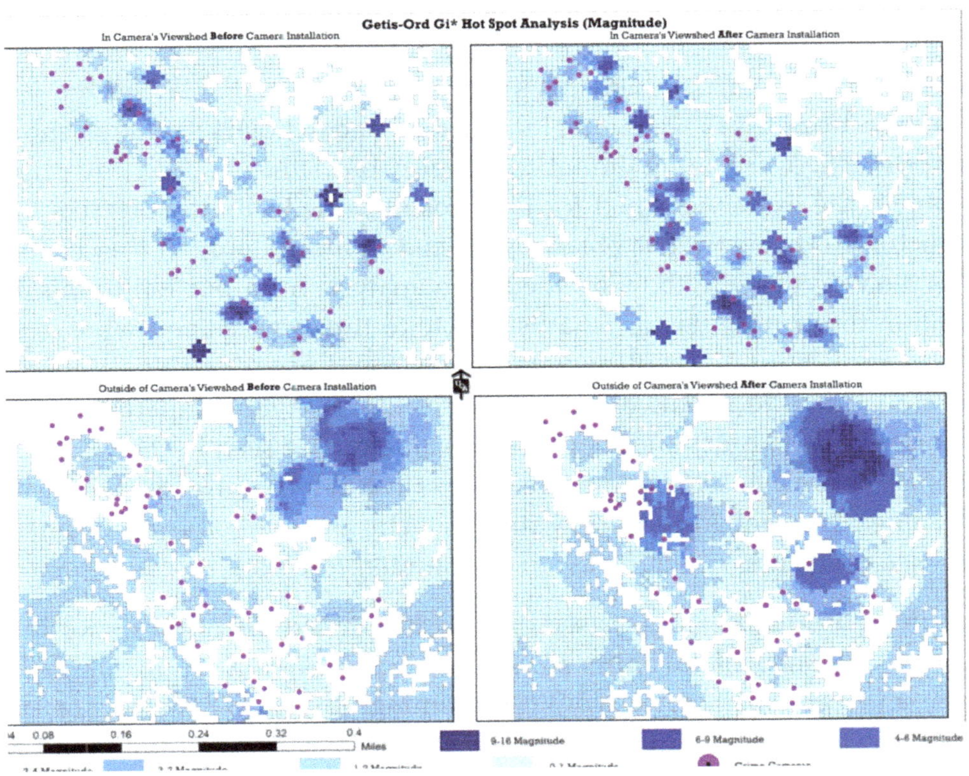

b)

Figure 6. (a) Getis-Ord Gits* optimized hot spot analysis comparison. (b) Getis-Ord Gits* optimized hot spot analysis magnitude comparison.

	Not In Viewshed		
	Before Install	After Install	
Average Magnitude	1.467597	1.416823	
		Change Rate:	-0.050774

	In Viewshed		
	Before Install	After Install	
Average Magnitude	1.028432	1.0678	
		Change Rate:	0.039368

Table 1. Magnitude change rate.

depicted may be due to the awareness of perpetrators of the location of CCTV units, and thus their blind spots, to commit a crime. The results below compare crime rates outside of the viewshed, before and after camera installation. Their findings gesture toward a good method to select new camera sites. These can be obtained by comparing maps of hotspots pre- and post-CCTV fitting, outside of the viewshed, as noticeable, as seen in **Figure 6(a)** and **(b)**.

9. Discussion and conclusion

As opposed to other similar studies, this study differs when displaying the evaluation of crime distribution from CCTV cameras. Its practical and scholarly value is shown through this unique approach, overcoming limitations in previous studies. Other investigations were not able to obtain the exact viewshed for their results, due to limitations like land use and road infrastructures. This work has obtained the exact view of sight of the crime cameras by using LiDAR data to calculate the viewshed. The analysis for a portion of this study is just contained within the boundaries of what is visible to the cameras and what is out of sight: a new practical application that police departments could use to analyze criminal hot spots. The use of LiDAR data to produce a viewshed of exactly what cameras see has not been undertaken prior to this study, and its practical value can be seen in the results. The results from WDQ analysis are exhibited as a decrease of crime in the control area, but an increase of crime in the camera's target area. Along with this tactic, the methods and results of previous studies were also utilized to conclude to a well-rounded result.

The WDQ analysis used buffer sizes based on the BPD procedure measures for camera line of sight. These measurements of the buffer sizes are based on camera vendors, local law enforcement officers, and camera monitor's typical zooming ability. When the hot spot analysis used the exact viewshed to calculate its results, its comparison with the WDQ was similar. This demonstrates that there is a clear benefit to installing crime cameras. It is worth noting that the hot spot analysis calculated the magnitude of the rate of crime. Thus, it can only give an idea of if there is displacement or diffusion of benefits taking place versus the WDQ analysis, as the WDQ results provide an exact measure to demonstrate both displacement and diffusion of benefits. Conversely, it could be argued that the hot spot analysis gives a better idea of criminal incidence patterns due to its recreation of the exact viewshed of the CCTV surveillance. Each method has its strengths and weaknesses, but together, they work in tandem to bolster each other. It should be pointed out that both results of WDQ and the hot spot analysis demonstrate an increase of crime post-installation within the parameters of CCTV surveillance. The analyses of both sets of results also corroborated the decrease of crime outside the camera's surveillance area: for WDQ, surveillance areas mean target area buffer (A) within .05 mile radius, but for the hot spot analysis, this represents its exact viewshed. This could indicate that both results had a diffusion of benefits.

With the use of the weight displacement quotient, it needs to be pointed out that the result measure should not be used as an exact quantification of displacement or diffusion of benefits, but it should be used to indicate the possible existence of these. The only way to obtain this exact measure would be through conducting extensive interviews with large numbers of local criminals. Overall, this type of research approach would be costly and error laden, potentially outweighing its practical benefits. Most research in this vein is with offenders who are incarcerated or on probation.

The crime data used in this research is restricted in interpreting and understanding the 'true' picture of crime. It must be understood that not every crime that occurs is reported then recorded, and that not every crime that is recorded is investigated. This means that caution should always be observed with interpreting and analyzing any crime data. Another issue with the crime data used in this research is that it was not possible to get the exact consistency from BPD for the data from 2003 to 2014. BPD could only supply data from 2009–present, due to a fire in their data facility in 2011. This data could have been extracted, but for an unrealistic cost. The 2003–2008 data were obtained through John Hopkins University's GIS data library: despite claiming that it originated from the BPD, the data differed slightly. With a few adjustments, however, it was successfully reconciled to BPDs'. This provided confidence in using it for this study, but it must be heeded as a limitation. Additionally, policing behavior over time can vary. A change before and after installation of cameras could lead to a change in crime distribution. A further variable to be considered is a possible shift in the demographics in Cherry Hill during this period. One more point of deliberation is over the possibility for human error when dealing with any large data set, such as the crime incident data. The complexity of crime must be re-emphasized, and other uncontrollable variables could play a part in the outcome of the results listed. If these limitations are understood, then they can be better accounted for.

Results from the viewshed and hot spot analysis show a decrease in magnitude outside the area of view of the CCTV. The magnitude of crime increased, however, after the installation of the cameras, in the line of the camera's visibility. This could be accounted for simply by the fact that after camera intervention, more criminal incidents are detectable, versus outside the viewshed where fewer incidents can be recorded. Whether this is directly due to camera footage, more victims coming forth, or police programs actively working in detected hot spots with CCTV is unclear, due to lack of acknowledgement in the crime incident data. Nonetheless, CCTV crime cameras have a small, desirable impact on crime, generally speaking, judging by the presence of diffusion of benefits. It has to be acknowledged that a successful CCCTV operation in a particular area's parameters may not be the same as in another crime fighting location. This is where the opportunity for the use of GIS within policing operation broadens. Research indicates that where people think they are being watched, crime reduces. GIS forecasts based on CCTV could forestall displacement and create a mobile and anticipatory police presence in crime hot spots, thus enhancing the capability of existing CCTV operations.

Baltimore patrons think that the city's public surveillance system prevents crime, and that it helps solve criminal activity that would have not been apprehended otherwise due to lack of identification, failure to recover the weapon, or even the absence of a witness willing to come forward. It has encouraged witnesses to cooperate with the police, where they would not otherwise. This ties in to theses previously proposed in this paper, about the elevated rates of general crime in hot spots post-installation of cameras. Although there are some complications with its usage, it has been observed that key investigations and prosecutions lie in the evidence provided by camera technology. It is important to view the evaluation of the crime surveillance cameras in the context of a larger community policing framework. These devices are not flawless, but just another—albeit useful—crime control and investigative tool. Cameras may make some difference, but this cannot be confidently stated. Ideally, this tool should be used as part of a process with all other law enforcement strategies. This technology is only as effective as the police program running it. GIS techniques merely enhance the CCTV's capabilities.

Author details

Brian Ways* and Brooks C. Pearson

*Address all correspondence to: briways@gmail.com

University of Central Arkansas, Conway, United States

References

[1] Simpson J, Weiner E. The Oxford English Dictionary. 2nd ed. England: Clarendon Press; 1989

[2] Farmer L. The New Oxford Companion to Law. England: Oxford University Press; 2008

[3] Skogan W. Community Policing: Common Impediments to Success: The Past, Present and Future. Washington, DC: The Annie E Casey Foundation; 2004

[4] La Vigne N, Lowry S, Markman J, Dwyer A. Evaluating the Use of Public Surveillance Cameras for Crime Control and Prevention. Washington, DC: The Urban Institute; 2011

[5] Nieto M. Public Video Surveillance: Is it An Effective Crime Prevention Tool? Sacramento: California Research Bureau; 1997. [Accessed: Sep 17, 2016]

[6] FBI: UCR. Crime in the United States. 2003. [Online]. Available from: https://ucr.fbi.gov/crime-in-the-u.s/2003 [Accessed: 2013]

[7] Federal Bureau of Investigations. FBI Releases 2009 Crime Statistics. 2009. [Online]. Available from: https://archives.fbi.gov/archives/news/pressrel/press-releases/fbi-releases-2009-crime-statistics [Accessed: Feb 18, 2015]

[8] Television Broadcast. National: Cable News Network. Turner Broadcasting System, a division of Time Warner. Mar 29, 1997

[9] Rosenberg T. Armed with Data, Fighting More than Crime. 2012. [Online]. Available from: https://opinionator.blogs.nytimes.com/2012/05/02/armed-with-data-fighting-more-than-crime/ [Accessed: Dec 2, 2014]

[10] National Ocean Service. What is LIDAR? 2013. [Online]. Available from: https://oceanservice.noaa.gov/facts/lidar.html [Accessed: Jun 4, 2013]

[11] Zhang K, Chen S, Whitman D, Shyu M, Yan J. A progressive morphological filter for removing nonground measurements from airborne LIDAR data. IEEE Transactions on Geoscience and Remote Sensing. 2003;**41**(4):872-882

[12] Bowers K, Johnson S. Measuring the geographical displacement and diffusion of benefit effects of crime prevention activity. Journal of Quantitative Criminology. 2003;**19**(3):275-301

[13] Lowry S, La Vigne N. Measuring potential diffusion of benefits and crime displacement near public surveillance systems. Geography and Public Safety. 2011;**3**(1):10

[14] Aimes A, Evans M, Fox L, Petteway R, Rutledge R. Neighborhood Health Profile: Cherry Hill. 2008. [Online]. Available from: https://www.baltimorehealth.org/dataresearch/ [Accessed: Sep 17, 2016]

[15] Burt J, Barber G, Rigby D. Elementary Statistics for Geographers. New York: Guilford Press; 2009

[16] Chang K. Introduction to Geographic Information Systems. Boston: McGraw-Hill; 2006

10

Absolute Density Measures Estimation Functions with Very High Resolution Satellite Images

Ana Cristina Gonçalves and Adélia M. O. Sousa

Abstract

Assessment and monitoring of forest structure is frequently done with absolute density measures with field forest inventory data and expansion methods. The development of basal area and the number of trees estimation functions with data derived from very high spatial resolution satellite images enable their short-term and cost-effective evaluation, allowing also the estimation for the area not requiring extrapolation methods. The functions of basal area and the number of trees per hectare are based on crown cover obtained with very high spatial resolution satellite images for two evergreen oaks and umbrella pine. The three tree species are especially important in the agroforestry systems of the Mediterranean region. The linear functions fitted for pure stands of the three species and mixed stands of cork and holm oak and of cork oak and umbrella pine showed a better performance for basal area than for the number of trees per hectare. The inclusion of dummy variables for species composition improved the accuracy of the functions.

Keywords: quickBird image, multi-resolution segmentation, crown horizontal projection, absolute density measures, regression

1. Introduction

Holm oak (*Quercus rotundifolia*), cork oak (*Quercus suber*), and umbrella pine (*Pinus pinea*) are three of the most representative species in agroforestry systems in the Mediterranean basin [1, 2]. Their stands have usually low density, open heterogeneous canopies and as main production bark for cork oak and fruit for all [3–5]. They also have other productions, such as agricultural crops, pasture and extensive grazing. Due to the climate variability of the Mediterranean region, these multiple use systems enable risk dispersion and the sustainability in time [1, 2, 6].

Absolute density measures, especially the number of trees and basal area, are referred in silviculture textbooks as a proxy for competition and growing space at tree level and thus for growing conditions at stand level [7–15]. These measures are reported as a single value per stand, as the sum of the tree values, reported for a standard area, usually the hectare [7, 13, 16–19], and have the advantage of enabling comparison between different stands [7–15].

The conventional approach to calculate the number of trees and basal area is using field plots of forest inventories, which have a statistical sampling design and an expansion factor when working on an area basis [20–25]. Though a high accuracy is attained at field inventory plots, they have disadvantages related to the area sampled and the extrapolation methods needed for the evaluation on an area basis.

Remote sensing-derived data have been used to estimate both the number of trees [26–33] and basal area [26, 29, 30, 32, 34–40]. The advantages are the estimation for small and large areas [25, 41–43]; enabling time series analysis in short time cycles [25, 42]; data collection at different scales, depending on the imagery spatial resolution [42, 43]; and enabling the evaluation of the area not requiring extrapolation methods [44–49]. The disadvantages are related to the imagery spatial resolution; accuracy decreases with the increase of the pixel size [50–52].

2. Materials and methods

2.1. Satellite images and inventory data

Two satellite images were acquired as four-band products (Blue, Green, Red and Near-Infrared), with the bands pan-sharpened with the UNB algorithm [53]. The image from QuickBird satellite (hereafter Mora) was collected in August 2006, has an area of 80 km^2 (central coordinate 8°4′53.98″W and 38°51′16.12″N) and a spatial resolution of 0.7 m. The image from WorldView-2 satellite (hereafter Alcácer) was collected in June 2011, has an area of 35 km^2 (central coordinate 8°40′28.20″W and 38°27′45.71″N) and has a spatial resolution of 0.5 m. The forest area is composed of holm oak and cork oak in the former and cork oak and umbrella pine in the latter.

The images were orthorectified, georeferenced and atmospherically corrected. Ortho-rectification was carried out with the rational polynomial coefficient model (RCP) using the image metadata and ASTER GDEM digital elevation model (resolution 30 m) in Envi 4.8 [54]. For georeferencing the images, field data points obtained with global navigation satellite system (GNSS) were used, which resulted in a root mean square error (RMSE) of 0.49 m for Mora and 0.30 m for Alcácer. Atmospheric correction was done using the image digital number for top of atmosphere (ToA) reflectance to convert it to soil level reflectance using dark-object subtraction method [55].

From each image a vegetation mask per specie by a set of sequential steps was derived, using object-oriented analysis [56, 57]. Contrast split segmentation based on the normalized difference vegetation index image (NDVI, [58] in eCognition, version 8.0.1 [59]) was used to isolate tree crowns, resulting in a vegetation mask. In the second step multi-resolution segmentation

[60] enabled the separation of the vegetation mask in different objects, using an algorithm with scale, shape and compactness as splitting parameters. The third step used the nearest neighbor algorithm to classify the objects per tree specie, in Weka 3.8.0 software [61], based on the descriptive statistics of the bands and on vegetation indices (NDVI, EVI, SAVI, and SR).

2.2. Inventory data

Each image was divided into a square grid, a function of their resolution, 65 × 65 pixels (45.5 × 45.5 m, 2070.25 m^2) for Mora and 90 × 90 pixels (45 × 45 m, 2025 m^2) for Alcácer. The grid area is similar to other studies with very high spatial resolution satellite images (e.g., [62]). Composition and crown cover (defined as the percent ratio between crown horizontal projection and grid area) were derived from the vegetation mask per specie using ArcGIS version 10 [63]. The grids were classified as monospecies if only one specie was present and multispecies otherwise. To design forest inventory, crown cover was grouped in three strata (10–30, 30–50 and >50%) and a random stratified sampling by proportional allocation was applied.

The field surveys took place in 2011 in Mora and 2013–2014 in Alcácer, measuring 17, 24 and 32 monospecies plots of holm oak (QR), cork oak (QS) and umbrella pine (PP), respectively, and 23 and 49 multispecies plots of holm oak and cork oak (QRQS) and cork oak and umbrella pine (QSPP), respectively. The total number of plots is 145. The main stand was defined as all individuals with diameter at breast height equal or larger than 6 cm for holm and cork oaks and 5 cm for umbrella pine. In each plot, for the main stand, diameter at breast height, total height and four crown radii (directions in north, south, east and west) were measured [16], as well as recorded tree location by GNSS. The number of trees per hectare (N) was defined as the sum of all individuals of the main stand per plot, scaled to a reference area, the hectare. Basal area per hectare (G) was defined as the sum of the sectional area of all the trees of the main stand in the plot, scaled to a reference area, the hectare [7, 16].

2.3. Statistical analysis

In the development of the functions for the number of trees and basal area per hectare, four types of functions were considered (**Figure 1**): (i) per composition classes (QR, QS, PP, QRQS and QSPP); (ii) not considering composition; (iii) considering crown cover per specie; and (iv) considering composition as dummy variables. Dummy variables were defined as a dichotomic variable where 1 refers to a specific composition (QR, QS, PP, QRQS or QSPP) and 0 to the other compositions. For (i) and (ii) the functions were fitted using ordinary least square linear regression through the origin (Eq. (1), where D is N or G, CC the crown cover and α the regression coefficient) and for (iii) and (iv) multiple linear regression with stepwise method and AIC as selection criteria (Eq. (2), where α and β are regression coefficients, d the dummy variables, i = QR, QS and PP and j = QR, QS, PP, QRQS and QSPP). For multiple regression, multicolinearity among explanatory variables was analyzed with the variance inflation factor (VIF), where no serious multicoliniarity is expected for VIF $\leq$ 10 [64, 65]. As suggested by several authors (e.g., [66, 67]) the sum of squares of the residuals (SQR), the coefficient of

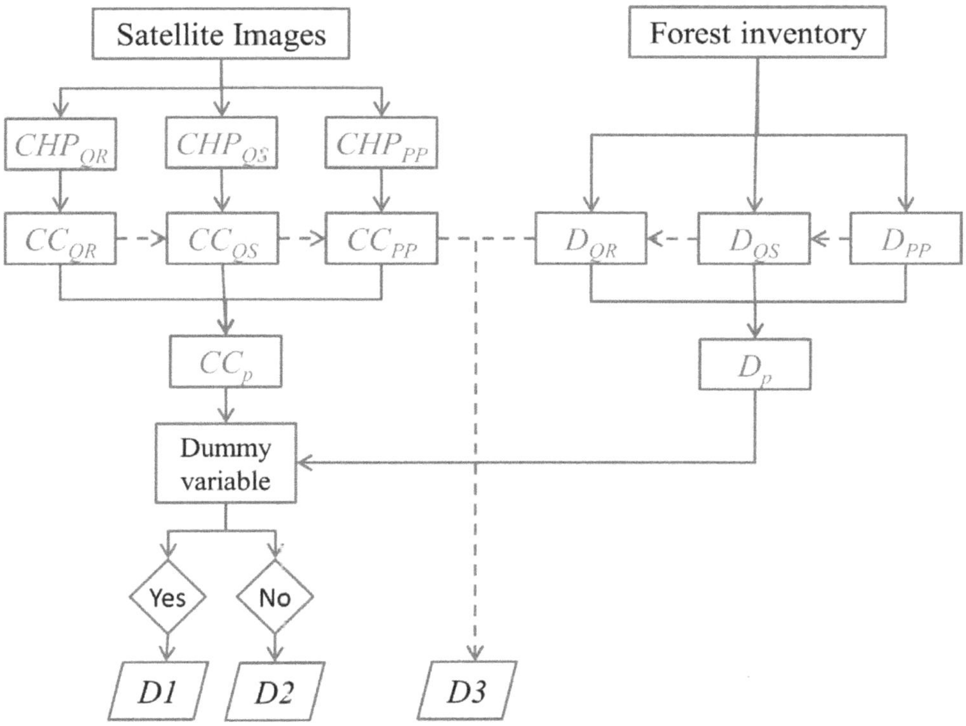

Figure 1. Model flow diagram (where *CHP* the crown horizontal projection in m², CC the crown cover, QR holm oak, QS cork oak, PP umbrella pine, and *p* the plot).

determination (R^2) and the adjusted coefficient of determination (R^2_{aj}) were used to evaluate the statistical properties of the models.

Whenever an independent data set is not available for model validation, re-sampling or jack-knifing methods are recommended. One of these methods uses PRESS residuals, calculated as the difference between the observed and the estimated value of a variable (in this study N or G), as a cross-validation statistic [68–70], which is considered also to improve reliability in the accuracy assessment [43]. PRESS residuals are attained by fitting the model iteratively k times with n-1 observations (where n is the total number of observations) guaranteeing their independence [68–70]. The validation of the models was tested for bias and precision—the former with the mean of the PRESS residuals (Eq. (3), where PRESSrm is the mean of the PRESS residuals, y the observed value, ŷ the estimated value and i the number of the observation) and the latter with the mean of the absolute PRESS residuals values (PRESSram, Eq. (4)). To the former was added the analysis of the 5th and 95th percentile of the PRESS residuals [70]. Heteroscedasticity associated with the error term was evaluated graphically with the plots of the studentized residuals and the estimated values and the normality of the studentized residuals with normal quantile-quantile plots and Shapiro Wilk normality test, for a probability level of 0.001 [68, 71]. To enable comparison with other studies (e.g., [36]), the relative root mean square error (RMSEr, Eq. (5)) was also computed. The statistical analysis was done using R statistical software [72].

$$D = \alpha \times CC \quad (1)$$

$$D = \alpha_i \times CC_i + \beta_j \times d_j \quad (2)$$

$$PRESS_{rm} = \frac{\sum_{i=1}^{n}(y_i - \hat{y}_{i,-1})}{n} \quad (3)$$

$$PRESS_{ram} = \frac{\sum_{i=1}^{n}\left|y_i - \hat{y}_{i,-1}\right|}{n} \quad (4)$$

$$RMSE_r = \frac{\sqrt{\frac{\sum_{i=1}^{n}(y_i - \hat{y}_i)^2}{n}}}{\bar{y}} \times 100 \quad (5)$$

3. Results and discussion

3.1. Satellite image analysis

The vegetation mask per species accuracy resulting from the multi-resolution and object-oriented classification, illustrated in **Figure 2**, depends on several factors, namely the forest tree density, the limits between forest trees and other land uses, image date, forest tree species reflectance and crown shape. Stands with low density and isolated forest trees, such as holm oak, cork oak and umbrella pine in agroforestry systems, enable the distinction and separation of the tree crowns with accuracy in very high spatial resolution satellite images (e.g., [62]), minimizing the confounding effects of shadow pixels which occur in tree crowns at closer spacings [73]. Conversely, the diffused boundaries between forest tree crowns and other land uses can decrease classification accuracy [74]. Thus image dates with the highest contrast between different land uses, especially between herbaceous vegetation and forest trees, such as the summer (June–August) in the Mediterranean region, produce the smallest confounding influence [75, 76]. The other two variables that influence accuracy of the vegetation mask per species are the reflectance similarity between species and crown shapes, the more similar the lower the accuracy. The forest areas of Mora and Alcácer are composed predominantly of isolated trees, though some clusters occur, and the images were acquired in the dry summer period (June and August), thus with high contrast between the forest trees and the other land uses, especially the forest understory that was composed of soil and dry herbaceous vegetation. Regarding reflectance and crown shape, holm oak and cork oak have larger similarities with wide ranges of reflectance and an irregular crown shape when compared to umbrella pine that has a narrower reflectance range and nearly circular crowns. Nonetheless, the spectral signatures of the tree species are different enough to enable a good classification [44, 46–49]. The accuracy evaluated with Kappa statistic [77, 78] encountered true positives for 90% of the holm oak and 87% of the cork oak for Mora and 98% of the umbrella pine and 74% of the cork oak for Alcácer.

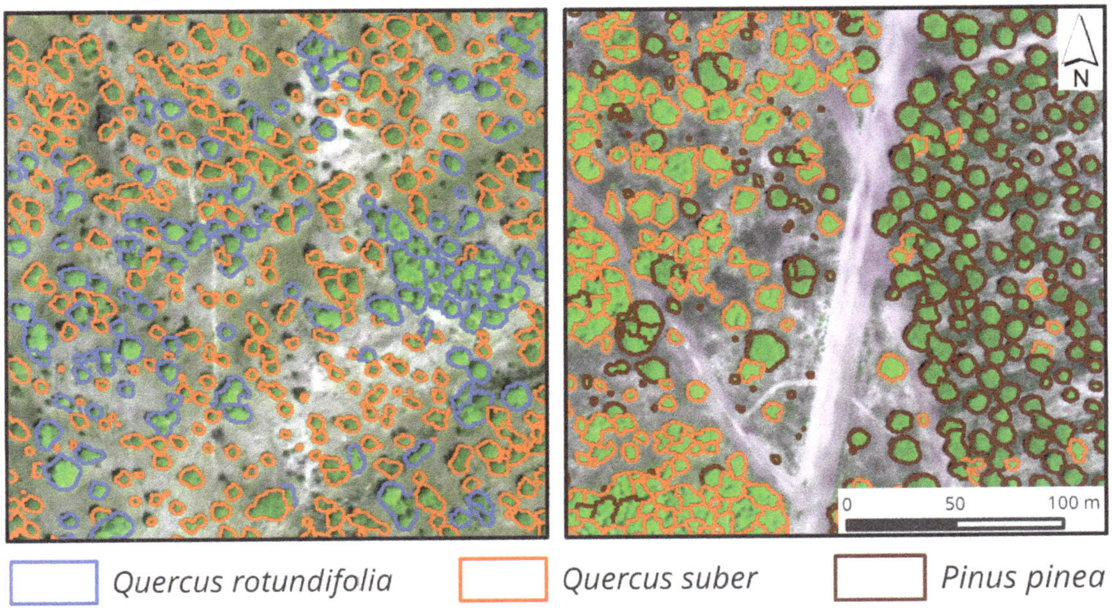

Figure 2. Illustration of vegetation mask for two small areas (left) Mora (cork and holm oak) and (right) Alcácer (cork oak and umbrella pine).

3.2. Number of trees and basal area functions

The variability for N is larger than for G and for QS, QRQS and QSPP than for QR and PP, denoted by the larger coefficient of variation (**Table 1** and **Figure 3**). The number of trees and their spatial distribution in the stand determine their growing space and the tree crown area is also a function of the species epinastic control [12] and spacing, with crown closure originating crown shyness [14], thus unbalancing the relation between crown cover and basal area and crown cover and number of trees. Holm oak, cork oak and umbrella pine have, especially in adult trees, weak epinastic control and are shade intolerant [5, 79], thus justifying the wide variability observed in the plots. However, positive correlations are observed for CC and G for all plots (0.72), stronger for QR (0.86) and QSPP (0.85) and weaker for QS and PP (0.66 each) and QRQS (0.62). These results are expected due to the almost direct relation between crown and stem diameters [80, 81]. Conversely, the correlation between CC and N is weaker when compared to the former. Positive correlations are found for all plots (0.47), stronger for QS (0.72), QR (0.57) and QSPP (0.54) and weaker for QRQS (0.36) and no correlation for PP (−0.04). The number of trees per hectare is frequently associated with the individual stems and the stage of development and the stand structure [7–15], giving rise to a higher variability in the number of trees (**Figure 4**).

In general, for the functions fitted for all composition classes (**Table 2**), there is an improvement in the statistical properties of the models from those not considering composition (N3 and G3) to those that included it as crown cover per specie (N2 and G2), with the best properties attained by those with composition as dummy variables (N1 and G1). For the number of trees, a reduction of −27.4% for SQR and an increase of 3.2% for R^2_{aj} from N3 to N1, a reduction of −18.0% for SQR and an increase of 2.1% for R^2_{aj} from N2 to N1 and a reduction

	Min	Max	Mean	SD	CV (%)	Min	Max	Mean	SD	CV (%)
	All					QR				
N	5	140	67	28.0	42.4	29	116	72	27	37.1
G (m^2 ha^{-1})	2.6	18.2	8.9	3.3	37.5	4.0	11.1	6.7	2.1	32.0
CC (%)	9.7	72.6	37.1	15.1	40.6	13.7	67.6	37.6	15.9	42.3
PHC (m^2)	195.8	1470.0	756.1	306.5	40.5	284.2	1399.0	777.7	329.2	42.3
	QS					PP				
N	5	135	60	33	54.2	5	119	72	23	32.4
G (m^2 ha^{-1})	5.2	15.4	9.1	2.8	30.7	3.8	15.9	9.5	2.6	27.5
CC (%)	13.4	70.5	35.3	14.0	39.7	13.0	66.7	40.5	17.1	42.3
PHC (m^2)	271.3	1460.2	723.9	293.0	40.5	263.5	1350.3	819.1	346.7	42.3
	QRQS					QSPP				
N	19	140	62	29	47.7	15	123	69	30	43.1
G (m^2 ha^{-1})	2.6	14.7	6.7	2.7	40.3	2.9	18.2	10.3	3.7	36.0
CC (%)	14.3	58.8	29.4	11.4	38.6	9.7	72.6	39.8	14.5	36.5
PHC (m^2)	296.9	1216.7	607.8	233.6	38.4	195.8	1470.0	804.9	293.9	36.5

Table 1. Descriptive statistics per composition classes (where N is the number of trees per hectare, G the basal area per hectare, CC the crown cover calculated with satellite data, PHC crown horizontal projection, max the maximum, min the minimum, SD the standard variation and CV the coefficient of variation).

of −8.0% for SQR and an increase of 1.0% for R^2_{aj} from N3 to N2 was observed. For basal area a reduction of −60.0% for SQR and an increase of 2.7% for R^2_{aj} from G3 to G1, a reduction of −53.5% for SQR and an increase of 2.5% for R^2_{aj} from G2 to G1 and a reduction of −4.3% for SQR and an increase of 0.2% for R^2_{aj} from G2 to G3 was observed. This is reflected in the better function performance of N1 or G1 when compared with N2 and N3 or G2 and G3. Bias is the lowest for N1 and G1 while precision is similar for all functions. The better statistical properties of N1 and G1 when compared with N3 and G3 could be, at least partially, explained by the inclusion of the composition variable in the functions as N and G have different behavior per composition classes (cf. **Table 2**). Comparing N1 and N2 or G1 and G2, the better performance of N1 and G1 can be explained by the relations between crown cover and the number of trees or basal area, for the different composition classes and probably by the influence of the clusters of mixed trees. When comparing N1 or G1, with the functions per composition class (Ni or Gi, i = 4,5,6,7,8), better model statistical properties are attained for N4 and N5 and G8 and G4, which correspond to the composition classes with smaller variability between CC and N and G (cf. **Table 2**), respectively. Nonetheless, in general, bias is larger for the number of trees than for the basal area. Similar results were attained for above-ground biomass functions with crown horizontal projection as an explanatory variable [44–46, 49]. The regression coefficients are presented in Eqs. (6)–(9) and **Table 3**.

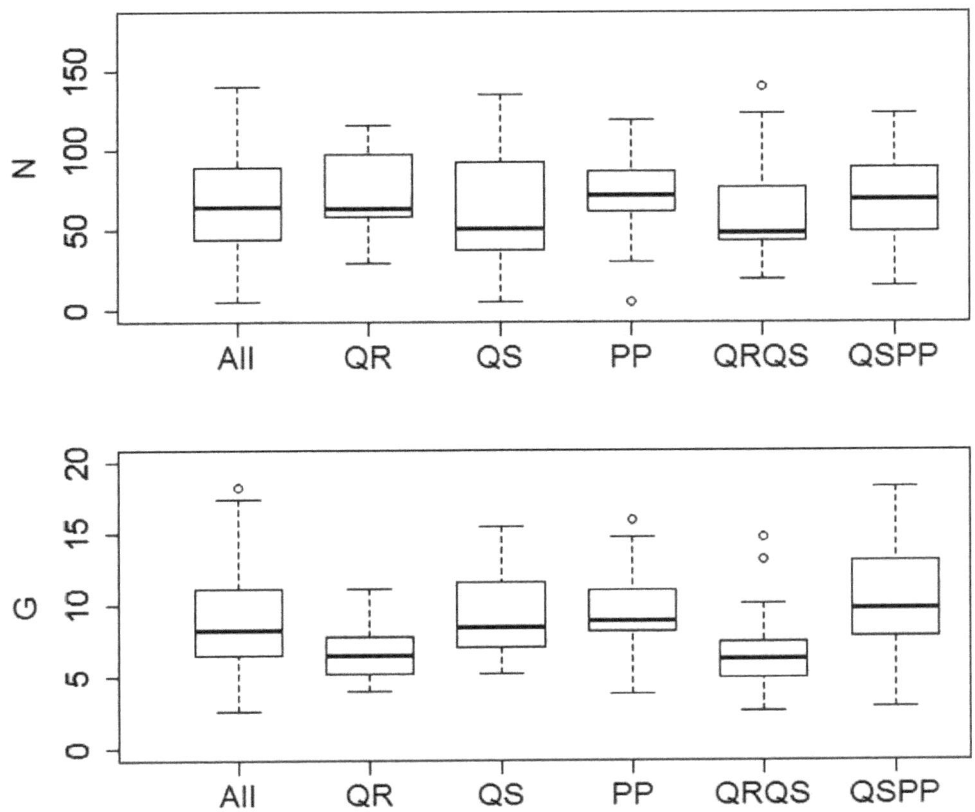

Figure 3. Boxplots of N and G for all plots and per composition class.

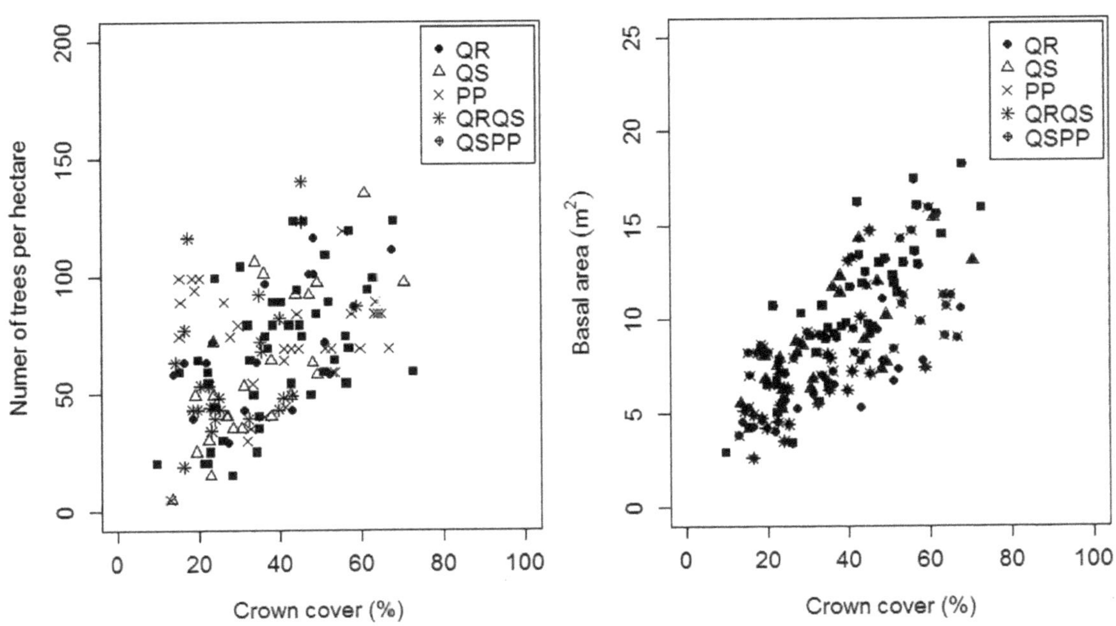

Figure 4. Crown cover and number of trees and basal area per composition class.

Model	Independent variables	SQR	R^2	R^2_{aj}	PRESSrm	PRESSram	Percentile		RMSEr
							5th	95th	
N									
N1	CC, dQR, dQS, dPP, dQRQS, dQSPP	89,148	0.884	0.879	0.002	0.828	−1.362	1.920	37.0
N2	CCQR, CCQS, CCPP	105,215	0.863	0.860	0.145	0.794	−1.178	1.966	40.1
N3	CC	113,595	0.852	0.851	0.172	0.802	−1.105	2.062	41.7
N4	CCQR	9870	0.900	0.893	0.196	0.912	−1.445	1.404	33.7
N5	CCQS	11,176	0.899	0.895	0.014	0.862	−1.146	1.873	35.9
N6	CCPP	37,293	0.795	0.788	0.300	0.776	−0.667	2.113	47.6
N7	CCQRQS	19,842	0.836	0.830	0.148	0.748	−1.303	1.873	44.6
N8	CCQSPP	31,382	0.878	0.875	0.111	0.833	−1.227	1.969	38.0
G									
G1	CC, dQR, dQS, dPP, dQRQS, dQSPP	585	0.955	0.953	0.001	0.794	−1.528	1.763	22.7
G2	CCQR, CCQS, CCPP	898	0.931	0.929	0.171	0.806	−1.690	1.729	28.1
G3	CC	936	0.928	0.927	0.162	0.810	−1.727	1.661	28.7
G4	CCQR	39	0.953	0.950	0.237	0.899	−1.301	1.602	22.8
G5	CCQS	142	0.935	0.932	0.202	0.883	−1.882	1.356	26.6
G6	CCPP	267	0.913	0.910	0.277	0.873	−1.304	1.627	30.5
G7	CCQRQS	136	0.898	0.894	0.136	0.805	−1.168	2.104	34.3
G8	CCQSPP	162	0.971	0.970	0.098	0.776	−1.432	1.654	18.1

Table 2. Statistical properties of the fitted functions (where CC is the crown cover, QR *Quercus rotundifolia,* QS *Querus suber,* PP Pinus pinea and d the dummy variable).

Model	α	Model	α
N3	1.68195	G3	0.22776
N4	1.7749	G4	0.16797
N5	1.7041	G5	0.2436
N6	1.5324	G6	0.21361
N7	1.9839	G7	0.21583
N8	1.65806	G8	0.255849

Table 3. Regression coefficients of the N and G functions.

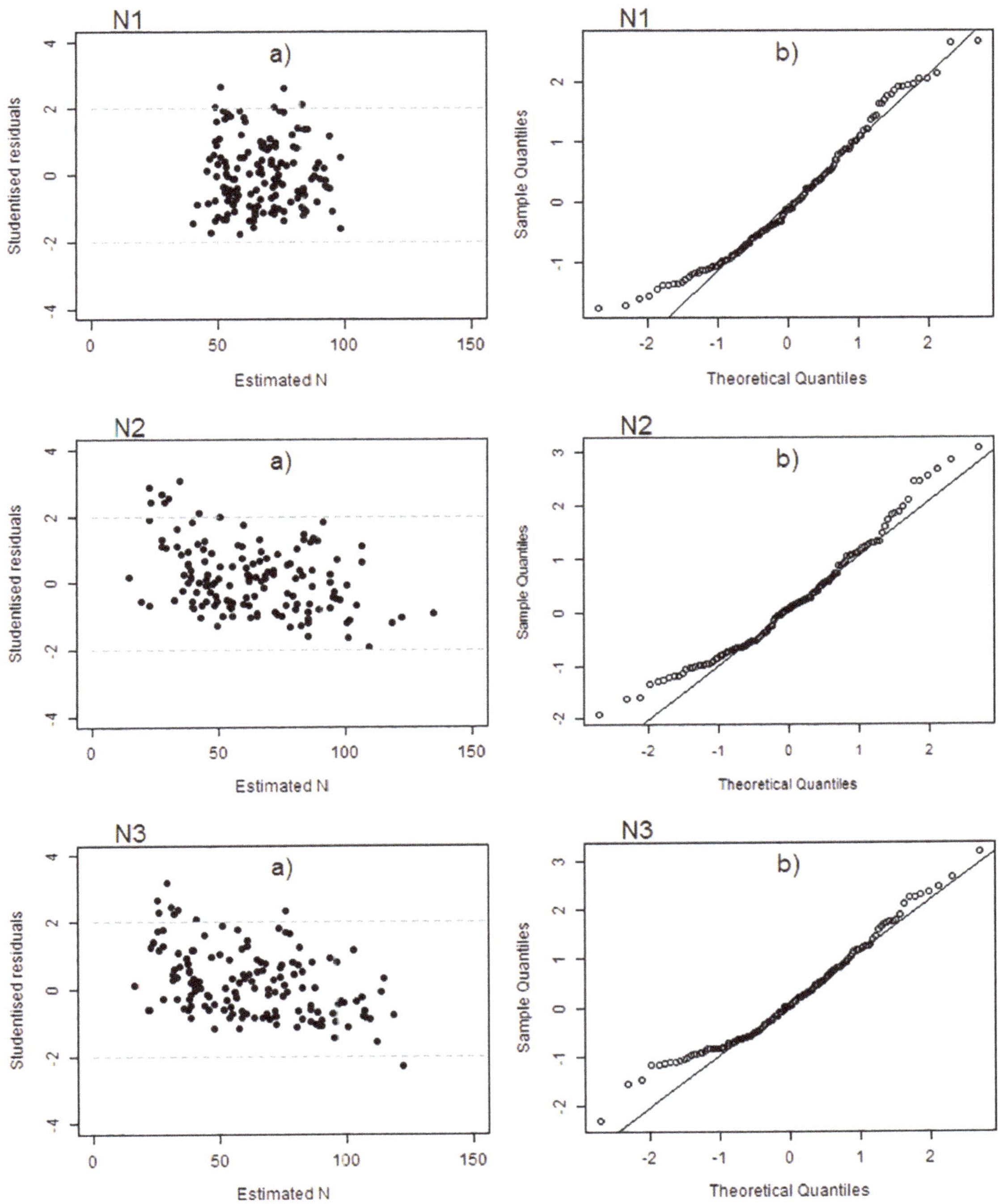

Figure 5. Studentized residuals (a) and normal probability (b) graphics for N1, N2, and N3.

$$N1 = 0.8982 \times CC + 37.7851 \times d_{QR} + 28.3933 \times d_{QS} + 35.4451 \times d_{PP} + 35.4373 \times d_{QRQS} + 33.1026 \times d_{QSPP} \quad (6)$$

$$N2 = 1.7334 \times CC_{QR} + 2.03536 \times CC_{QS} + 1.50642 \times CC_{PP} \quad (7)$$

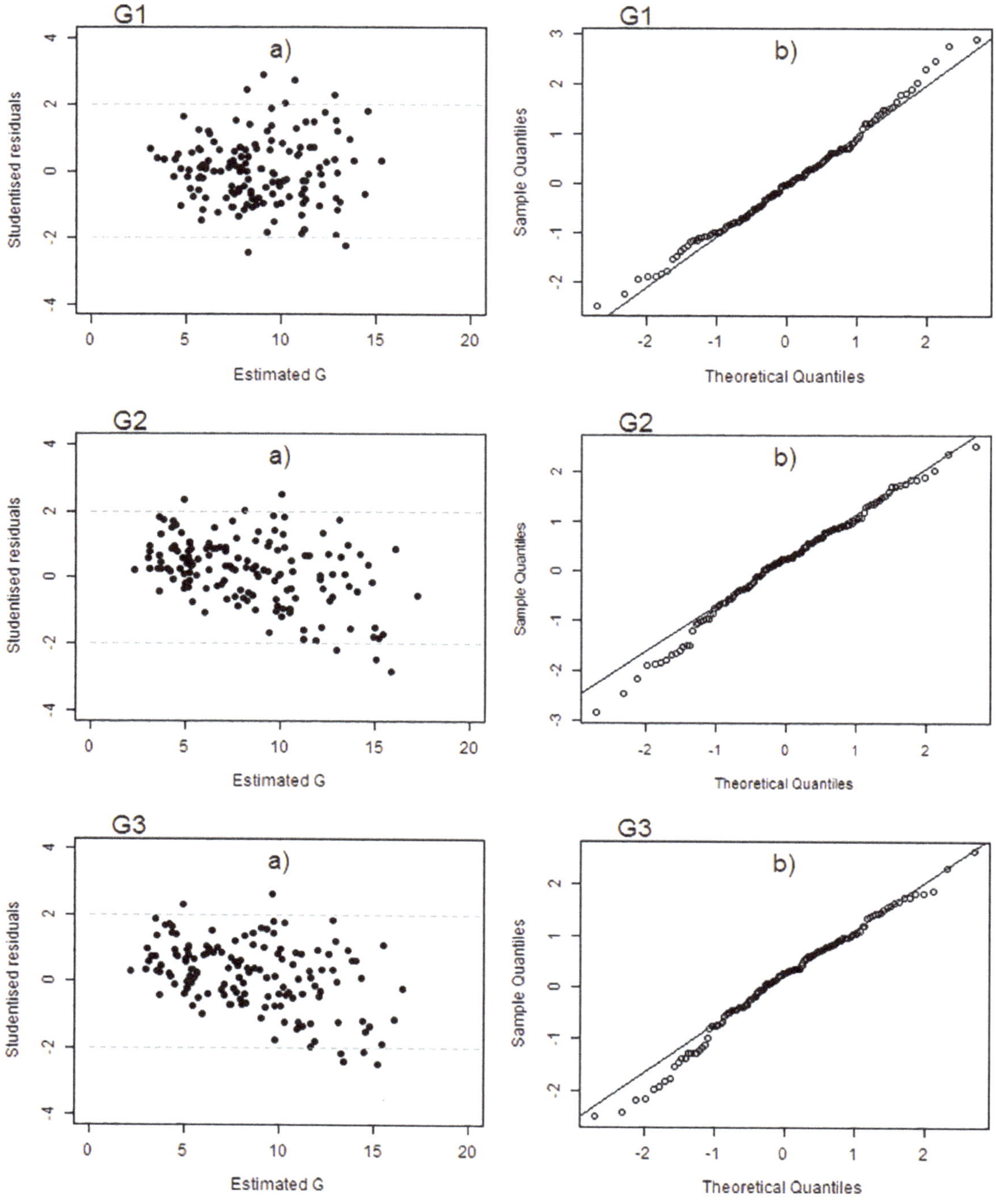

Figure 6. Studentized residuals (a) and normal probability (b) graphics for G1, G2, and G3.

$$G1 = 0.14992 \times CC + 1.04399 \times d_{QR} + 3.84375 \times d_{QS} + 3.39116 \times d_{PP} + 2.25474 \times d_{QRQS} + 4.37992 \times d_{QSPP} \tag{8}$$

$$G2 = 0.19954 \times CC_{QR} + 0.2234 \times CC_{QS} + 0.237722 \times CC_{PP} \tag{9}$$

The coefficient of determination of the functions to estimate the number of trees per hectare with explanatory variables derived from satellite data is rather variable from 0.38 to 0.88 [26, 29, 30, 32]; an even wider variation is found for the functions to estimate the basal area per hectare, from 0.07 to 0.92 [26, 29, 30, 32, 35, 40]. In general, the functions for N and G estimation developed in this study have higher R^2 than those of the aforementioned studies. RMSEr decreases from N3 and G3 to N1 and G1 (**Table 2**) and is smaller than that attained value for the basal area as studied by reference Hyvönen et al. [36] (46–53%) and Günlü et al. [40] (29–34%).

For N1, N2, G1 and G2, no serious multicoliniarity is expected as VIF for all explanatory variables is smaller than 7.7. The studentized residuals of N1, N2, N3, G1, G2 and G3 do not show systematic variations; the normal probability graphs approach the straight line (**Figures 5** and **6**), thought with deviations in the extremes, which is corroborated by the residuals not meeting normality (all, $p < 0.001$). Similar results were attained for N3–N8 and G4–G8.

4. Conclusions

The vegetation mask per species from very high-resolution satellite images enables deriving accurately crown cover. In spite of the similarities between holm oak, cork oak and umbrella pine, the methods of multi-segmentation and object-oriented classification allow their distinction with accuracy.

The functions with the best performances for the number of trees per hectare and basal area per hectare for the three species were attained with the explanatory variable crown cover and composition classes as dummy variables.

The agroforestry systems in general, and those of holm oak, cork oak and umbrella pine in particular, have high spatial variability which makes these functions well suited to evaluate and monitor their stands in space and time.

Acknowledgements

The authors would like to thank all the forest producers for permission to undertake forest inventory; Paulo Mesquita for image processing; and Carla Coelho, David Gomes and Pedro Antunes for their fieldwork. The work was financed by Programa Operativo de Cooperação Transfronteiriço Espanha–Portugal (POCTEP), project Altercexa–Medidas de Adaptación y Mitigación del Cambio Climático a Través del Impulso de las Energías Alternativas en Centro, Alentejo y Extremadura (Refª 0317_Altercexa_I_4_E and 0406_ALTERCEXA_II_4_E), TrustEE–innovative market based Trust for Energy Efficiency investments in industry (Project ID: H2020–696140) and National Funds through FCT–Foundation for Science and Technology under the Project UID/AGR/00115/2013.

Conflict of interest

Both authors declare that there is no conflict of interest.

Author details

Ana Cristina Gonçalves* and Adélia M. O. Sousa

*Address all correspondence to: acag@uevora.pt

Department of Rural Engineering, School of Sciences and Technology, Institute of Mediterranean Agricultural and Environmental Sciences (ICAAM), Institute of Research and Advanced Information (IIFA), University of Évora, Évora, Portugal

References

[1] Nerlich K, Graeff-Hönninger S, Claupein W. Agroforestry in Europe: A review of the disappearance of traditional systems and development of modern agroforestry practices, with emphasis on experiences in Germany. Agroforestry Systems. 2013;**87**:475-492

[2] Eichhorn MP, Paris P, Herzog F, Incoll LD, Liagre F, Mantzanas K, Mayus M, Moreno G, Papanastasis VP, Pilbeam DJ, Pisanelli A, Dupraz C. Silvoarable systems in Europe—Past, present and future prospects. Agroforestry Systems. 2006;**67**:29-50

[3] Gonçalves AC. Multi-species stand classification: Definition and perspectives. In: Chakravarty S, Shukla G, editors. Forest Ecology and Conservation. Rijeka: INTECH; 2017. pp. 4-23

[4] Gonçalves AC, Afonso A, Pereira DG, Pinheiro A. Influence of umbrella pine (*Pinus pinea* L.) stand type and tree characteristics on cone production. Agroforestry Systems. 2017a;**91**:1019-1030

[5] Correia AV, Oliveira AC. Principais espécies florestais com interesse para Portugal: zonas de influência mediterrânica. [Main forest species with interest for Portugal: Zones of Mediterranean influence]. Lisboa: Direcção-Geral das Florestas. Estudos e Informação, 1999. p. 318

[6] Jose S, Gillespie AR, Pallardy SG. Interspecific interactions in temperate agroforestry. Agroforestry Systems. 2004;**61**:237-255

[7] Assmann E, editor. The Principles of Forest Yield Study. Oxford: Pergamon Press; 1970. p. 506

[8] Boudru M. Forêt et Sylviculture. Le traitement des forêts [Forest and Silviculture. The Treatment of Forests]. Tome 2. Gembloux: Presses Agronomiques de Gembloux; 1989. p. 344. [in French]

[9] Matthews JD. Silvicultural Systems. Oxford: Claredon Press; 1989. p. 284

[10] Lanier L, Badré M, Delabraze P, Dubourdieu J, Flammarion JP. Précis de sylviculture [Compendium of Silviculture]. Nancy: ENGREF; 1986. p. 468. [in French]

[11] Schütz JP. Sylviculture 1. Principes d'éducation des forêts [Silviculture 1. Principals of Tending Forests]. Lausanne: Presses Polytechniques et Universitaires Romandes; 1990. p. 245. [in French]

[12] Oliver CD, Larson BC. Editors. Forest Stand Dynamics. Update Editions. New York: John Wiley & sons, Inc; 1996. p. 544

[13] Smith DM, Larson BC, Kelty MJ, Ashton PMS. The Practice of Silviculture. Applied Forest Ecology. 9th ed. New York: John Wiley & Sons, Inc; 1997. p. 560

[14] Schütz JP. Sylviculture 2. La gestion des forêts irrégulières et mélangées. Collection Gérer L'environement, n° 13. Lausanne: Presses Polytechniques et Universitaires Romandes; 1997. p. 178. [in French]

[15] O'Hara KL. Multiaged Silviculture: Managing for Complex Forest Stand Structures. Oxford: Oxford University Press; 2014. p. 213

[16] Avery TE, Burkhart HE, editors. Forest Measurements. 4th ed. New York: Macgraw-Hill Inc; 1994. p. 480

[17] Philip MS. Measuring Trees and Forests. 2nd ed. Cambridge: CAB International; 1994. p. 310

[18] Av L, Akça A, editors. Forest Msensuration. Sringer. Dordrecht: Sringer; 2009. p. 383

[19] West PW. Tree and Forest Measurement. 2nd ed. Dordrecht: Sringer; 2009. p. 190

[20] Brown S, Gillespie AJR, Lugo AE. Biomass estimation methods for tropical forests with applications to forest inventory data. Forest Science. 1989;**35**:881-902

[21] Gillespie AJR, Brown S, Lugo AE. Tropical forest biomass estimation from truncated stand tables. Forest Ecology and Management. 1992;**48**:69-87

[22] Houghton RA, Lawrence KT, Hackler JL, Brown S. The spatial distribution of forest biomass in the Brazilian Amazon: A comparison of estimates. Global Change Biology. 2001;**7**:731-746

[23] Tomppo E, Haakana M, Katila M, Peräsaari J. Multi-Source National Forest Inventory—Methods and Applications. Managing Forest Ecosystems 18. Dordrecht: Springer Science+ Business Media; 2008. p. 373

[24] Vidal C, Lanz A, Tomppo E, Schadauer K, Gschwantner T, di Cosmo L, Robert N. Establishing forest inventory reference definitions for forest and growing stock: A study towards common reporting. Silva Fennica. 2008;**42**(2):247-266

[25] McRoberts R, Tomppo E, Naesset E. Advanced and emerging issues on national forest inventories. Scandinavian Journal of Forest Research. 2010;**25**:368-381

[26] Hudak AT, Crookston NL, Evans JS, Falkowski MJ, Smith AM, Gessler PE, Morgan P. Regression modeling and mapping of coniferous forest basal area and tree density from discrete-return LiDAR and multispectral satellite data. Canadian Journal of Remote Sensing. 2006;**32**:126-138

[27] Hirata Y, Tabuchi R, Patanaponpaiboon P, Pougparn S, Yoneda R, Fujioka Y. Estimation of aboveground biomass in mangrove forest using high-resolution satellite data. Journal of Forest Research. 2014;**19**:34-41

[28] Sedano F, Goméz D, Gong P, Biging G. Tree density estimation in a tropical woodland ecosystem with multiangular MISR an MODIS data. Remote Sensing of Environment. 2008;**112**:2523-2537

[29] Kayitakire F, Hamel C, Defourny P. Retrieving forest structure variables based on image texture analysis and IKONOS-2 imagery. Remote Sensing of Environment. 2006; **102**:390-401

[30] Özdemir I, Karnieli A. Predicting forest structural parameters using the image texture derived from WorldView-2 multispectral imagery in a dryland forest, Israel. International Journal of Applied Earth Observation and Geoinformation. 2011;**13**(5):701-710

[31] Mohammadi J, Joibary SS, Yaghmaee F, Mahiny AS. Modeling forest stand volume and tree density using Landsat ETM + data. International Journal of Remote Sensing. 2010;**31**(11):2959-2975

[32] Gómez C, Wulder MA, Montes F, Delgado JA. Modelling forest structural parameters in Mediterranean pines of Central Spain using QuickBird-2 imagery and classification and regression tree analysis (CART). Remote Sensing. 2012;**4**(1):135-159

[33] Kahriman A, Günlü A, Karahalil U. Estimation of crown closure and tree density using landsat TM satellite images in mixed forest stands. Journal of Indian Society of Remote Sensing. 2014;**42**(3):559-567

[34] Meng Q, Cieszewski C, Madden M. Large area forest inventory using Landsat ETM+: A geostatistical approach. ISPRS Journal of Photogrammetry and Remote Sensing. 2009;**64**:27-36

[35] Hyyppä J, Hyyppä H, Inkinen M, Engdahl M, Linko S, Zhu Y-H. Accuracy comparison of various remote sensing data sources in the retrieval of forest stand attributes. Forest Ecology and Management. 2000;**128**:109-120

[36] Hyvönen P, Pekkarinen A, Tuominen S. Segment-level stand inventory for forest management. Scandinavian Journal of Forest Research. 2005;**20**(1):75-84

[37] Poulain M, Peña M, Schmidt A, Schmidt H, Schulte A. Relationships between forest variables and remote sensing data in a Nothofagus pumilio forest. Geocarto International. 2010;**25**:25-43

[38] Gebreslasie MT, Ahmed FB, van Aardt JAN. Predicting forest structural attributes using ancillary data and ASTER satellite data. International Journal of Applied Earth Observation and Geoinformation 2010;**12**:S23-S26

[39] Jung M, Tautenhahn S, Wirth C, Kattge J. Estimating basal area of spruce and fir in post-fire residual stands in Central Siberia using Quickbird, feature selection, and random forests. Procedia Computer Science. 2013;**18**:2386-2395

[40] Günlü A, Ercanlı I, Sönmez T, Başkent EZ. Prediction of some stand parameters using pan-sharpened IKONOS satellite image. European Journal of Remote Sensing. 2014;**47**(1):329-342

[41] Greenberg JA, Dobrowski SZ, Ustion SL. Shadow allometry: Estimating tree structural parameters using hyperspatial image analysis. Remote Sensing of Environment. 2005;**97**:15-25

[42] Eisfelder C, Kuenzer C, Dech S. Derivation of biomass information for semi-arid areas using remote-sensing data. International Journal of Remote Sensing. 2012;**33**(9):2937-2984

[43] Lu D, Chen Q, Wang G, Liu L, Li G, Moran EA. A survey of remote sensing-based aboveground biomass estimation methods in forest ecosystems. International Journal of Digital Earth. 2016;**9**(1):63-105

[44] Sousa AMO, Gonçalves AC, Mesquita P, Marques da Silva JR. Biomass estimation with high resolution satellite images: A case study of *Quercus rotunifolia*. ISPRS Journal of Photogrametric and Remote Sensing. 2015;**101**:69-79

[45] Gonçalves AC, Sousa AMO, Mesquita PG. Estimation and dynamics of above ground biomass with very high resolution satellite images in *Pinus pinaster* stands. Biomass and Bioenergy. 2017b;**106**:146-154

[46] Gonçalves AC, Sousa AMO, Silva JRM. Pinus pinea above ground biomass estimation with very high spatial resolution satellite images. In: Carraquinho I, Correia AC, Mutke S, editors. Mediterraneanpine Nuts From Forest and Plantations. Vol. 122. Options Mediterranées. 2017. pp. 49-54

[47] Macedo FL, Sousa AMO, Gonçalves AC, Silva HR, Rodrigues RAF. Estimativa do volume de madeira para *Eucalyptus* sp. com imagens de satélite de alta resolução espacial. Scientia Forestalis. 2017;**45**(114):237-247

[48] Malico I, Gonçalves AC, Sousa AMO. Assessment of the availability of forest biomass for biofuels production in Southwestern Portugal. In: Defect and Diffusion Forum. Vol. 371. 2017. pp. 121-127

[49] Sousa AMO, Gonçalves AC, Silva JRM. Above ground biomass estimation with high spatial resolution satellite images. In: Tumuluru JS, editor. Biomass Volume Estimation and Valorization for Energy. Rijeka: InTech; 2017. pp. 47-70

[50] Xie Y, Sha Z, Yu M. Remote sensing imagery in vegetation mapping: A review. Journal of Plant Ecology. 2008;**1**:9-23

[51] Boyle SA, Kennedy CM, Torres J, Colman K, Pérez-Estigarribia PE, de la Sancha NU. High-resolution satellite imagery is an important yet underutilized resource in conservation biology. PLoS One. 2014;**9**(1):e86908

[52] As-syakur RA, Osawa T, Adnyana IWS. Medium spatial resolution satellite imagery to estimate gross primary production in an urban area. Remote Sensing. 2010;**2**:1496-1507

[53] Zhang Y, Mishra RK. From UNB PanSharp to Fuze go—The success behind the pan-sharpening algorithm. International Journal of Image Data Fusion. 2013;**5**:39-53

[54] Envi. Reference Guide—Exelis Visual Information Solutions. Boulder, Colorado: Exelis Visual Information Solutions [Internet]. 2009. Available from: http://www.exelisvis.com/portals/0/pdfs/envi/envi_zoom_user_guide.pdf [Accessed: November 27, 2012]

[55] Chavez PS Jr. An improved dark-object subtraction technique for atmospheric scattering correction of multispectral data. Remote Sensing of Environment. 1988;**24**(3):459-479

[56] Blaschke T. Object based image analysis for remote sensing. ISPRS Journal of Photogrammetry & Remote Sensing. 2010;**65**:2-16

[57] Blaschke T, Hay GJ, Kelly M, Lang S, Hofmann P, Addink E, Queiroz Feitosa R, van der Meer F, van der Werff H, van Coillie F, Tiede D. Geographic object-based image analysis—Towards a new paradigm. ISPRS Journal of Photogrammetry & Remote Sensing. 2014;**87**:180-191

[58] Rouse JW, Haas RH, Schell JA, Deering DW. Monitoring vegetation systems in the Great Plains with ERTS. In: 3rd ERTS Symposium, NASA; 1973. SP-351 I. pp. 309-317

[59] Definiens Imaging. eCognition Developer 8.0.1 Reference Book [Internet]. 2010. Available from: http://www.definiens.com [Accessed: October 23, 2012]

[60] Baatz M, Schäpe A. Multiresolution segmentation: an optimization approach for high quality multi-scale image segmentation. In: Angew. Geogr. Informationsverarbeitung XII. Beiträge zum Agit. Salzbg. Karlsruhe: Herbert Wichmann Verlag; 2000. pp. 12-23

[61] Eibe F, Hall MA, Witten IH. The WEKA Workbench. Online Appendix for Data Mining: Practical Machine Learning Tools and Techniques. 4th ed. Cambridge (USA): Morgan Kaufmann; http://www.cs.waikato.ac.nz/ml/weka/Witten_et_al_2016_appendix.pdf

[62] Ozdemir I. Estimating stem volume by tree crown area and tree shadow area extracted from pan# sharpened Quickbird imagery in open Crimean juniper forests. International Journal of Remote Sensing. 2008;**29**(19):5643-5655

[63] Esri. ArcGIS Desktop: Release 10. Redlands, CA: Environmental Systems Research Institute [Internet]. 2010. Available from: http://www.esri.com [Accessed: January 23, 2013]

[64] Legendre P, Legendre L, editors. Numerical Ecology. 3th ed. Vol. 24. Amsterdam: Elsevier Science BV; 2012. p. 1006

[65] Sheather SJ, editor. A Modern Approach to Regression with R. New York: Springer Texts in Statistics; 2009. p. 393

[66] Burkhart HE, Tomé M, editors. Modelling Forest Trees and Stands. Dordrecht: Springer Science+Business Media; 2012. p. 457

[67] Pretzsch H, editor. Forest Dynamics, Growth and Yield: From Measurement to Model. Berlin Heidelberg: Springer-Verlag; 2009. p. 664

[68] Myers RH, editor. Classical and Modern Regression with Applications. Chicago: Duxbury Press; 1986. p. 488

[69] Clutter JL, Fortson JC, Pienaar LV, Briester GH, Bailey RL, editors. Timber Managment: A Quantitative Approach. New York: John Wiley & Sons, Inc; 1983. p. 333

[70] Paulo JA, Palma JHN, Gomes AA, Faias SP, Tomé J, Tomé M. Predicting site index from climate and soil variables for cork oak (*Quercus suber* L.) stands in Portugal. New Forests. 2015;**46**:293-307

[71] Montgomery DC, Peck EA. Introduction to Linear Regression Analysis. New York: Wiley; 1982

[72] R Development Core Team. R: A language and environment for statistical computing. R Foundation for Statistical Computing [Internet]. 2017. Available from: http://www.R-project.org [Accessed: December 06, 2017]

[73] Ke Y, Quackenbush LJ. A review of methods for automatic individual tree-crown detection and delineation from passive remote sensing. International Journal of Remote Sensing. 2011;**32**(17):4725-4747

[74] Massada AB, Kent R, Blank L, Perevolotsky A, Hadar L, Carmel Y. Automated segmentation of vegetation structure units in a Mediterranean landscape. International Journal of Remote Sensing. 2012;**33**(2):346-364

[75] Nguyen HC, Jung J, Lee J, Choi S-U, Hong S-Y, Heo J. Optimal atmospheric correlation for above-ground forest biomass estimation with the ETM+ remote sensor. Sensors. 2015;**15**:18865-18886

[76] Wu W, De Pauw E, Helldén U. Assessing woody biomass in African tropical savannahs by multiscale remote sensing. International Journal Remote Sensing. 2013;**34**(13):4525-4549

[77] Stehman SV. Estimating the kappa coefficient and its variance under stratified random sampling. Photogrammetric Engineering & Remote Sensing. 1996;**62**:401-407

[78] Congalton RG, Oderwald RG, Mead RA. Assessing Landsat classification accuracy using discrete multivariate statistical techniques. Photogrammetric Engineering and Remote Sensing. 1983;**49**:1671-1678

[79] Mutke S, Calama R, González-Martínez SC, Montero G, Gordo J, Bono D, Gil L. Mediterranean stone pine: Botany and horticulture. Horticultural Reviews. 2012;**39**: 153-201

[80] Hemery GE, Savill PS, Pryor SN. Applications of the crown diameter–stem diameter relationship for different species of broadleaved trees. Forest Ecology and Management. 2005;**215**:285-294

[81] Verma NK, Lamb DW, Reid N, Wilson B. An allometric model for estimating DBH of isolated and clustered eucalyptus trees from measurements of crown projection area. Forest Ecology and Management. 2014;**326**:125-132

11

One World, One Health Challenge: The Holistic Understanding of Rickettsiosis Integrating Multi-Criteria Analysis Techniques and Spatial Statistics

Diego Montenegro, Ana Paula da Cunha, Ingrid Machado, Liliane Duraes, Stefan Vilges de Oliveira, Marcel Pedroso, Gilberto S. Gazêta and Reginaldo P. Brazil

Abstract

Among the human diseases caused by etiological agents transmitted by ticks in Brazil, the most prevalent and public health interest is that one caused by species of *Rickettsia*—Spotted Fever (SF). We applied the concept "One World, One Health" to achieve a proper understanding of SF and determine risk scenarios for human infection by pathogenic *Rickettsia* in the state of Rio de Janeiro (RJ). Multi-criteria decision analysis and spatial statistics were performed on data encompassing epidemiology, health care, biotic determents, and socioeconomic and demographic variables. The construction of multi-criteria descriptors used 33 indicators ordered in 12 sub criteria of 5 major categories: public health, environmental health, acarology, veterinary health, and microbiology. SF happened nonrandomly in RJ and the risk was heterogeneous in the weight of indicators contribution mainly associated with the criteria acarology (35.11%), microbiology (33.25%), and veterinary medicine (23.96%). Spatially, and according to the PROMETHEE algorithm, ticks from *Amblyomma sculptum* class and dogs are determinants for the occurrence of human cases of SF in RJ. To define indicators of SF and plan health actions, the "One World, One Health" concept proved to be applicable at three levels: (i) the local or possible areas of infection, (ii) the population or municipality, and (iii) the ecosystem or state. The model of study is flexible according to the reality of the endemic areas and also demonstrates its applicability from a national to a local (home) scale.

Keywords: one world, one health, tick-borne diseases, risk, spotted fever, multi-criteria decision analyses

1. Introduction

A total of 1415 species of infectious agents have been associated with human diseases throughout the world. From these, about 60% (868) are considered zoonosis [1], representing 75% of emerging diseases and 80% of agents with bioterrorist potential [2]. These high prevalence have led to new considerations regarding understanding pathogens that have negative impacts on public and veterinary health [2].

"One World, One Health" is a holistic and interdisciplinary vision that integrates ecosystem or environmental health, animal health, public health and food safety at local, country and global scales [3–7].

The concept, which has many historical backgrounds, was presented in 2004 by the Wildlife Conservation Society [8]. It is now widely accepted, incorporated, and disseminated by the World Health Organization (WHO), the World Organization for Animal Health (OIE), and the Food and Agriculture Organization of the United Nations (FAO). The first joint efforts of this vision gave rise to the Global Early Warning System for Major Animal Diseases including Zoonosis – GLEWS [9].

Several interdisciplinary, cooperative, and holistic studies have been developed toward understanding and combating a wide variety of pathogenic diseases [3, 5–7, 10–17].

In the present work, we consider the concept *"One World, One Health" sensu lato*; that is, as synonymous with "One Medicine", "One Health", Universal Health and Risk, and Disease Ecology and Eco-epidemiology [3, 5, 6, 10, 16, 17] for the prevention, surveillance, and control of infectious diseases of zoonotic origin.

Among zoonosis, rickettsiosis are produced by Gram-negative proteobacteria of the genus *Rickettsia* that are transmitted mainly by ticks to mammalian hosts and, accidentally, to man [18–21].

In Brazil, among the diseases caused by etiological agents transmitted by ticks, the most prevalent and of most interest to public health are Brazilian spotted fever (FMB) produced by *Rickettsia rickettsii*, and caused by other species of *Rickettsia* that are considered as emerging (species of the spotted fever group of *Rickettsia*-SFGR) such as *Rickettsia* strain Atlantic rainforest [22–24]. However, for the purposes of the present study all these types of spotted fever will be treated collectively as spotted fever (SF).

In the transmission cycle of *Rickettsia*, different co-specific relationships between ticks and vertebrate hosts have been identified that involve geographical, ecological, and evolutionary processes [19, 25–27].

Recent studies [24, 27–29] indicate that the dynamics of SF in the endemic states of Brazil is of a seasonal nature, with the highest incidence rates in the second part of the year (June to October). This seasonality has been associated with the population dynamics of species of ticks of the genus *Amblyomma*, but particularly high densities of larvae and nymphs of *Amblyomma sculptum* [22, 30–34]. This tick is considered the main vector of *R. rickettsii* among the human population of Brazil [22, 23, 27].

However, there are a number of species of ticks that could be considered reservoirs and vectors of pathogenic *Rickettsia* and some mammalian species that could function as amplifiers and reservoirs of these bioagents. These elements are potentially responsible for the occurrence and maintenance of epizootic cycles and the emergence and/or reemergence of epidemic outbreaks of SF in Brazil. Although they are main determinants of bacterial transmission cycles, they are not significant hazards due to the establishment of vulnerability and risk areas, and surveillance programs, as well as efforts to control human outbreaks, as demonstrated by the findings of a number of studies [25, 35–39].

An integrated approach is a promising strategy for understanding the dynamics of SF and determining risk scenarios for infection of humans by pathogenic *Rickettsia* because it would result in the establishment of SF surveillance and environmental control programs. This is precisely what we aim to achieve in the present work by employing the concept of *"One World, One Health"*. With the establishment of a clear risk scenario, clinically suspected cases in high risk areas may be treated early [40, 41].

2. Materials and methods

2.1. Study area

The state of Rio de Janeiro is located in the eastern portion of Brazil's Southeast Region and occupies an area of 43,777.954 km^2 divided into 92 municipalities. It is the fourth smallest state (by area) in Brazil, yet has the highest population density (365.23 inhabitants/km^2) with an estimated population of 16,640,00 inhabitants. It is also the most urbanized state in the country, with 97% of the population living in cities [42].

2.2. Epidemiological data and health care

The epidemiological data presented here was obtained from the Sistema de Informação de Agravos de Notificação—SINAN (Notifiable Diseases Information System) [43] and provided by the Secretaria de Estado de Saúde do Rio de Janeiro—SES/RJ (State Secretary Health of Rio de Janeiro). The data comprised cases confirmed by laboratory tests (PCR or serology) and/or clinical and epidemiological nexus of SF between 2007 and 2016.

These data were made available with the protection of the identity of the patients; therefore, information such as names or addresses cannot be displayed at any time to comply with national ethical regulations of Brazil [44].

To assess access to health care, the following municipal indicators were used: number of health professionals (HP) and number of hospital beds (HB) per 100 thousand inhabitants. For these indicators, we used the average of the information from the years 2010 and 2016 obtained from the Secretaria de Atenção em Saúde-SAS (Health Care Secretary) of the Ministério da Saúde do Brasil (Ministry of Health-MH) [45].

2.3. Biotic determinants

Information regarding potential vectors of SF and hosts of the tick was obtained from the database of the Laboratório de Referência Nacional em Vetores das Riquetsioses (LIRN—Labortory of the National Reference of Rickettsial Vectors), built from samples received and analyzed within the workflow of the Rede Nacional de Vigilância de Ambiental para Febre Maculosa e outras Rickettsioses (National Network for Environmental Monitoring for spotted fever and other rickettsial diseases) of MH from 2007 to 2016. The samples were collected in sampling units (specimens originating from the same host or environment), by cloth-dragging, visual searches on hosts and inspection of plant litter and abiotic surfaces by state and municipal health teams from Rio de Janeiro for environmental surveillance and investigation of cases of SF.

Ticks were identified using dichotomous keys, while species of *Rickettsia* infecting ticks were identified using PCR [see in detail in Montenegro et al. [27]].

To characterize conditions of climatic suitability throughout the study region, potential distribution models for *A. sculptum, Amblyomma autreolatum,* and *Amblyomma ovale,* the main vectors of SF in Brazil [22], were produced using the Maxent algorithm version 3.2.1 [46]. For this, we used seven climatic variables that we considered crucial to the biology of the species and which had already been evaluated for collinearity by principal component analysis (PCA) in Niche Software version 3.0: annual mean temperature, mean diurnal temperature range, maximum temperature in the warmest month, minimum temperature in the coldest month, annual precipitation, and precipitation in the wettest and driest months [37, 38, 47, 48].

To generate the final models, we selected the logistic values for the final output of the 25 bootstrap interactions. The mean number of repetitions was selected as the final model and converted into binary models to generate adequacy maps that were overlaid as a shape-file to subtract information about adequacy in relation to the municipality of occurrence of these vectors (**Figure 1**).

2.4. Socioeconomic and demographic data

Data for socioeconomic and demographic indicators are being used to understand the dynamics of spotted fever [49–51]. It was obtained from Brazilian census data acquired by the Instituto Brasileiro de Geografia e Estatística-IBGE [52]. The following indicators were used for the analysis: Municipal Human Development Index (MHDI), GINI Index (IG), Demographic Density (DD), Land Use Municipality (LUM), Percentage of Poverty in the Rural Area (PPRA), Degree of Urbanization (DU), and the Percentage of the Agricultural Contribution to the Municipal GDP (GDP-AM).

2.5. Analysis of data

The first phase of analysis evaluated whether human infections with SF-causing pathogens occur randomly with respect to time. Three statistical tests were used in statistical software [53], with at least two coincident tests with confidence levels of 95% determining randomness, or not, of human cases of SF.

In the second phase of analysis, two multi-criteria decision analyses (MDA) were performed, for the risk of human infection with pathogenic tick-borne *Rickettsia*. Multi-criteria decision analysis is a method that approaches a certain object from different characteristics that influence the occurrence of a given problem. The method has become increasingly used for understanding of events of interest in public health [54].

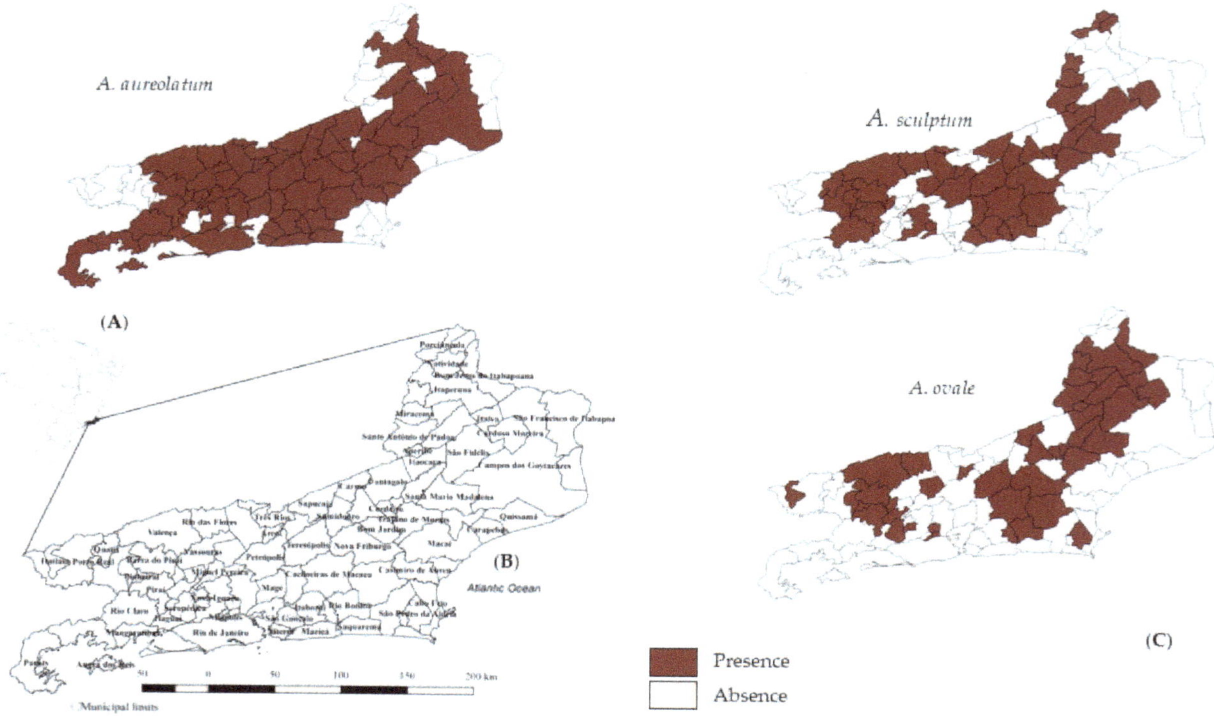

Figure 1. Location of the state of Rio de Janeiro, Brazil (A), its municipalities (B) and potential distribution models for *A. sculptum*, *A. aureolatum*, and *A. ovale* by Maxent algorithm (C).

The following phases of MDA were defined: Structuring Phase, Modelling Phase and Evaluation Phase. The objective of the Structuring Phase is to understand the decision tree formed by the determinant variables of SF. The Modelling Phase mathematically expresses human risk to SF and the Evaluation Phase presents the results of the model.

2.5.1. *MDA structuring phase*

To construct multi-criteria descriptors and decision-making processes, three criteria comprising 13 subcriteria, which in turn include 34 indicators of similar nature and that are considered determinants of SF, were used (**Box 1**).

2.5.2. *MDA modelling phase*

For modelling MDA, numerical values of criteria, subcriteria, and indicators were incorporated into the decision-making model by employing peer-to-peer comparison. This comparison method attributes a preferential value to two criteria, two sub-criteria or two indicators, which are graphically compared to one another using the D-sight program [55]. Theoretical weights, ranging from 1 to 100, for the components of the decision-making model were generated using the PROMETHHE method (preference ranking method for enrichment evaluation) through peer-to-peer comparison [56].

Criteria	Subcriteria	Indicators	Sources
Public Health	Health care	HP	SAS
		HB	
		Ratio HP/HB	
	Social	MHDI	IBGE
		GI	
		PPRA	
Environmental Health	Land use	LUM	IBGE
		DU	
		GDP-AM	
	Demographic	DD	
	Vector fitness (presence of vectors in the environment (not hosts))	Modeled presence	LIRN
		Real presence	LIRN
Acarology	Tick vector (proven *R. rickettsii* transmission)	*A. aureolatum*	LIRN
		A. sculptum	
		A. ovale	
		R. sanguineus	
	Tick carrier synanthropic (infected with *Rickettsia* but without proof of transmission to hosts and associated with anthropized environments)	*A. dubitatum*	
		D. nitens	
		Rh. microplus	
		A. longirostre	
	Tick wild (infected or not with *Rickettsia*)	*Ha. leporispalustris*	
		Ornithodoros sp.	
Veterinary health	Reservoir animals (proven infection of tick)	*C. familiares*	LIRN
		H. hydrochaeris	
	Host animals (has not been shown to function as a reservoir)	*E. caballus*	
		B. taurus	
		F. catus	
		E. asinus	
		Mula	
	Parasitism in humans	*H. sapiens*	
Microbiology	Etiological agent	*R. rickettsii*	LIRN and Montenegro et al. [27]
		Spotted fever group of *Rickettsia* – SFGR	
		Rickettsia	

Box 1. Criteria, sub-criteria, and indicators of multi-criteria decision analysis (MDA).

It should be noted that the PROMETHEE method was developed in order to help the individual or collective decision maker. These methods serve to solve problems by selecting, or making arrangements of, possible alternatives (municipality, options, and shares), subject to an assessment of various of criteria (variables, qualitative and quantitative indices, indicators, attributes, and any criteria with numerical or nominal values), which may be in conflict with each other, and seek to simultaneously satisfy different solutions for decision making [56, 57].

The third phase of analysis involved using the scores resulting from the PROMETHEE model for spatial statistical analysis. Spatial dependence of risk for each area assessed was analyzed using the Moran Local Index—LISA (Anselin 2010), with 9999 permutations, no spatial autocorrelation and statistical significance (LISASig) considered to be greater than 0.05. All spatial statistical analyses were performed using the program TerraView [58].

Finally, using data points from data on probable infection sites for confirmed SF cases (using addresses reported in the epidemiological form with Google Maps), we sought to focus on areas with the highest probability of human SF infection within the biomes, and the greatest spatial association with hosts and ectoparasites in RJ, using the program QGis [59].

2.5.3. MDA evaluation phase

This is the phase where the results of the modelling are presented

3. Results

Human infections with pathogenic *Rickettsia* in RJ occur mainly between the months of March–April and August–October, with it being more pronounced in the latter (**Figure 2**).

Two of the three tests employed showed that the temporal dynamics of SF cases is not a sequence of temporally random events. The first test counted the number of times the data sequence was above and below the median and determined that the sample size was too small to perform a valid test (it was inconclusive).

The second test counted the number of times a sequence went up or down (number of executions up and down = 46, expected number of runs = 77, and Z statistic = 6.77, Pv = 1.3^{-11}) and the third was based on the sum of squares of the autocorrelation coefficients—*Box Pierce test* (based on the first 24 autocorrelation coefficients, Z = 37.3, Pv = 0.040).

Multi-criteria modelling found the risk to tick-borne transmission of spotted fever group of *Rickettsia* in Rio de Janeiro to be heterogeneous with regard to the contribution of the weights of the two scenarios examined. The main sources of variation are in the contributions of the criteria acarology (35.11%), microbiology (33.25%), and veterinary medicine (23.96%) (**Table 1**). The five criteria are made up of 13 subcriteria, and 21 of the original 34 priority indicators and variables, contributed at least 2% to the two models of MDA.

With the exception of environmental health, all criteria contributed equally to the determination of risk scenarios for SF; the greatest influence was by acarology, with 27.3% of the total

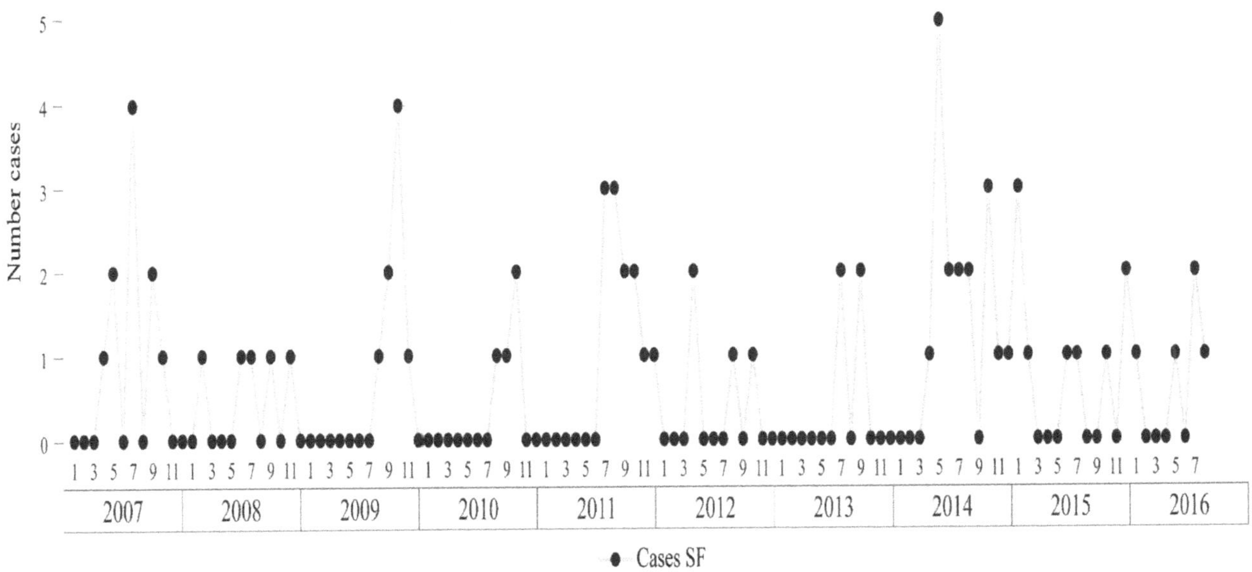

Figure 2. Temporal (monthly) dynamics of human infection with tick-borne pathogenic *Rickettsia* that causes spotted fever in the state of Rio de Janeiro.

weight of the model. On the other hand, the risk scenario was found to be more associated with acarology and microbiology (**Table 1**). The five criteria are made up of 13 subcriteria, and 21 of the original 34 priority indicators and variables; the indicators contributed at least 2% to the two models of MDA (**Table 1**).

According to the weight of each indicator, the largest (5%) absolute value generated by the PROMETHEE algorithm for the occurrence of infection with SFGR is the criteria associated with the subcriteria health care (rate HP/HB), reservoir animals (*C. familiares*), human parasitism, tick vector (*A. sculptum*), and etiological agent (*R. rickettsii* and SFGR) (**Table 1**).

The multi-criteria evaluation of the results of the PROMETHEE method found heterogeneous risk scores for the 92 municipalities of the state of RJ. Based on the scores, 25% (23/92) of the territories are located in the highest level (4th quartile) of risk for SF (**Figure 3a**).

The LISA technique identified 48.9% (45/92) of the municipalities as having high spatial dependence, with 19.6% (18/92) are in the category High (High-High) and 29.3% (27/92) in the category Low (Low-Low). For the same categories of spatial dependence mentioned for risk were found (**Figure 3b**). Statistically significant spatial correlation was found in 13.0% (12/92) of the municipalities (LISASig ≤5%) (**Figure 3c**).

The risk model was sensitive to identifying areas with human SF infection (**Figure 3d**).

Finally, it was possible to georeference 50% (39/78) of the confirmed cases of SF with probable areas of infection (PAIs) in RJ. The PAIs are located mainly in the areas where the Atlantic Forest biome and anthropic areas intersect (**Figure 4**). The ticks *A. sculptum* and *R. sanguineus* were the most frequent and abundant species in the PAIs, while dogs and the environment are the most significant host and site of infestation, respectively, with the most ticks occurring in places where humans acquire rickettsias.

Criterion	Subcriterion	Indicators	Theoretical weight (%)	Absolute weight (%)—PROMETHEE
Public Health			**18.16**	
	Health care		75	
		Rate HP/HB	50	6.81
		HP	25	3.4
		HB	25	3.4
	Social		25	
		PPRA	33.33	1.51
		GI	33.33	1.51
		MHDI	33.33	1.51
Environmental Health			**11.98**	
	Land use		29.7	
		GDP-AM	42.86	1.52
		DU	28.57	1.02
		LUM	28.57	1.02
	Demographic		16.34	
		DD	100	1.96
	Vector fitness		53.96	
		Real presence	66.67	4.31
		Modeled presence	33.33	2.16
Veterinary Health			**23.96**	
	Reservoir animals		31.89	
		H. hydrochaeris	33.33	2.55
		C. familiares	66.67	5.09
	Host animals		22.11	
		Mula	9.59	0.51
		E. asinus	9.59	0.51
		F. catus	14.53	0.77
		B. taurus	30.46	1.61
		E. caballus	35.83	1.9
	Parasitism in humans		46	
		H. sapiens	100	11.02
Acarology			**35.11**	
	Tick vector		52.78	
		R. sanguineus	13.16	2.44
		A. aureolatum	25.23	4.68
		A. sculptum	42.43	7.86

Criterion	Subcriterion	Indicators	Theoretical weight (%)	Absolute weight (%)—PROMETHEE
		A. ovale	19.17	3.55
	Tick carrier			
		Rh. microplus	33.33	3.89
		D. nitens	33.33	3.89
		A. dubitatum	33.33	3.89
	Tick wild		13.96	
		Ornithodoros sp.	33.33	1.63
		Ha. leporispalustris	33.33	1.63
		A. longirostre	33.33	1.63
Microbiology			**33.25**	
	Etiological agent		100	
		Rickettsia	9.53	1.03
		SFGR	24.99	2.7
		R-rickettsii	65.48	7.07

Table 1. Weights of the criteria used for determining vulnerability for tick-borne transmission of the spotted-fever group of ***Rickettsia*** to humans.

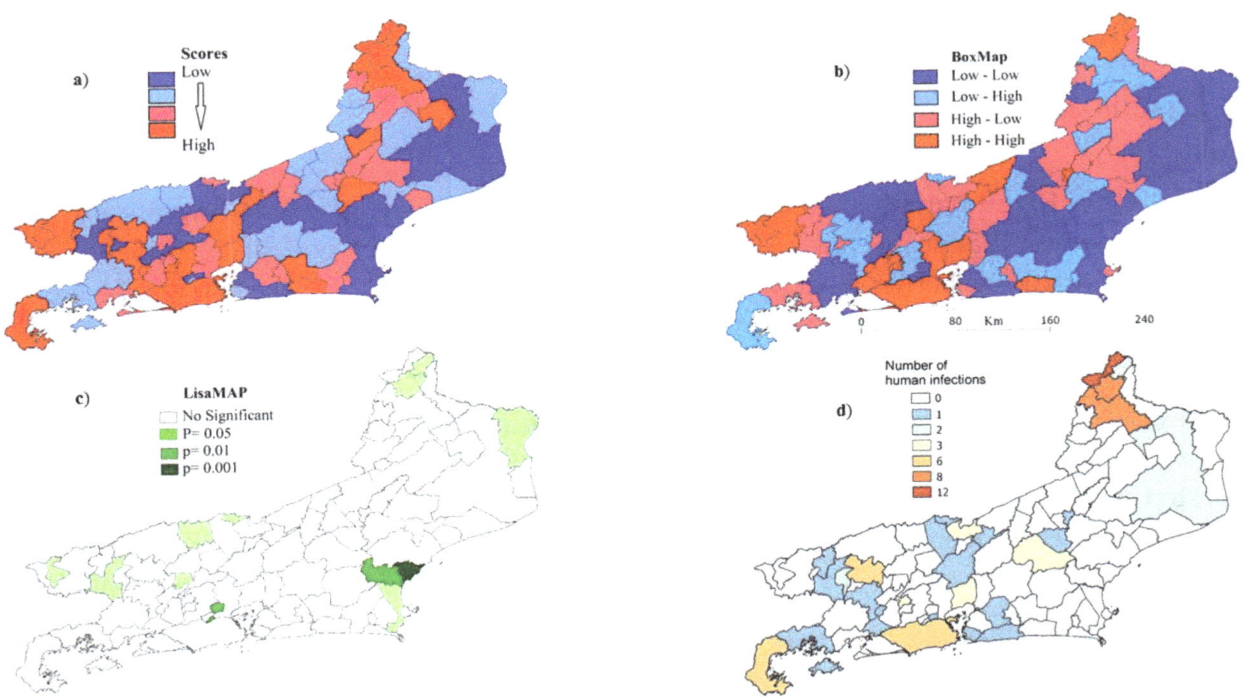

Figure 3. Spatial vulnerability for SF in the state of Rio de Janeiro. The letter **a** corresponds to the range of absolute values, **b** to spatial dependence, **c** to the statistical significance of spatial correlation and **d** to the number of cases with local infection.

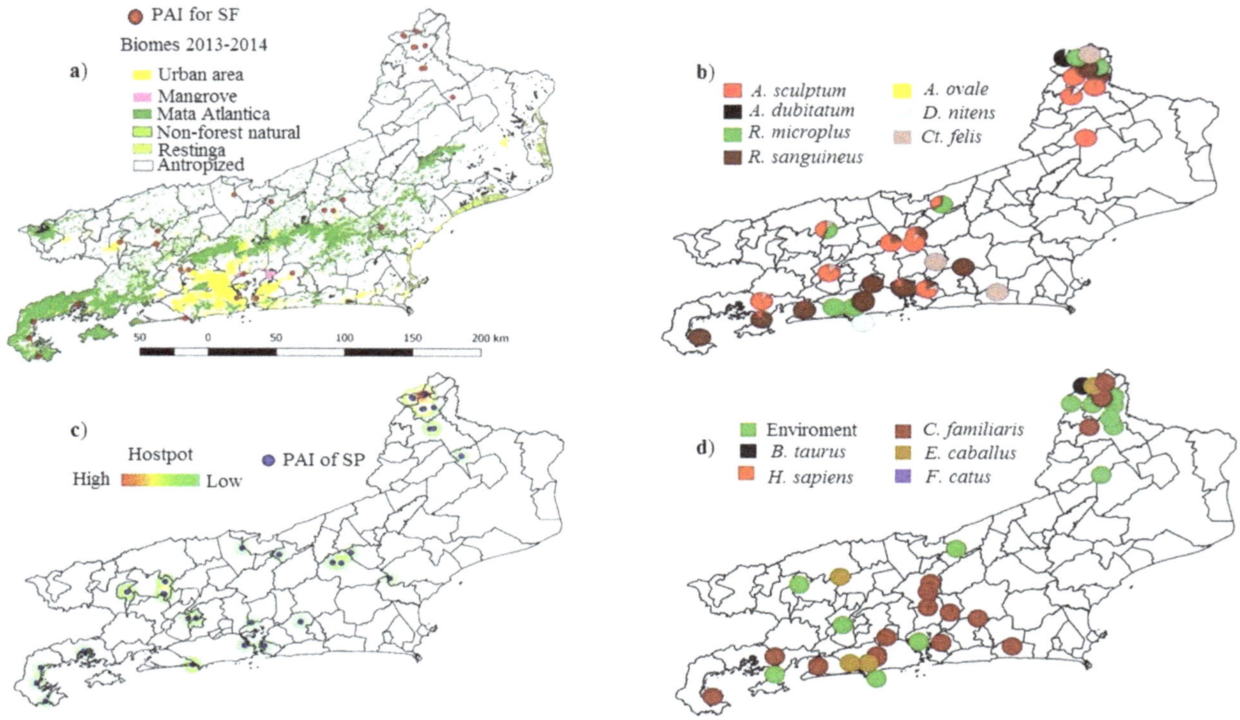

Figure 4. Distributions of biological and environmental determinants of probable areas of infection (PAIs) of pathogenic *Rickettsia* in the state of Rio de Janeiro, 2007–2016: (a) PAIs relative to the biomes of RJ, (b) tick species, (c) hostpots for SF with 5Km radius, and (d) host and environment infested with ticks.

The municipalities of Porciúncula, Natividade, and Itaperuna are located in a hostpot for human acquisition of bacteria (**Figure 4**).

4. Discussion

In the state of RJ, human infections with pathogenic *Rickettsia* are an event of significant public health interest that do not happen randomly; there is seasonality with occurrence between March and April and between August to October, as in all endemic states of Brazil [24, 28, 29, 60]. This understanding requires that collective health actions be deployed before, during, and after these periods, in order to combat the scenario of high mortality in RJ [27, 60].

There exists a spatial association between increased frequency and abundance of the tick species *A. sculptum* and *R. sanguineus*, the presence of dogs and the environment (grass and vegetation) infested. At least spatially, and according to the PROMETHEE algorithm, *A. sculptum* and dogs are essential determinants for the occurrence of human cases of SF in RJ. However, we emphasize the need for focused studies on the participation of *R. sanguineus* in outbreaks in RJ [27].

This tick (*R. sanguineus*) functions as reservoir and vector among hosts, keeping other *Rickettsia* wild strains brought to the peridomicile environment by other species of ticks that parasitize dogs (*A. aureolatum*), as happens in the state of São Paulo, Brazil [61].

Recent studies of SF epidemiology in RJ [60] have found that 69% (72/104) of SF cases confirmed in the last 10 years were in urban areas and the rate of lethality for people infected in the peri-urban environment is greater (86.6%, 13/15). Thus, in addition to solving this problem, it is also necessary to determine the degree to which the dog tick participates in urban outbreaks and what strains of *Rickettsia* it amplifies.

The application of the concept of "*One World, One Health*" in the present study addresses three scales for defining indicators and health actin plans: (i) the individual or PAIs; (ii) the population or municipality; and (iii) The ecosystem or state. The first level includes hotspots for human acquisition of the bacteria (**Figure 4c**). These same areas possess the greatest number of biological elements in the PAI and thus are the places where advertising posters should be placed announcing the risk of SF acquisition [41].

Note that health care represents a considerable weight in the categorization of risk (13.61%), indicating that intersectoral municipal policies (second level) could have an impact on this indicator. Ensuring that the number of health care sites, the number of health professionals and, above all, the actions of epidemiological surveillance and health care, including treatment of suspected patients are sufficient to reverse the lethality coefficients in the state, even without confirmatory evidence [40, 41].

The third level serves to define and prioritize indicators in public health and environmental health, as well as interdisciplinary, to intervene and monitor in the state of RJ. The adequacy of resources and the development of pilot eco-epidemiological studies at a sub-regional level, following the patterns of risk and spatial dependence (**Figure 3a** and **b**), are also recommended.

Despite the fact that climate data were used for modelling suitability for the presence of the principal vectors in Brazil, these variables (temperature, humidity, rainfall, elevation, etc.) did not permit a local evaluation, as recommended [26, 37, 38, 47, 48, 62–66]. This limitation is associated with the low number of georeferenced cases (50% of the cases confirmed with PAIs were in RJ) and the lack of a necessary control of georeferenced unconfirmed cases of SF.

We urgently recommend that all cases (suspected and confirmed) be georeferenced. For this purpose, we recommend using the free, Internet-based GPS application essentials, which also functions from satellites without the need of an internet connection (http://www.gpsessentials.com/).

Although this is not the first interdisciplinary effort to integrate different determinants for understanding SF [22, 61], it is the first in Brazil to use multi-criteria analysis with mathematical algorithms applied to that pathology.

The present work uses the PROMETHEE algorithm to transform qualitative perceptions to quantitative values of the different dangers and threats that comprise the socio-environmental risk of human infection and defines the spatial dependence for infection with SF. Therefore, in addition to being a model study for cases where weights of the contributions of each element of the transmission chain are flexible according to the realities of the endemic areas, this study also demonstrates applicability from a national to local (domicile) scale.

Acknowledgements

We thank the Secretarias Municipais e Estaduais de Saúde (Secretary Municipal and of State of Health) of RJ for logistic and administrative support in acquiring information. This chapter is part of the doctoral thesis of D. Montenegro and was supported by the PhD scholarship program funded by Coordenação de Aperfeiçoamento de Pessoal de Nível Superior (Capes—Brasil sem Miséria) /FIOCRUZ.

Conflict of interest

The authors declare that we have prepared the manuscript in accordance with the standards of the journal, possess exclusive responsibility for the accuracy and correctness of the contents of the submitted article and declare that we have no conflicts of interest. The opinions expressed by authors contributing to this journal do not necessarily reflect the opinions of the journal or the institutions with which the authors are affiliated.

Author contributions

DM—contributed to the concept, design, and application of MDA techniques and spatial statistics; APC and MP—contributed to the design and application of MDA techniques; SVO made the climate suitability modelling for vector; IBM, LD, SVO, GSG and RPB—contributed to research project design and concept, data gathering and interpretation of results. All authors contributed to critically revising for important intellectual content, final approval of the version to be published, and all are in agreement to be accountable for all aspects of the work in ensuring that questions related to the accuracy or integrity of any part of the work are appropriately investigated and resolved.

Author details

Diego Montenegro[1,2,3,6*], Ana Paula da Cunha[4], Ingrid Machado[2], Liliane Duraes[2], Stefan Vilges de Oliveira[5], Marcel Pedroso[4], Gilberto S. Gazêta[2,5*] and Reginaldo P. Brazil[1]

*Address all correspondence to: dc.montenegro85@gmail.com and gsgazeta@ioc.fiocruz.br

1 Laboratório de Doenças Parasitária, Instituto Oswaldo Cruz/Fundação Oswaldo Cruz, Rio de Janeiro, Brazil

2 Laboratório de Referência Nacional em Vetores das Riquetsioses—Secretaria de Vigilância em Saúde/Ministério da Saúde, Instituto Oswaldo Cruz/Fundação Oswaldo Cruz, Rio de Janeiro, Brazil

3 Fundación Chilloa/Santa Marta, Colombia

4 Instituto de Comunicação e Informação Científica e Tecnologia em Saúde-ICICT/Fiocruz, Rio de Janeiro, Brazil

5 Secretaria de Vigilância em Saúde—Ministério da Saúde, Brasilia, Brazil

6 Research Group in Evolution, Systematics and Ecology Molecular, Universidad del Magdalena, Santa Marta, Colombia

References

[1] Taylor LH, Latham SM, Woolhouse MEJ. Risk factors for human disease emergence. Philosophical Transactions of the Royal Society B. 2001;**356**(1411):983-989. DOI: 10.1098/rstb.2001.0888

[2] OIE. One Health "at a glance" [Internet]. Paris, France: OIE—World Organisation for Animal Health; 2015. p. 2 Available from: http://www.oie.int/en/for-the-media/onehealth/#oe_mainContent

[3] Mwangi W, De Figueiredo P, Criscitiello MF. One health: Addressing global challenges at the Nexus of human, animal, and environmental health. PLoS Pathogens. 2016 Sep;**12**(9):e1005731. Available from: http://www.ncbi.nlm.nih.gov/pubmed/27631500

[4] Frank D. One world, one health, one medicine. The Canadian Veterinary Journal. 2008 Nov;**49**(11):1063-1065. Available from: http://www.ncbi.nlm.nih.gov/pubmed/19183729

[5] Lerner H, Berg C. The concept of health in one health and some practical implications for research and education: What is one health? Infection Ecology & Epidemiology. 2015;**5**:25300. Available from: http://www.ncbi.nlm.nih.gov/pubmed/25660757

[6] Osburn B, Scott C, Gibbs P. One world--one medicine--one health: emerging veterinary challenges and opportunities. Rev Sci Tech. 2009 Aug;**28**(2):481-486. Available from: http://www.ncbi.nlm.nih.gov/pubmed/20128454

[7] Calistri P, Iannetti S, Danzetta ML, Narcisi V, Cito F, Di Sabatino D, et al. The components of "one world - one health" approach. Transboundary and Emerging Diseases. 2013:4-13. Available from: http://www.ncbi.nlm.nih.gov/pubmed/24589096

[8] Cook RA, Karesh WB, Osofsky SA. The Manhattan Principles on "One World, One Health" [Internet]. September 2004 Symposium. New York: Wildlife Conservation Society; 2004. p. 1-1. Available from: http://www.oneworldonehealth.org/sept2004/owoh_sept04.html

[9] FAO-OIE-WHO. GLEWS, the Global Early Warning System [Internet]. Paris, France: GLEWA; 2013. Available from: http://www.glews.net/

[10] Chomel BB, Osburn BI. Zoological medicine and public health. Journal of Veterinary Medical Education. 2006;**33**(3):346-351. Available from: http://www.ncbi.nlm.nih.gov/pubmed/17035205

[11] Cunningham AA, Scoones I, JLN W. One health for a changing world: New perspectives from Africa. Philosophical Transactions of the Royal Society of London. Series B. 2017;**372**: 1725. Available from: http://rstb.royalsocietypublishing.org/content/372/1725/20160162

[12] Regier Y, O'Rourke F, Kempf VAJ. Bartonella spp. - a chance to establish one health concepts in veterinary and human medicine. Parasites & Vectors. 2016 Dec 10;**9**(1):261. Available from: http://parasitesandvectors.biomedcentral.com/articles/10.1186/s13071-016-1546-x

[13] Purohit MR, Chandran S, Shah H, Diwan V, Tamhankar AJ, Lundborg CS. Antibiotic resistance in an indian rural community: A "one-health" observational study on commensal

coliform from humans, animals, and water. International Journal of Environmental Research and Public Health. 2017 Apr 6;**14**(4):386. Available from: http://www.mdpi.com/1660-4601/14/4/386

[14] Smith J, Taylor EM, Kingsley P. One world-one health and neglected zoonotic disease: Elimination, emergence and emergency in Uganda. Social Science & Medicine. 2015 Mar;**129**: 12-19. Available from: http://linkinghub.elsevier.com/retrieve/pii/S0277953614004122

[15] Singh BB, Gajadhar AA. Role of India's wildlife in the emergence and re-emergence of zoonotic pathogens, risk factors and public health implications. Acta Tropica. 2014 Oct;**138**:67-77. Available from: http://www.ncbi.nlm.nih.gov/pubmed/24983511

[16] Brown T. "Vulnerability is universal": Considering the place of "security" and "vulnerability" within contemporary global health discourse. Social Science & Medicine. 2011 Feb;**72**(3):319-326. Available from: http://www.ncbi.nlm.nih.gov/pubmed/20947233

[17] Gibbs EPJ, Anderson TC. One world - one health' and the global challenge of epidemic diseases of viral aetiology. Veterinaria Italiana. 2009;**45**(1):35-44. Available from: http://www.ncbi.nlm.nih.gov/pubmed/20391388

[18] Hoogstraal H. Ticks in relation to human diseases. Annual Review of Microbiology. 1967;**12**:377-420

[19] Hoogstraal H, Aeschlimann A. Tick-host specificity. Bull la Société Entomol Suisse. 1982;**55**:5-32. Available from: http://doc.rero.ch/record/19829

[20] Walker DH. Rickettsiae. In: Medical Microbiology [Internet]. Galveston: University of Texas Medical Branch; 1996. Available from: http://www.ncbi.nlm.nih.gov/pubmed/21413251

[21] Travassos J. Resumo dos trabalhos ecologicos e experimentais sobre Tifo exantematicos de São Paulo dessenvolvidos por Juaquim Travassos. Biol e Hyg. 1937;**9**(1):61-73

[22] Szabó MPJ, Pinter A, Labruna MB. Ecology, biology and distribution of spotted-fever tick vectors in Brazil. Frontiers in Cellular and Infection Microbiology. 2013;**3**:27. Available from: http://www.scopus.com/inward/record.url?eid=2-s2.0-84891426655&partnerID=tZOtx3y1

[23] Brasil. Febre Maculosa Brasileira. Penna G, Teixeira M, Costa M, Pereira S, Carmo E, do Nascimento E, editors. Guia de Vigilância Epidemiológica [Internet]. 7th ed. Brasilia DF: Ministério da Saúde/Secretaria de Vigilância em Saúde/Departamento de Vigilância Epidemiológica; 2009. p. 1-14. Available from: http://bvsms.saude.gov.br/bvs/publicacoes/guia_vigilancia_epidemiologica_7ed.pdf

[24] Brasil. Febre Maculosa Brasileira e Outras Riquetsioses. In: da Silva J, editor. Guia de Vigilância em Saúde [Internet]. 1st ed. Brasilia DF: Secretaria de Vigilância em Saúde/Ministerio da Saúde; 2014. p. 445-54. Available from: http://portalsaude.saude.gov.br/images/pdf/2014/novembro/27/guia-vigilancia-saude-linkado-27-11-14.pdf

[25] de Souza CE, Pinter A, Donalisio MR. Risk factors associated with the transmission of Brazilian spotted fever in the Piracicaba river basin, state of São Paulo, Brazil. Revista

da Sociedade Brasileira de Medicina Tropical. 2015;**48**(1):11-17. Available from: http://www.ncbi.nlm.nih.gov/pubmed/25860458

[26] Brites-Neto J, Nieri-Bastos FA, Brasil J, Duarte KMR, Martins TF, Verissimo CJ, et al. Environmental infestation and rickettsial infection in ticks in an area endemic for Brazilian spotted fever. Revista Brasileira de Parasitologia Veterinária. 2013;**22**(3): 367-372. Available from: www.cbpv.com.br/rbpv

[27] Montenegro DC, Bitencourth K, de Oliveira SV, Borsoi AP, Cardoso KM, Sousa MSB, et al. Frontiers in Microbiology. 2017;**8**:505. Available from: http://journal.frontiersin.org/article/10.3389/fmicb.2017.00505/full

[28] de Oliveira S, Guimarães JN, Reckziegel GC, da Neves BMC, de Araújo-Vilges KM, Fonseca LX, et al. An update on the epidemiological situation of spotted fever in Brazil. Journal of Venomous Animals and Toxins Including Tropical Diseases. 2016 Dec 22;**22**(1): 22. Available from: http://jvat.biomedcentral.com/articles/10.1186/s40409-016-0077-4

[29] Katz G, de Neves VLFC, Angerami RN, do Nascimento EMM, Colombo S. Situação epidemiológica e importância da febre maculosa no Estado de São Paulo. BEPA, Bol. epidemiol. paul. 2009;**6**(69):4-13. Available from: http://periodicos.ses.sp.bvs.br/scielo.php?script=sci_abstract&pid=S1806-42722009000900001&lng=pt&nrm=iso&tlng=pt

[30] Labruna MB, Kasai N, Ferreira F, Faccini JLH, Gennari SM. Seasonal dynamics of ticks (Acari: Ixodidae) on horses in the state of São Paulo, Brazil. Veterinary Parasitology. 2002 Apr 19;**105**(1):65-77. Available from: http://www.ncbi.nlm.nih.gov/pubmed/11879967

[31] Oliveira PR, Borges LM, Lopes CM, Leite RC. Population dynamics of the free-living stages of Amblyomma cajennense (Fabricius, 1787) (Acari: Ixodidae) on pastures of Pedro Leopoldo, Minas Gerais state, Brazil. Veterinary Parasitology. 2000 Oct 20;**92**(4):295-301. Available from: http://www.ncbi.nlm.nih.gov/pubmed/10996741

[32] Oliveira PR, Borges LMF, Leite RC, Freitas CMV. Seasonal dynamics of the cayenne tick, Amblyomma cajennense on horses in Brazil. Medical and Veterinary Entomology. 2003 Dec;**17**(4):412-416. Available from: http://www.ncbi.nlm.nih.gov/pubmed/14651655

[33] Toledo R, Tamekuni K, Haydu V, Vidotto O. Dinamica sazonal de carrapatos do genero Amblyomma (Acari: Ixodidae) em um parque urbano da cidade de Londrina. Revista Brasileira de Parasitologia Veterinária (Brazilian Journal of Veterinary Parasitology). 2008;**17**(1):50-54

[34] de Lemos ER, Machado RD, Coura JR, Guimarães MA, Freire NM, Amorim M, et al. Epidemiological aspects of the Brazilian spotted fever: Seasonal activity of ticks collected in an endemic area in São Paulo, Brazil. Revista da Sociedade Brasileira de Medicina Tropical. 1997;**30**(3):181-185. Available from: http://www.scielo.br/pdf/rsbmt/v30n3/0725.pdf

[35] Childs JE, Paddock CD. Passive surveillance as an instrument to identify risk factors for fatal Rocky Mountain spotted fever: Is there more to learn? The American Journal of Tropical Medicine and Hygiene. 2002;**66**(5):450-457

[36] Silveira A. Ecosystems characterization with potential risk for infestation by ticks and rickettsial transmission to humans in the Rio de Janeiro state. Dissertação. 2010:50. http://www.dominiopublico.gov.br/pesquisa/DetalheObraForm.do?select_action=&co_obra=179198

[37] Estrada-Peña A, Gray JS, Kahl O, Lane RS, Nijhof AM. Research on the ecology of ticks and tick-borne pathogens--methodological principles and caveats. Frontiers in Cellular and Infection Microbiology. 2013 Jan;**3**:29. Available from: http://www.pubmedcentral.nih.gov/articlerender.fcgi?artid=3737478&tool=pmcentrez&rendertype=abstract

[38] Estrada-Peña A, Alexander N, Wint GRW. Perspectives on modelling the distribution of ticks for large areas: So far so good? Parasites & Vectors. 2016;**9**(1):179. Available from: http://www.ncbi.nlm.nih.gov/pubmed/27030357

[39] Barros-Silva PMR, Fonseca LX, Carneiro ME, de Vilges KMA, de Oliveira SV, Gurgel-Gonçalves R. Occupational risk of spotted fever: An evaluation of knowledge, attitudes and prevention practices among veterinary medicine students. Rev patol trop. 2014:389-397. Available from: http://www.revistas.ufg.br/index.php/iptsp/article/view/33597/17784

[40] Valbuena G. Fiebres que no deberían matar. Biomédica. 2007;**27**(3):321-324

[41] Pinter A, França A, de Souza C, Sabbo C, do Nascimento E, dos Santos F, et al. Febre Maculosa Brasileira. Vol. 8, Suplemento Bepa. Sao Paulo: Centro de produção e divulgação científica CCD/SES-SP; 2011. 33p

[42] IBGE. Demografia do estado do Rio de Janeiro. Brasilia D.F, Brazil: Instituto Brasileiro de Geografia e Estatistica; 2016. Available from: http://www.ibge.gov.br/estadosat/perfil.php?sigla=rj

[43] Brasil. FEBRE MACULOSA—Casos confirmados notificados no Sistema de Informação de Agravos de Notificação—Rio de Janeiro [Internet]. Brasilia D.F, Brazil: Brasil or Ministerio da Saúde do Brasil; 2016. Available from: http://tabnet.datasus.gov.br/cgi/deftohtm.exe?sinannet/cnv/febremaculosarj.def

[44] Brasil. Resolução 466/12 do Conselho Nacional de Saúde/MS Sobre Diretrizes e Normas Regulamentadoras de Pesquisa envolvendo seres humanos [Internet]. Brasil: Diário Oficial da União, Seção 1, Página 21082, Coluna 2, Resolução No 466, de 12 de Dezembro de 2012 2012 p. 5. Available from: http://bvsms.saude.gov.br/bvs/saudelegis/cns/2013/res0466_12_12_2012.html

[45] Ministerio da Saúde. Dados e Indicadores da Saúde [Internet]. MS. 2013. Available from: http://portalms.saude.gov.br/dados-e-indicadores-da-saude

[46] Elith J, Phillips SJ, Hastie T, Dudík M, Chee YE, Yates CJ. A statistical explanation of MaxEnt for ecologists. Diversity and Distributions. 2011 Jan 1;**17**(1):43-57. Available from: http://doi.wiley.com/10.1111/j.1472-4642.2010.00725.x

[47] Barbieri JM, Da Rocha CMBM, Bruhn FRP, Cardoso DL, Pinter A, Labruna MB. Altitudinal assessment of Amblyomma aureolatum and Amblyomma ovale (Acari: Ixodidae), vectors of spotted fever group Rickettsiosis in the state of São Paulo, Brazil. Journal of Medical Entomology. 2015 Sep;**52**(5):1170-1174. Available from: http://www.ncbi.nlm.nih.gov/pubmed/26336213

[48] Eremeeva ME, Dasch GA. Challenges posed by tick-borne rickettsiae: Eco-epidemiology and public health implications. Frontiers in Public Health. 2015;**3**:55. Available from: http://www.pubmedcentral.nih.gov/articlerender.fcgi?artid=4404743&tool=pmcentrez&rendertype=abstract

[49] Kaplan JE, Newhouse VF. Occurrence of Rocky Mountain spotted fever in relation to climatic, geophysical, and ecologic variables. The American Journal of Tropical Medicine and Hygiene. 1984 Nov;**33**(6):1281-1282. Available from: http://www.ncbi.nlm.nih.gov/pubmed/6507738

[50] Newhouse VF, Choi K, Holman RC, Thacker SB, D'Angelo LJ, Smith JD. Rocky Mountain spotted fever in Georgia, 1961-75: Analysis of social and environmental factors affecting occurrence. Public Health Reports. 1979;**101**(4):419-428. Available from: http://www.ncbi.nlm.nih.gov/pubmed/3090609

[51] Raghavan RK, Goodin DG, Neises D, Anderson GA, Ganta RR. Hierarchical Bayesian spatio-temporal analysis of climatic and socio-economic determinants of rocky mountain spotted fever. PLoS One. 2016;**11**(3):e0150180. Available from: http://www.ncbi.nlm.nih.gov/pubmed/26942604

[52] IBGE. Instituto Brasileiro de Geografia e Estatistica [Internet]. IBGE. 2010. Available from: https://www.ibge.gov.br/

[53] Statpoint Technologies I. STATGRAPHICS® Centurion [Internet]. Virginia, Estados Unidos: Statgraphics Technologies, Inc; 2006. Available from: http://www.statgraphics.com/

[54] Montenegro D, da Cunha AP, Ladeia-Andrade S, Vera M, Pedroso M, Junqueira A, et al. Multi-criteria decision analysis and spatial statistic: An approach to determining human vulnerability to vector transmission of Trypanosoma cruzi. Memórias do Instituto Oswaldo Cruz. 2017 Oct;**112**(10):709-718. Available from: http://www.scielo.br/scielo.php?script=sci_arttext&pid=S0074-02762017001000709&lng=en&tlng=en

[55] Hayez Q, De Smet Y, Bonney J. D-sight: A new decision making software to address multi-criteria problems. International Journal of Decision Support Systems. 2012 Jan;**4**(4):1-23. Available from: http://services.igi-global.com/resolvedoi/resolve.aspx?doi=10.4018/jdsst.2012100101

[56] Brans JP, Mareschal B. The PROMCALC & GAIA decision support system for multicriteria decision aid. Decision Support Systems. 1994 Nov;**12**(4-5):297-310. Available from: http://linkinghub.elsevier.com/retrieve/pii/0167923694900485

[57] Hayez Q, Mareschal B, De Smet Y. New GAIA visualization methods. In: Proceedings of the International Conference on Information Visualisation [Internet]. IEEE; 2009. p. 247-51. Available from: http://ieeexplore.ieee.org/lpdocs/epic03/wrapper.htm?arnumber=5190781

[58] INPE. TerraView [Internet]. São José dos Campos: SP: Instituto Nacional de Pesquisas Espaciais; 2010. Available from: http://www.dpi.inpe.br/terraview/index.php

[59] QGIS Development Team. QGIS Geographic Information System [Internet]. Open Source Geospatial Foundation Project; 2016. Available from: http://www.qgis.org/

[60] Montenegro D, de Mello FL, Giordano Dias CM, Almeida P, Araújo M, Magalhães MA, et al. Evaluating the surveillance system for spotted fever in Brazil using machine-learning techniques. Frontiers in Public Health. 2017 Nov 30;**5**:323. Available from: http://journal.frontiersin.org/article/10.3389/fpubh.2017.00323/full

[61] Mitsumori A, Pinter A, Nunes A, Marcusso C, Costa C, Moises D, et al. In: Boulos M, editor. A febre Maculosa Braileira na Região Metropolitana de São Paulo. Vol. 13. São Paulo, SP, Brasil: Boletim Epidemiológico Paulista- BEPA; 2016. 53p

[62] Milagres BS, Padilha AF, Barcelos RM, Gomes GG, Montandon CE, Pena DCH, et al. Rickettsia in synanthropic and domestic animals and their hosts from two areas of low endemicity for Brazilian spotted fever in the eastern region of Minas Gerais, Brazil. The American Journal of Tropical Medicine and Hygiene. 2010 Dec 6;**83**(6):1305-1307. Available from: http://www.ajtmh.org/content/83/6/1305

[63] Labruna MB. Ecology of rickettsia in South America. Annals of the New York Academy of Sciences. 2009;**1166**:156-166

[64] Ogrzewalska M, Uezu A, Jenkins CN, Labruna MB. Effect of forest fragmentation on tick infestations of birds and tick infection rates by rickettsia in the Atlantic Forest of Brazil. EcoHealth. 2011;**8**(3):320-331

[65] Londoño AF, Acevedo-Gutiérrez LY, Contreras VM, Marín DM, Díaz FJ, Valbuena G, et al. Ecological endemicity for rickettsia in Colombia. F1000Research. 2014 Jun 16;**5**:1. Available from: http://f1000research.com/posters/1095698

[66] Labruna MB, Kerber CE, Ferreira F, Faccini JL, De Waal DT, Gennari SM. Risk factors to tick infestations and their occurrence on horses in the state of São Paulo, Brazil. Veterinary Parasitology. 2001 May 9;**97**(1):1-14. Available from: http://www.ncbi.nlm.nih.gov/pubmed/11337122

12

The Use of Photos of the Social Networks in Shaping a New Tourist Destination: Analysis of Clusters in a GIS Environment

Hélder Tiago da Silva Lopes,
Paula Cristina Almeida Cadima Remoaldo and
Vitor Ribeiro

Abstract

The use of new photo-sharing services in social networks has favoured a perception of the interests of locals and visitors. The photos presented in these networks are geocoded by the users, residents or visitors, allowing extensive databases to be obtained. The research that was conducted between 2015 and 2016 followed an essentially quantitative approach. Based on the georeferenced photos on social networks, the main distribution patterns of places of interest to tourists, visitors and residents were analysed in a rural emergent tourist destination in northeastern Portugal. We used geographical information systems (GISs) to apply various spatial and statistical analysis techniques. One the main conclusions was that there is a high number of natural and cultural heritage locations with tourism potential, and, in some cases, their accessibility standards make them favourable destinations for tourists.

Keywords: rural tourism, spatial analysis, accessibility, photo-sharing services, GIS

1. Introduction

The development of new methods of analysis of the images of various destinations has accompanied the emergence of extensive research that uses the method of analysing photos [1–5]. Another fact is due to the massive use of the Internet, which is increasingly introducing significant changes in the way people interact in society [6, 7]. This new dimension of communication has made the expression of multiple cultural values possible [8]. Because of the evolution of the Internet and the advent of smart technology, the proliferation of the number of photos registered per trip has occurred in recent decades [9]. However, the habit of sharing them with co-workers, family and friends, as well as in the broader media, comes as a result of online sharing, often in real time and synchronously on various websites, such as Flickr, Picasa, Facebook, Panoramio and Pinterest [4, 10–15].

Images have assumed a fundamental role in the dissemination of the dimensions and amplitudes of tourist spaces. The sharing of photos of a place through social networks has an increasingly strong power in promoting tourism, where the dissemination of the tourist image has an effective and lasting role. Different social media platforms have been accommodating new features, and coupled with their simplicity of handling and attractiveness, they are competing increasingly with traditional travel guides or leaflets because they allow people to post photographs on the Internet immediately after they are taken [4, 12, 13, 16–21]. However, it should be noted that these platforms still tend to be used by the younger age groups and by groups with higher socioeconomic characteristics.

Conversely, research centred on the analysis of online images of rural destinations has not been explored extensively [22–24]. Increasingly, urban dwellers' interest in staying overnight and visiting rural areas shows that it is essential to market and promote rural destinations online [18, 25–29].

With the evolution of geographic information technologies, it is possible to analyse distribution patterns (intensity, concentration and dispersion) of tourism resources in different spaces through density maps, central trend measures or with the use of indicators of distribution patterns (e.g. Getis-Ord General G or Moran's Index).

Based on these assumptions, the main objectives of this research are (i) to identify the spatial distribution patterns of the photos of visitors and residents; (ii) to characterise the different "looks" of the tourist destination and (iii) to contribute to the development of a tourist image that is closer to the interests of visitors and locals. This study complements other approaches undertaken in the territory, resulting mainly from an exploratory analysis of existing and potential tourism resources, the results of the focus group that was conducted, and the results of a self-administered survey [30, 31].

This chapter consists of four sections. After the introductory section, the second section presents the main methods and sources used to acquire the data that were used. Section 3 is a summary of the main results that were achieved, and it highlights the main potentialities of the analysis of the images and how they can contribute to the segmentation of the visitors. Section 4 presents our discussions. Section 5 presents the conclusions concerning the main results, proposes some challenges for future research and identifies the main limitations that were intrinsic to the study.

2. Materials and methods used in the study area

2.1. Geographical context

The case study was based on the municipality of Boticas, which is located in the Nomenclature of Territorial Units for Statistics (NUTS) III of Alto Tâmega in the northeast of Continental Portugal.

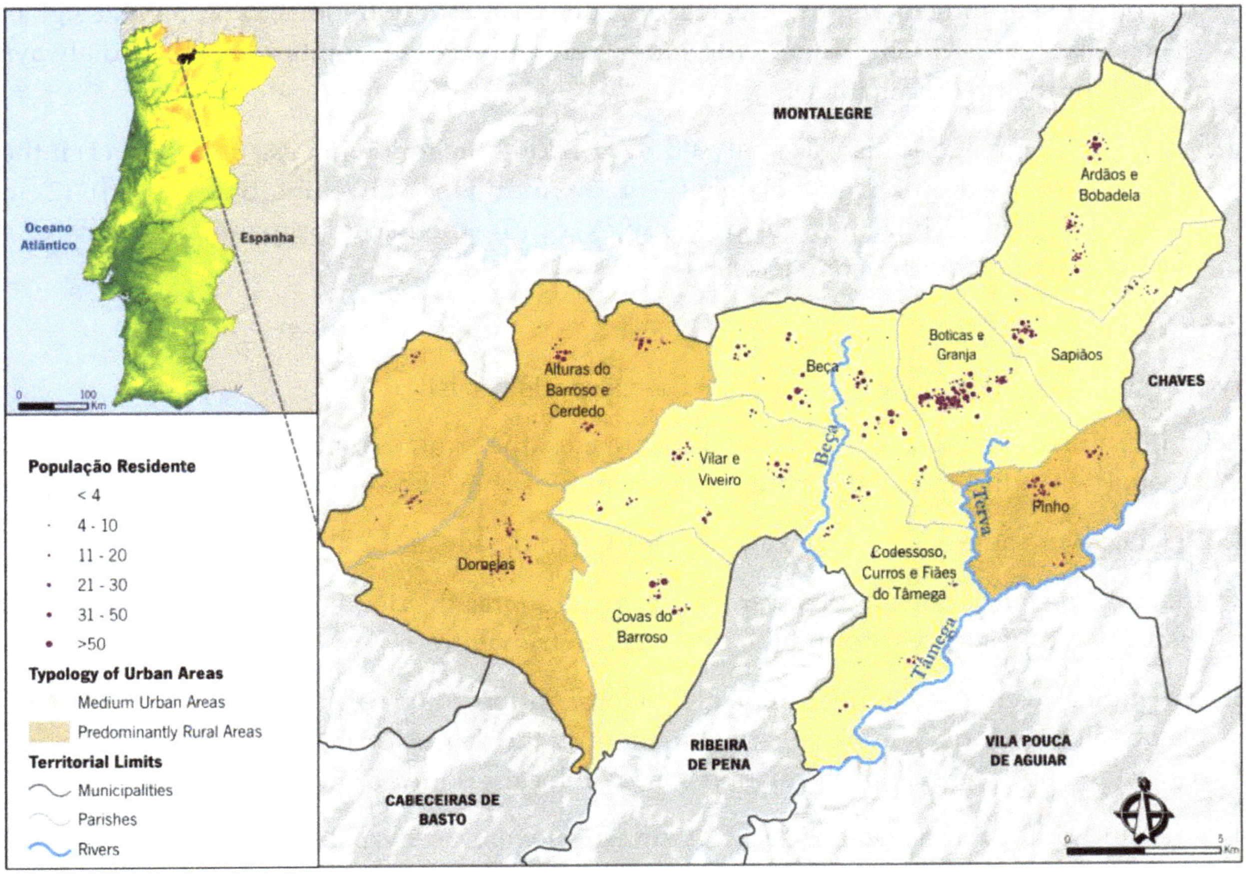

Figure 1. Geographical context of the municipality of Boticas: Population by subsection, TIPAU and MDT. Source: Own elaboration (based on the Census, 2011; TIPAU 2014 (INE) and Atlas of the Environment).

According to the Typology of Urban Areas (TIPAU) of 2014, seven of the parishes are considered Medium Urban Areas (MUA) and three as Predominantly Rural Areas (PRA) [32]. The area is subdivided into 10 parishes, with an area of 322 km^2 (**Figure 1**). In 2011 (last census), the population of the municipality of Boticas was 5750, of which 1510 residents (26.3% of the population) lived in Boticas and Granja. In fact, this municipality, like other territories located in the interior of the country, has lost its vitality. One of the municipality's main potentials is the use of its endogenous resources to attract tourists, associated economic activities and to enhance its population.

2.2. Materials and methods

The database used in this research was constructed with data from Panoramio's photo platform, Google Earth. The period from January 2005 to March 2016 was used, and 728 photos were used. The photos were analysed by two quantitative methods, that is, (1) univariate and multivariate statistical analysis using the SPSS 22.0 statistical package and (2) geospatial analysis using the ArcGIS 10.3 package. The data were grouped into several typologies considering the assumptions defined in previous studies [3–5]:

i. Four categories—(1) built heritage (e.g. monuments, historic buildings, churches and typical homes), (2) nature (landscape, fauna and flora), (3) culture (combining expressions of local culture, traditions or festivals) and (4) tourist services (comprising all forms of tourism equipment and infrastructures, such as accommodations, catering and signs). The situations in which images with more than one set of elements were verified always opted for the predominant set.

ii. Zoom of the image—this was analysed by checking the following assumptions: (1) if the image is focused on a single element (e.g. a window of a dwelling or a church), (2) in its context (a church in the housing complex) or (3) whether it is a panoramic view or a scenery (e.g. a mountain or a river).

iii. Presence of people.

iv. Origin—whether the image originates from locals or visitors.

After defining the cataloguing criteria of the images, the main assumptions that are inherent to the analysis of the spatial distribution of the photos should be emphasised:

1. the data were aggregated in hexagons with 150 m of side and 300 m of diameter;
2. standard distance calculations were performed to infer the degree of concentration or the dispersion of resources around the mean geometric centre;
3. two indices were used to determine global localization patterns, that is, Getis-Ord General G and Global Moran's I. These served to identify the degree of agglomeration of high and low values and the spatial correlation based on local resources and attribute values;
4. Anselin Local Moran's statistics (LISA statistics) were used, making it possible to determine emerging local trends for the intensification of Boticas' tourism.

These typologies were the basis of the identification of clusters. This technique allows the identification of groups with high homogeneity within the group as well as intergroup heterogeneity [3]. The k-means method was used considering the following assumptions: (i) the choice of a sample was made based on a sample in each group, from a group of groups (2–5); (ii) minimum loss of information through the merging of the elements and (iii) instead of considering the categories *per se*, items were used to segment the groups. After obtaining the clusters, an ANOVA test was performed to determine the differences between the groups.

3. Results

The destination image can be promoted based on the sets of photos that residents and visitors take, considering their ability to become souvenirs, postcards or tourists' objects. Photo density analysis of Sightsmaps (http://www.sightsmap.com/, developed by [33]) clearly showed a density of photos in more urbanised areas (e.g. Oporto Metropolitan Area (OMP), Braga, Guimarães, Viana do Castelo, Chaves and Bragança) in the course of the Douro River (in Portugal, Barca d'Alva to the mouth between Porto and VN Gaia) or in the Peneda and Gerês National Park (PNPG), which was different from what occurs in territories such as Boticas.

In the municipality of Boticas, as with the other municipalities of Alto Tâmega (with the exception of Chaves), there is a density of photos that were much smaller and circumscribed to the central area of the municipality. **Figure 2** summarises the distribution of the photographic

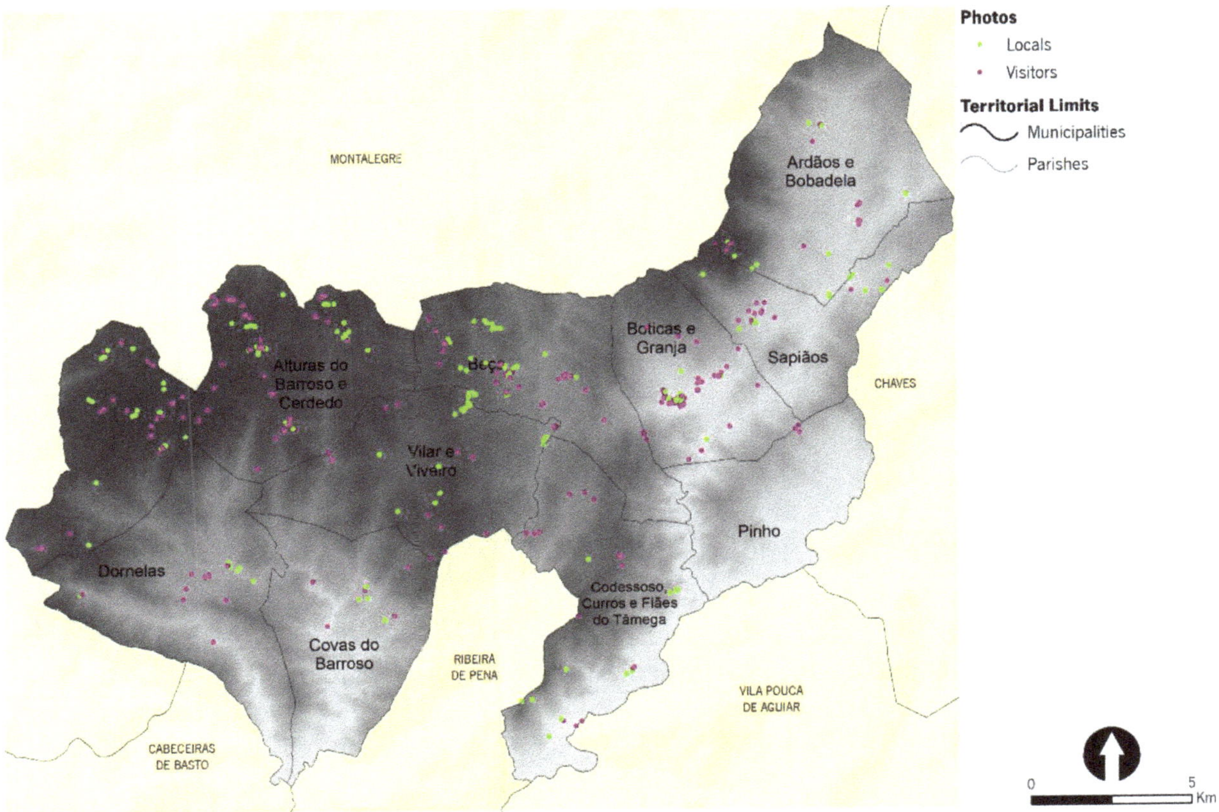

Figure 2. Geographic distribution of photos of visitors and residents. Source: Panoramio photos, collected during the period from January 2003 through March 2015.

series, differing according to whether they were taken by residents or visitors. In general, there is a concentration of photos in the northwest and central areas of the municipality of Boticas, although most of the photos are distributed somewhat throughout the municipality, mainly along the main axes of the road network.

However, it is in the parish of Boticas and Granja that there was a greater density of photos, which was due to the amount of equipment available to visitors. For this reason, in all parishes, there are more photos of visitors than residents (258 for locals and 470 for visitors—**Table 1**).

The maximum of photos recorded per hexagon for this group was 55 photos (**Table 2**).

The grouping of the photos in hexagons shows a concentration of photos to the north of the municipality, and the areas with the lower densities of photos are located to the south, especially in the parish of Pinho (three photos—**Figure 3**).

To identify the clusters, the G-statistics were analysed, and the Moran index was used. The G-statistics indicated that there was a tendency for the concentration of values (high clusters), with high levels of significance (p-value < 0.01). Likewise, the Moran index indicated a very strong spatial correlation for the formation of spatial agglomerates (p-value < 0.01, **Figure 4**).

The Anselin local Moran's statistic was calculated with the intention of mapping the presence of these groups, and it was found that the number of clusters with High-High (HH) values was not very significant. Concerning the distribution of these clusters, there was a spatial

Parish	Locals		Visitors		Total	
	No.	%	No.	%	No.	%
Alturas do Barroso and Cerdedo	85	32.9	105	22.3	190	26.1
Dornelas	7	2.7	16	3.4	23	3.2
Ardãos and Bobadela	15	5.8	15	3.2	30	4.1
Boticas and Granja	49	19.0	201	42.8	250	34.3
Beça	38	14.7	72	15.3	110	15.1
Pinho	0	0.0	3	0.6	3	0.4
Codessoso, Curros and Fiães do Tâmega	15	5.8	14	3.0	29	4.0
Covas do Barroso	10	3.9	9	1.9	19	2.6
Vilar and Viveiro	28	10.9	24	5.1	52	7.1
Sapiãos	11	4.3	11	2.3	22	3.0
Total	258	100	470	100	728	100

Source: Panoramio photos, collected for the period from January 2003 through March 2015.

Table 1. Distribution of photos by parish (in number and percent).

	Number of photos	D.P.	Maximum of photos per hexagon
Total	728	1.56	66
Locals	238	0.43	17
Visitors	490	1.22	55

Source: Panoramio photos, collected for the period from January 2003 through March 2015.

Table 2. Statistics of photos taken by visitors and locals.

agglomeration in several villages typical of the municipality (e.g. Vilarinho Sêco, Coimbró), in places of landscape-natural interest (e.g. Carvalhelhos, Mosteiró, Vilarinho de Mó) and in the village centre, either by locals or visitors. There are some differences between places photographed by locals and visitors. The former has High-High (HH) clusters in Coimbró, Serra do Barroso Wind Farm and in Fiães do Tâmega, while the latter have an agglomeration in Sapiãos (e.g. anthropomorphic graves, some typical houses, river beach), which are not present in the photographs made by the residents. It should be noted, however, that there were small clusters expressed in Low-High values, which correspond to atypical values.

Figure 6 presents the typologies created for the photos. The concentration of photos is strongly associated with two types of characteristics, that is, built-heritage (57%) and natural elements (35%). Tourism services (5%) and local culture (4%, **Figure 5A**) appear less often.

The distances from which the images were photographed were very different. In some cases, the person taking the photos focused only on the details, while others sought to take photographs of the entire scene. Therefore, we chose to classify the photographs into three subgroups, as shown in **Figure 5(B)**. The results allowed us to conclude that 57% of the sites/

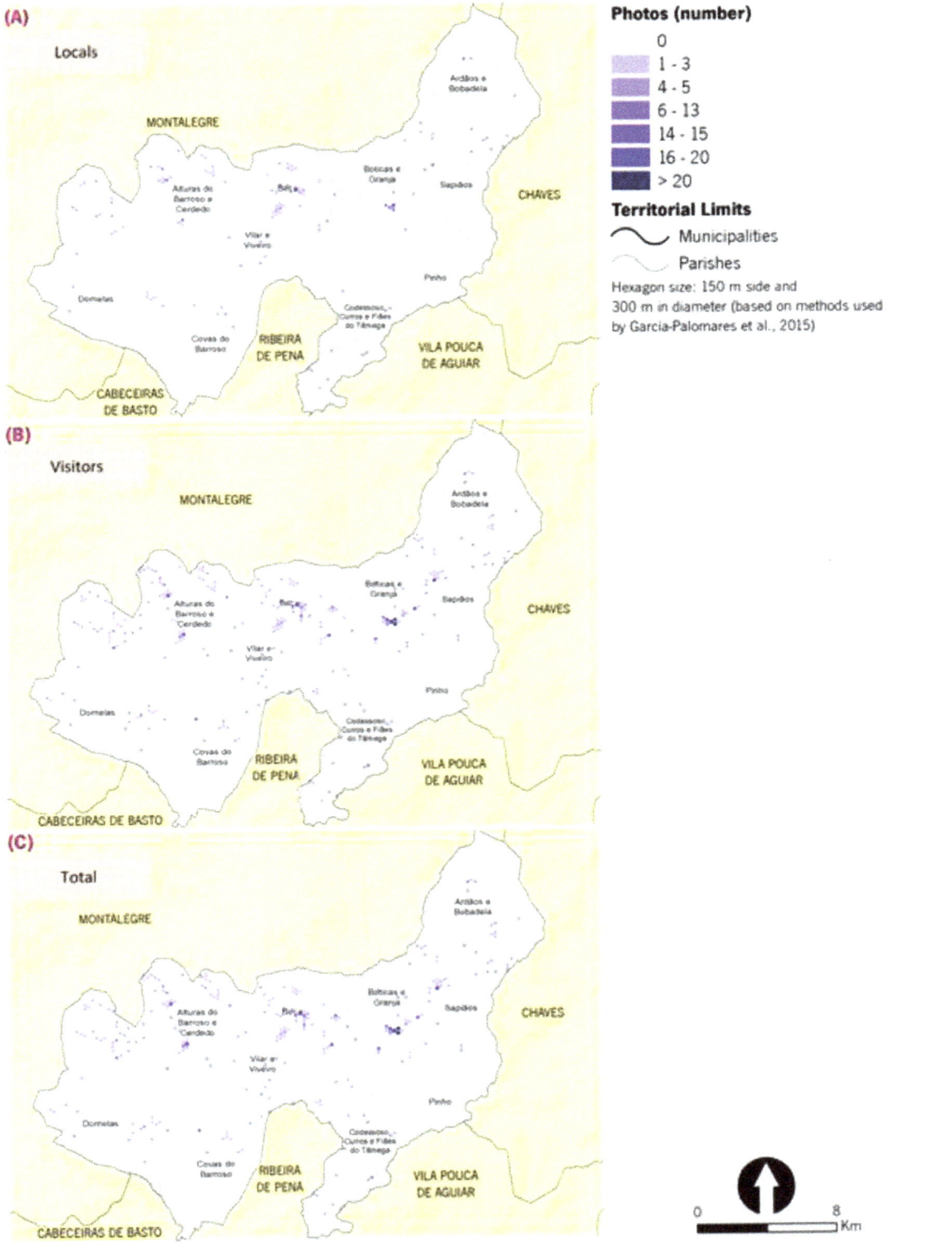

Figure 3. Density of photos: (A) locals, (B) visitors, (C) total. Source: Own elaboration, based on Panoramio photos for the period from January 2003 through March 2015.

visitors photographed elements in the context, which reflected the demand of visitors and residents to capture what is within a context, so that it can be recognised easily. Another element used for analysis was the degree of the presence of people in the photos (**Figure 5C**).

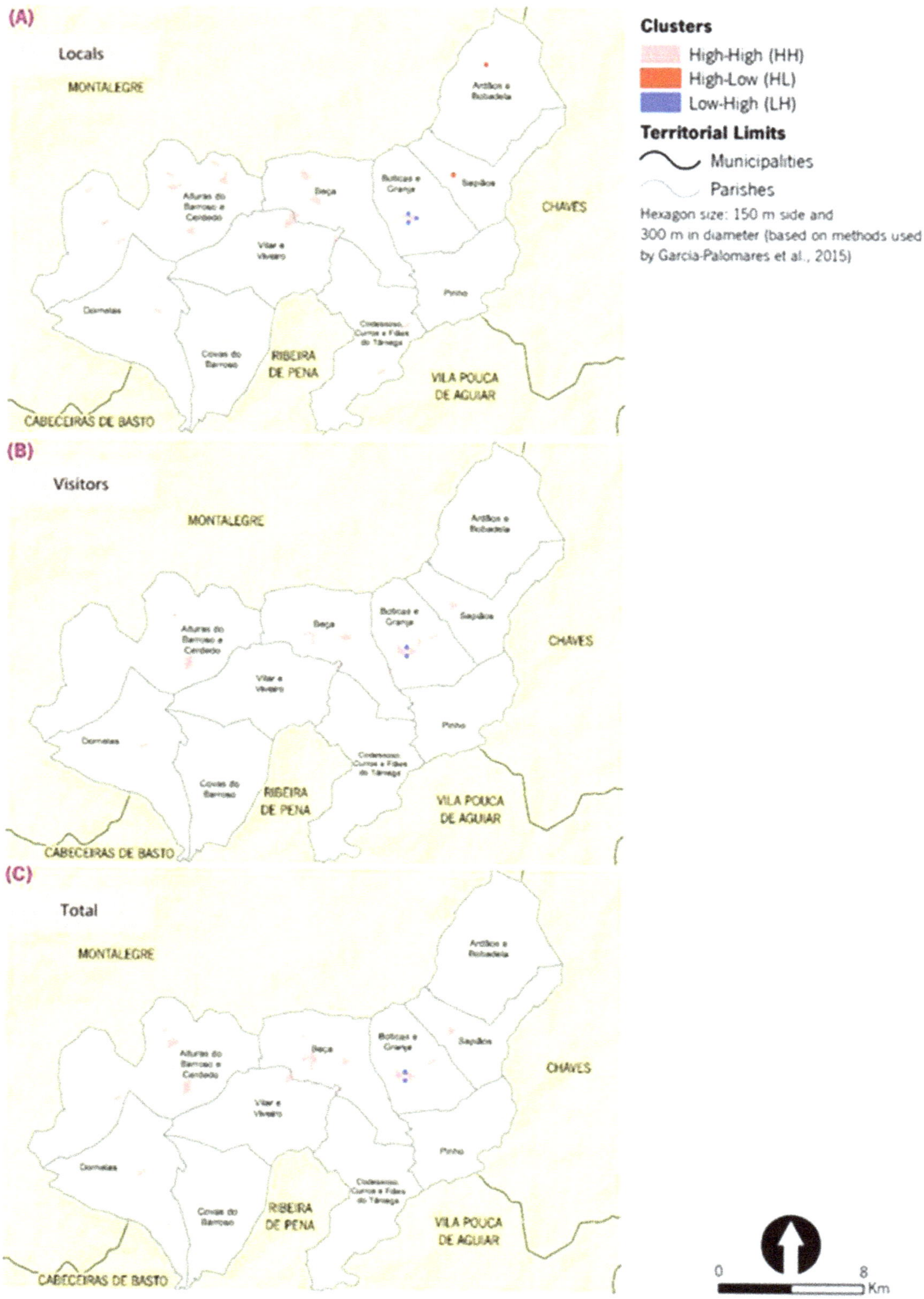

Figure 4. Anselin local Moran's statistics. Source: Own elaboration, based on Panoramio photos for the period from January 2003 through March 2015.

The most common circumstance for visitors and residents was to take photos without the presence of people (93%; **Figure 5D**), and only 7% of the photos included people. In addition, 67.3% of the photos were taken by visitors, showing their renewed interest, which is perfectly

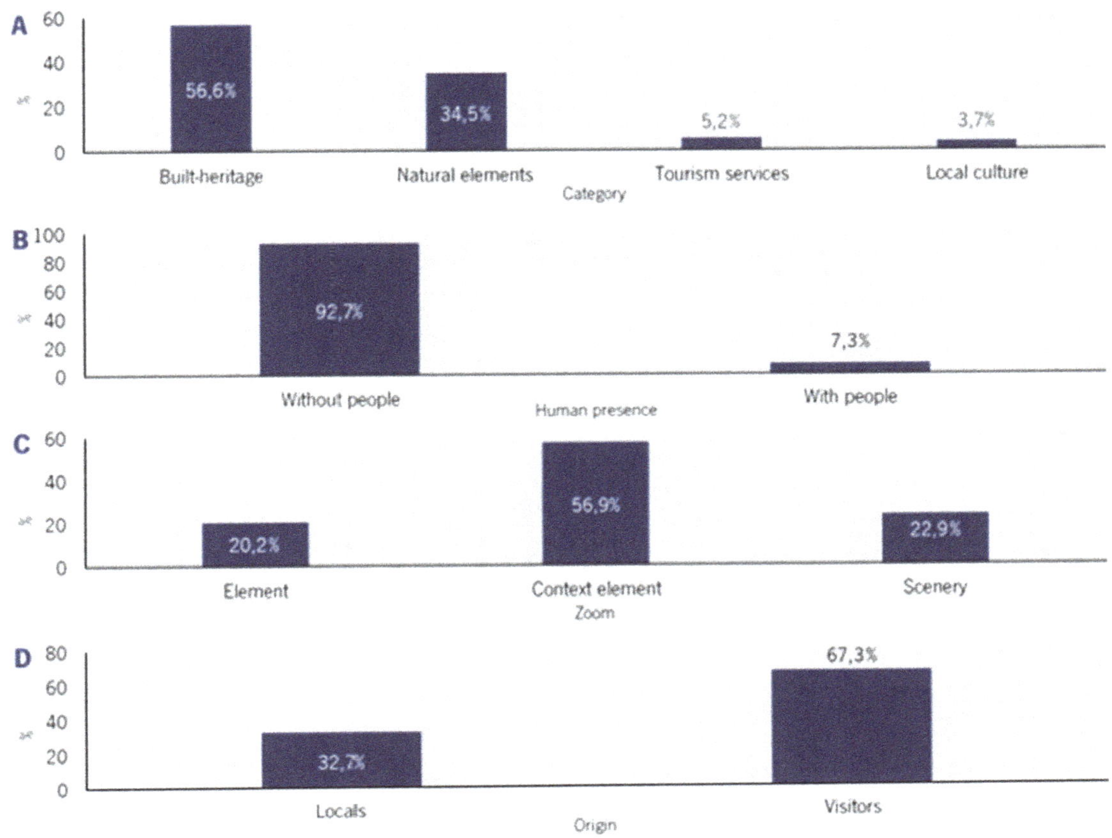

Figure 5. Typologies of photos (%): (A) category, (B) human presence, (C) zoom, and (D) origin. Source: Own elaboration, based on Panoramio photos for the period from January 2003 through March 2015.

understandable. In fact, it is surprising that there were a significant number of locals placing Panoramio photos, which may indicate the recognition of the natural and built heritage that exists in the municipality.

Figure 6 represents the temporal distribution of the photos taken by visitors and locals. It should be noted that the figures do not refer to all of the photos sent to the Panoramio platform over the review period; rather, they refer only to the photos that contain information concerning the dates they were taken.

There were 528 photos with year-to-year, intra-annual and intra-day information that were included among the 729 previously classified photos. In fact, the year with the highest number of photographs taken by visitors was 2014, while in the case of locals, the highest number of photographs taken occurred in previous years, that is, between 2006 and 2008 (**Figure 6A**). Thus, in cumulative terms, there has been a redoubled interest in taking photographs by visitors in recent years. However, when analysing the values recorded monthly in the period from 2013 through 2015, there is a seasonality of the photo records, with a concentration in August (**Figure 6B**). Concerning the hourly range, these photographs generally were taken between 10:00 am and 5:00 pm (**Figure 6C**).

It should be noted that residents contributed most to post-5:00 pm photographs, which may be due to the return of excursions to the place of departure after an overnight visit to the Boticas municipality, indicating that the municipality had visitors but not tourists. There are some records

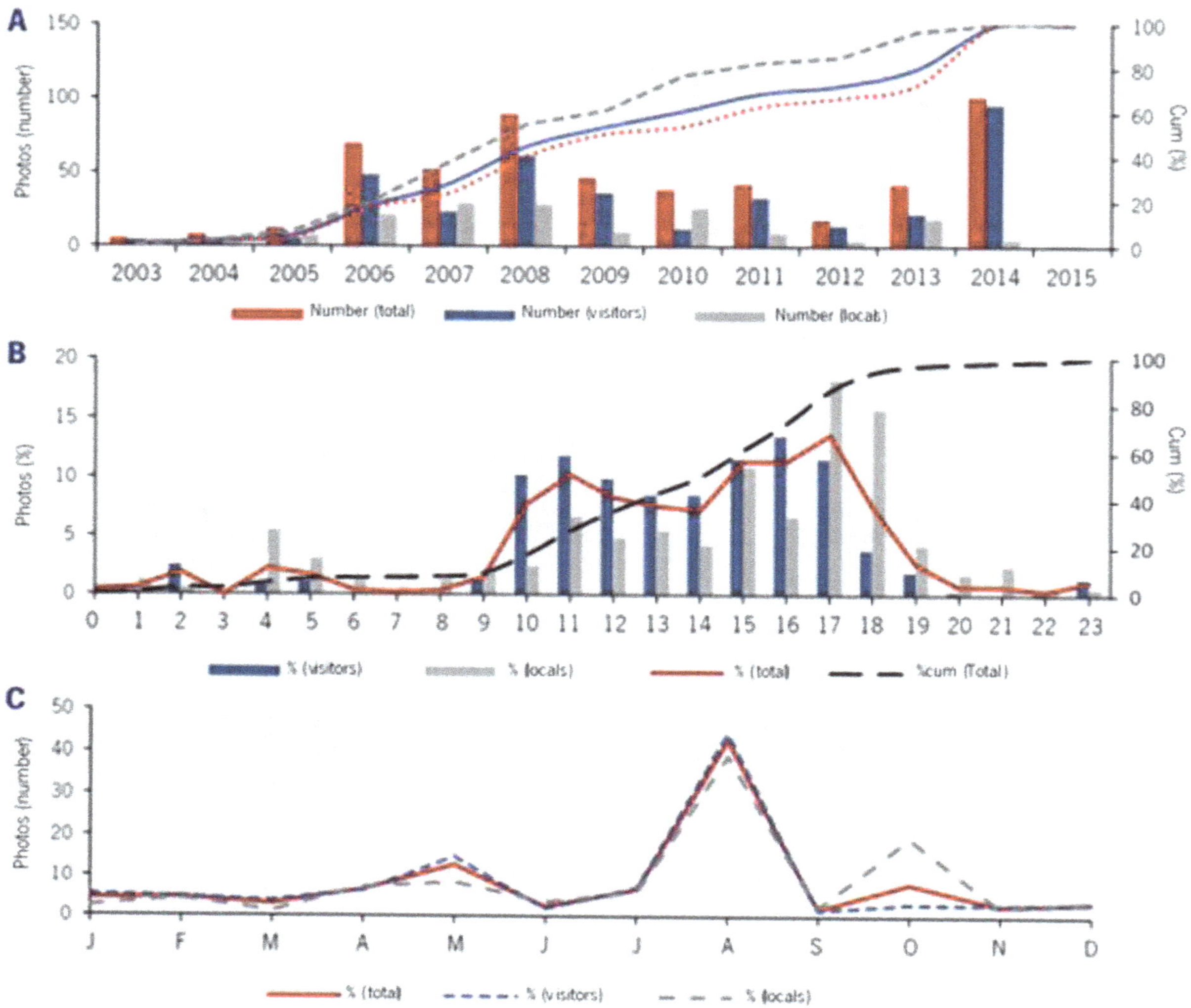

Figure 6. Temporal patterns of locals' and visitors' photos: (A) photos taken per year with registration of the number of photos and cumulative percentage (% cum) for locals, visitors and both; (B) photos taken per month with percent (%) recorded for locals, visitors and both; (C) photos taken per hour with percentage (%) for locals, visitors and both and cumulative percentage (% cum). Source: Panoramio photos collected for the period from January 2003 through March 2015.

during the night period, but they should be viewed with caution because there may have been errors associated with the instruments' recordings of the time when the photographs were taken.

An analysis also was conducted for the areas with the highest concentration of photos (i.e. the areas classified in **Figure 7** with High-High (HH) clusters).

It should be noted that these concentration areas received a larger number of photos than the others, and it was decided to compare the evolutionary rhythms by year and month with respect to the total records of photos in the municipality.

Although there was a trend toward the concentration of photographs in some areas of the Boticas municipality, as was previously indicated, this centralization only began to occur in 2014 (**Figure 7A**). Even so, these photos were taken mostly in the summer months, with emphasis on the month of August, which should result from visitors to the country (**Figure 7B**).

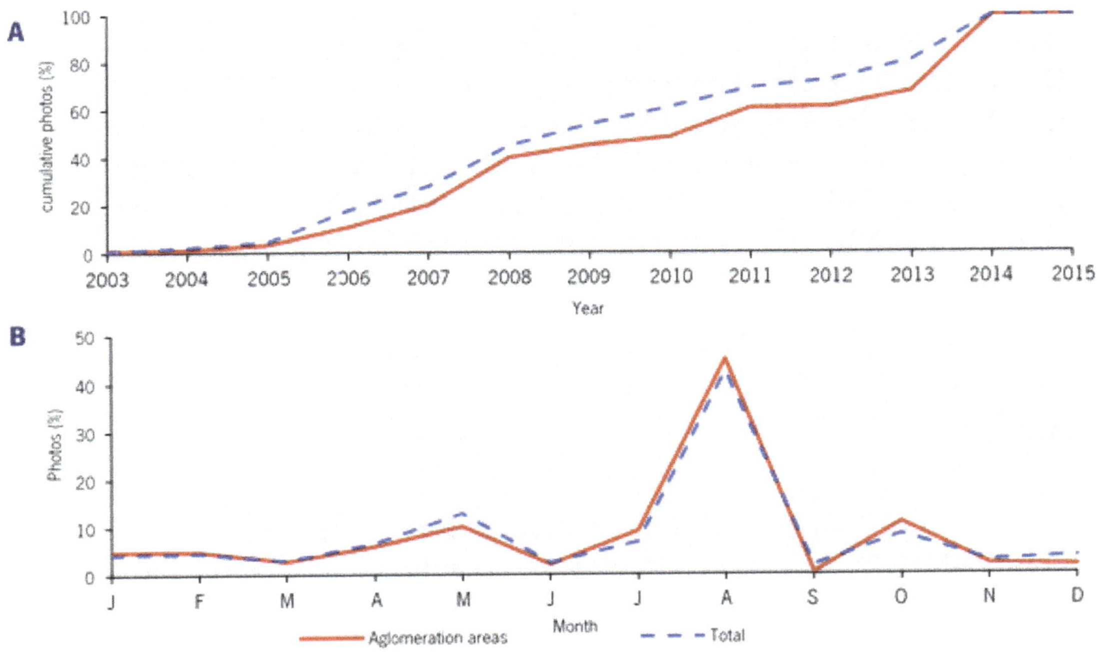

Figure 7. Time patterns in photos in agglomeration areas (high-high (HH) clusters) in comparison with all photographs: (A) cumulative percentage (% cum) of captured photos per year from 2003 through 2015; (B) percentage (%) of photos per month. Source: Own elaboration, based on Panoramio photos for the period from January 2003 through March 2015.

	Cluster A (n = 250)	Cluster B (n = 325)	Cluster C (n = 153)	F-ratio	p-value
Built heritage	0	89.8	78.4	4698.348	00.000*
Local culture	0	2.5	12.4		
Nature	100	0	0.7		
Tourist services	0	7.7	8.5		
Element	10	29.2	17.6	163.294	00.000*
Element in context	34.4	64.9	76.5		
Scenario	55.6	5.8	5.9		
Without people	96	93.8	85	3.008	00.050**
With people	4	6.2	15		
Residents	34	0	100	94.401	00.000*
Visitors	66	100	0		

*$p < 0.01$.
**$p \leq 0.05$.
Source: Own elaboration, based on Panoramio photos for the period from January 2003 through March 2015.

Table 3. Analysis of the clusters of the photos.

Table 3 summarises the results of grouping items into clusters. It was observed that the characteristics of the photographs indicated that there were two types of patterns, that is, tourists and residents. Note that, during the analysis, it was verified that all the items presented a p-value of ≤0.05. The ANOVA test result also showed that the variables included in the model were sufficiently different for their grouping, and, thus, three groups were created. Cluster A (n = 250) essentially presents photographs of nature (100%), and the cluster is associated with photos that focus on scenarios (55.6%) and includes the visitors who most often take this type of photos. Cluster B represents only the class of visitors, highlighting the elements in the context (64.9%) and the built heritage category (89.8%). Cluster C represents the least expressive group and incorporates photographs that focus on elements located in the context (76.5%) and in the category of the built heritage (78.4%). Clusters B and C contain essentially photographs that do not include people.

4. Discussion

As demonstrated during the study, Boticas is a tourist destination with seasonal demand, particularly during the summer period (**Figure 8**), a fact substantiated by the intersection of the number of visitors registered in Interactive Shop Porto and North Tourism and the number of photographs available in Panoramio. In any case, it is important to mention that some of the rural tourism houses in the municipality have a higher number of visitors during the winter period, as was evident from direct observation and discussions with these local agents.

The relationship between the photos that were taken and the promotion conveyed by various means of communication does not always present a direct relationship. In fact, although the Nadir Afonso Arts Center and the Boticas Nature and Biodiversity Park present a significant number of photographs and are key tourist resources for the promotion of the territory, there are other resources that the municipality presents that are underutilised due to the lack of the dissemination of information about their existence.

When the number of photographs is considered with respect to the tourist resources identified in the territory (**Figure 9**), it was verified that there are villages in which some of the traditional customs have been preserved, for example, Vilarinho Sêco and Coimbró. Although such considerations also are woven by residents and other stakeholders (i.e. [23, 30, 31]) when

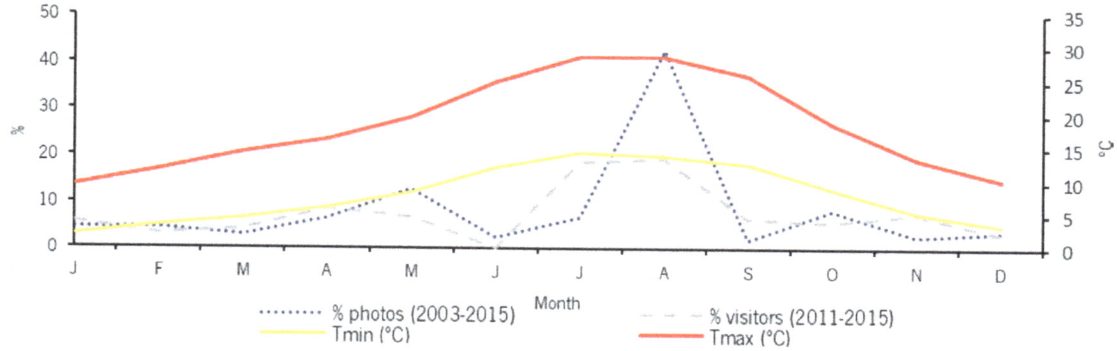

Figure 8. Photos recorded in the Panoramio (from 2003 through 2015) and visitors (from 2011 through 2015) in comparison with the minimum and maximum temperatures registered in the climatological normal for the period from 1971 through 2000 in Vila Real. Source: Own elaboration, based on Panoramio photos for the period from January 2003 through March 2015, in data from the Interactive Tourism Store and http://www.ipma.pt/pt/oclima/normais.clima/1981-2010/023.

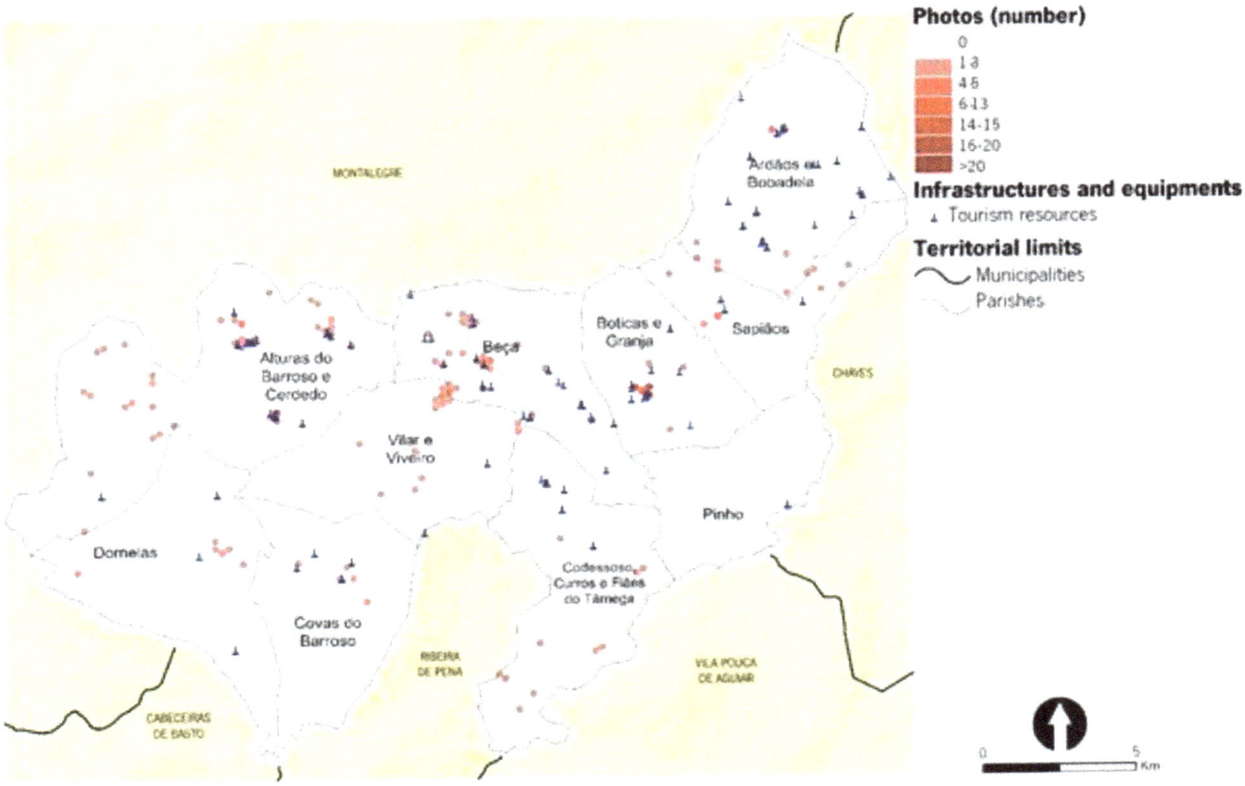

Figure 9. Relationship between the number of photographs and tourist resources identified in Boticas. Source: Photographs of Panoramio, collected for the period from January of 2003 through March 2015 and in the tourist resources referenced in http://patrimonio.cm-boticas.pt and in loco.

they identify the valued tourist attributes or those with high potential for appreciation, they are still not as effective in the tourism promotion strategy.

The results also show that visitors and local people more often have photographed patrimonial and natural elements. However, it should be noted that this difference was more significant in the case of the local people. Nevertheless, it is important to mention that these evidences present a high degree of similarity with the results obtained by [3]. The perceived image of Boticas is very much related to the built heritage rather than to nature, although the differences are not so significant when compared with the categories of tourist services and local culture. This was a surprising discovery, and, if it is true that this municipality has a unique built-heritage (especially churches and museums), there also were some atrocities committed during the patrimonial recovery processes with profound changes to the structures of churches and chapels. However, the relevance of nature, especially on the part of the visitors, is consistent with the reading made by the various stakeholders, considering the magnificence of the natural ecosystem, that is, the mountains and valleys, of this territory.

Another element of analysis that allows us to infer the image of visitors is associated with the degree to which people are present in the photographs. While in some European destinations the captured heritage photographs do not have individuals present, in others they appear more frequently, especially in exotic or allocentric destinations. Photographs with the presence of visitors occur commonly in destinations where there is a degree of interaction between the receiving community and visitors [3].

A careful analysis of the photographs allowed us to infer that those taken in the Boticas territory were largely free of the presence of both local people and visitors. This situation is not uncommon, as was evident in the investigations conducted by Girona [3, 34, 35] and Vale de Boí [3]. One of the reasons that support the absence of people in the photos is the ideal of individual consumption of places, which is supported by the assertion that the presence of people in the photographs could represent a kind of alteration of reality [36].

Nevertheless, in most cases, tourists who visit a destination such as Boticas seek to eliminate any vestige of humanity to avoid detracting attention away from the element or the landscape, for example, empty churches or the splendour of the landscape. These photographs usually also have a smaller focus, where the elements in the context (56.9%) are the most photographed, which may be due to the search for a broader framework than they are visiting (global vision of the place), to facilitate their later recognition.

In fact, visitors' behaviours may be framed in accordance with the photographs that are taken in the sense that the length of stay in the municipality is very small. Most of the photos with a visiting author occur between 10:00 A.M. and 4:00 P.M, which may denote (i) the absence of tourist facilities that are open for a longer time period, (ii) the absence of other interesting activities in the municipality, (iii) the absence of catering services that are open until later hours during the weekly period and (iv) the lack of desirable hotel accommodations that meet visitors' expectations.

Considering these elements, it would seem reasonable to start some efforts to counteract this trend and to create a precise and accurate image of this territory. This could be done in the form of tourism products, based, for example, on nature, health and well-being. This approach could attract visitors who have an interest in these locations, and it may allow the maintenance of sustainable tourism practices.

5. Conclusion

The photos taken by locals and visitors show that there are certain observable distribution patterns and that these patterns have some similarities and some significant differences. In fact, visitors take photos less concentrated in certain housing areas than in locals, which shows a certain predisposition to value elements of the intangible heritage. It should be noted that 34.5% of the photos that were taken were associated with the natural component.

In some cases, there is a certain disarticulation between the promotion of tourism conducted by the municipality and the image of the destination based on photographs. In this way, there are some clues that can be put forward for future work, namely (i) assessing the seasonality of photographs taken by residents and visitors and, thus, which potential sites to promote and at which time of the year; (ii) verification of the distribution of the photographs at the supra-municipal scale, given that this sample is incipient and that it is possible, based on photographs standards on a broader scale, to develop common strategies with other adjacent territories; and (iii) objectives, goals and strategies that can be outlined in the short term and medium term for the development of tourism.

Such research has some limitations. While it is critical to take into consideration that residents or visitors only share the photos they deem to be most relevant, only some groups are frequent users of these platforms. Nonetheless, these approaches must be used in combination with other approaches, such as surveys, focus groups and interviews, to determine important information concerning the use of the territory's resources from the perspective of enhancing tourism.

Acknowledgements

This work has the financial support of the Project Lab2PT—Landscapes, Heritage and Territory Laboratory—AUR/04509 and FCT through national funds and when applicable of the FEDER cofinancing, in the aim of the new partnership agreement PT2020 and COMPETE2020—POCI 01 0145 FEDER 007528.

Conflict of interest

The authors declare that there is no conflict of interest.

Author details

Hélder Tiago da Silva Lopes[1,2]*, Paula Cristina Almeida Cadima Remoaldo[1] and Vitor Ribeiro[1]

*Address all correspondence to: htsltiago@gmail.com

1 Laboratory of Landscape, Heritage and Territory, Department of Geography, University of Minho, Braga, Portugal

2 Group of Climatology, Water Research Institute, Department of Geography, University of Barcelona, Spain

References

[1] Camprubí R, Guia J, Comas J. The new role of tourists in destination image formation. Current Issues in Tourism. 2013;**16**:203-209

[2] Choi S, Lehto XY, Morrison AM. Destination image representation on the web: Content analysis of Macau travel related websites. Tourism Management. 2007;**28**:118-129

[3] Donaire JA, Camprubí R, Galí N. Tourist clusters from Flickr travel photography. Tourism Management Perspectives. 2014;**11**:26-33

[4] García-Palomares JC, Gutiérrez J, Mínguez C. Identification of tourist hot spots based on social networks: A comparative analysis of European metropolises using photo-sharing services and GIS. Applied Geography. 2015;**63**:408-417

[5] Santos J. The role of tour operators' promotional material in the formation of destination image and consumer expectations: The case of the People's republic of China. Journal of Vacation Marketing. 1998;**4**:282-297

[6] Buhalis D, Law R. Progress in information technology and tourism management: 20 years on and 10 years after the internet—The state of eTourism research. Tourism Management. 2008;**29**:609-623

[7] Myhill M, Shoebridge M, Snook L. Virtual research environments–a Web 2.0 cookbook? Library Hi Tech. 2009;**27**:228-238

[8] Castells M, Espanha R. A era da informação: economia, sociedade e cultura. Paz e terra. 1999

[9] Donaire JA, Galí N. La imagen turística de Barcelona en la comunidad de flickr. Cuadernos de Turismo. 2011:291-303

[10] Kavitha S, Jobi V, Rajeswari S. Tourism recommendation using social media profiles. In: Artificial Intelligence and Evolutionary Computations in Engineering Systems. Springer; 2017. pp. 243-253

[11] Kisilevich S, Krstajic M, Keim D, et al. Event-based analysis of people's activities and behavior using Flickr and Panoramio geotagged photo collections. In: Information Visualisation (IV), 2010 14th International Conference. IEEE; 2010. pp. 289-296

[12] Lo IS, McKercher B, Lo A, et al. Tourism and online photography. Tourism Management. 2011;**32**:725-731

[13] Munar AM, Jacobsen JKS. Motivations for sharing tourism experiences through social media. Tourism Management. 2014;**43**:46-54

[14] Yang L, Wu L, Liu Y, et al. Quantifying tourist behavior patterns by travel motifs and geo-tagged photos from Flickr. ISPRS International Journal of Geo-Information. 2017;**6**:345

[15] Wood SA, Guerry AD, Silver JM, et al. Using social media to quantify nature-based tourism and recreation. Scientific Reports. 2013;**3**:2976

[16] Fotis J, Buhalis D, Rossides N. Social Media Use and Impact during the Holiday Travel Planning Process. Springer-Verlag; 2012

[17] Jenkins O. Photography and travel brochures: The circle of representation. Tourism Geographies. 2003;**5**:305-328

[18] Munar AM. Social media strategies and destination management. Scandinavian Journal of Hospitality and Tourism. 2012;**12**:101-120

[19] Leung D, Law R, Van Hoof H, et al. Social media in tourism and hospitality: A literature review. Journal of Travel & Tourism Marketing. 2013;**30**:3-22

[20] Gomez B. La relación clima-turismo: Consideraciones básicas en los fundamentos teóricos y prácticos. Investig Geográficas. 1999:21-34

[21] Xiang Z, Gretzel U. Role of social media in online travel information search. Tourism Management. 2010;**31**:179-188

[22] Arriaza M, Cañas-Ortega JF, Cañas-Madueño JA, et al. Assessing the visual quality of rural landscapes. Landscape and Urban Planning. 2004;**69**:115-125

[23] Lopes HT da S. O turismo como alavanca do desenvolvimento de áreas rurais: o caso de estudo do município de Boticas. University of Minho; 2016

[24] Pérez JG. Ascertaining landscape perceptions and preferences with pair-wise photographs: Planning rural tourism in Extremadura, Spain. Landscape Research. 2002;**27**:297-308

[25] Haven-Tang C, Sedgley D. Partnership working in enhancing the destination brand of rural areas: A case study of made in Monmouthshire, Wales, UK. J Destin Mark Manag. 2014;**3**:59-67

[26] Hashim NH, Murphy J, Hashim NM. Islam and online imagery on Malaysian tourist destination websites. Journal of Computer and Communications. 2007;**12**:1082-1102

[27] Juravle AIG, Sasu C, Terec-Vlad L. The destination image of Bucovina among Romanian tourists. Journal of Cross Cultural Management. 2016;**10**:139-149

[28] Polo Peña AI, Frías Jamilena DM, Rodríguez Molina MÁ, et al. Online marketing strategy and market segmentation in the Spanish rural accommodation sector. Journal of Travel Research. 2016;**55**:362-379

[29] Zhou L. Online rural destination images: Tourism and rurality. J Destin Mark Manag. 2014;**3**:227-240

[30] Lopes H, Remoaldo P, Ribeiro V, et al. The creation of a new tourist destination in low density areas: The Boticas case. J Spat Organ Dyn. 2016;**4**:118-131

[31] Remoaldo P, Freitas I, Matos O, et al. The planning of tourism on rural areas: The stakeholders' perceptions of the Boticas municipality (Northeastern Portugal). Eur Countrys. 2017;**9**:504-525

[32] INE. Tipologia de Áreas Urbanas. 2014. Available from: http://smi.ine.pt/Versao/Detalhes/3485%3E [Accessed: February 15, 2018]

[33] Tammet T, Luberg A, Järv P. Sightsmap: Crowd-sourced popularity of the world places. In: Information and Communication Technologies in Tourism 2013. Springer; 2013. pp. 314-325

[34] Galí N. La humanización de las imágenes emitidas por la publicidad de los destinos turísticos monumentales: el Caso de Girona. PASOS Rev Tur y Patrim Cult;**3**

[35] Galí N, Donaire JA. The social construction of the image of Girona: A methodological approach. Tourism Management. 2005;**26**:777-785

[36] Urry J. Tourist Gaze: Leisure and Travel in Contemporary Societies (Theory, Culture & Society). Sage Publications; 1990

Permissions

All chapters in this book were first published in SAMP, by InTech Open; hereby published with permission under the Creative Commons Attribution License or equivalent. Every chapter published in this book has been scrutinized by our experts. Their significance has been extensively debated. The topics covered herein carry significant findings which will fuel the growth of the discipline. They may even be implemented as practical applications or may be referred to as a beginning point for another development.

The contributors of this book come from diverse backgrounds, making this book a truly international effort. This book will bring forth new frontiers with its revolutionizing research information and detailed analysis of the nascent developments around the world.

We would like to thank all the contributing authors for lending their expertise to make the book truly unique. They have played a crucial role in the development of this book. Without their invaluable contributions this book wouldn't have been possible. They have made vital efforts to compile up to date information on the varied aspects of this subject to make this book a valuable addition to the collection of many professionals and students.

This book was conceptualized with the vision of imparting up-to-date information and advanced data in this field. To ensure the same, a matchless editorial board was set up. Every individual on the board went through rigorous rounds of assessment to prove their worth. After which they invested a large part of their time researching and compiling the most relevant data for our readers.

The editorial board has been involved in producing this book since its inception. They have spent rigorous hours researching and exploring the diverse topics which have resulted in the successful publishing of this book. They have passed on their knowledge of decades through this book. To expedite this challenging task, the publisher supported the team at every step. A small team of assistant editors was also appointed to further simplify the editing procedure and attain best results for the readers.

Apart from the editorial board, the designing team has also invested a significant amount of their time in understanding the subject and creating the most relevant covers. They scrutinized every image to scout for the most suitable representation of the subject and create an appropriate cover for the book.

The publishing team has been an ardent support to the editorial, designing and production team. Their endless efforts to recruit the best for this project, has resulted in the accomplishment of this book. They are a veteran in the field of academics and their pool of knowledge is as vast as their experience in printing. Their expertise and guidance has proved useful at every step. Their uncompromising quality standards have made this book an exceptional effort. Their encouragement from time to time has been an inspiration for everyone.

The publisher and the editorial board hope that this book will prove to be a valuable piece of knowledge for researchers, students, practitioners and scholars across the globe.

List of Contributors

Khalid Al-Ahmadi
National Centre for Remote Sensing Technology, Space and Aeronautics Research Institute, King Abdulaziz City for Science and Technology, Riyadh, Saudi Arabia

Andreas Koch
University of Salzburg, Salzburg, Austria

Andreas Rienow and Nora Jennifer Schneevoigt
Geomatics Research Group, Department of Geography, Ruhr-University Bochum, Germany

Frank Thonfeld
Remote Sensing Research Group, Department of Geography, University of Bonn, Germany

José António Tenedório
Interdisciplinary Centre of Social Sciences (CICS. NOVA), Faculty of Social Sciences and Humanities (NOVA FCSH), Universidade NOVA de Lisboa, Lisbon, Portugal

Jorge Rocha
Universidade de Lisboa, Lisbon, Portugal

Toshiaki Ichinose
National Institute for Environmental Studies/Nagoya University, Tsukuba/Nagoya, Japan

John J. Williams and Fred Bidandi
University of the Western Cape, Cape Town, South Africa

Norman Schofield
Department of Political Science, Washington University, Saint Louis, USA

Valentina Svalova
Sergeev Institute of Environmental Geoscience, Russian Academy of Sciences, Moscow, Russia

Brian Ways and Brooks C. Pearson
University of Central Arkansas, Conway, United States

Ana Cristina Gonçalves and Adélia M. O. Sousa
Department of Rural Engineering, School of Sciences and Technology, Institute of Mediterranean Agricultural and Environmental Sciences (ICAAM), Institute of Research and Advanced Information (IIFA), University of Évora, Évora, Portugal

Diego Montenegro
Laboratório de Doenças Parasitária, Instituto Oswaldo Cruz/Fundação Oswaldo Cruz, Rio de Janeiro, Brazil
Laboratório de Referência Nacional em Vetores das Riquetsioses—Secretaria de Vigilância em Saúde/ Ministério da Saúde, Instituto Oswaldo Cruz/ Fundação Oswaldo Cruz, Rio de Janeiro, Brazil
Fundación Chilloa/Santa Marta, Colombia
Research Group in Evolution, Systematics and Ecology Molecular, Universidad del Magdalena, Santa Marta, Colombia

Ana Paula da Cunha and Marcel Pedroso
Instituto de Comunicação em Informação Científica e Tecnologia em Saúde-ICICT/Fiocruz, Rio de Janeiro, Brazil

Ingrid Machado and Liliane Duraes
Laboratório de Referência Nacional em Vetores das Riquetsioses—Secretaria de Vigilância em Saúde/ Ministério da Saúde, Instituto Oswaldo Cruz/ Fundação Oswaldo Cruz,Rio de Janeiro, Brazil

Stefan Vilges de Oliveira
Secretaria de Vigilância em Saúde—Ministério da Saúde, Brasilia, Brazil

Gilberto S. Gazêta
Laboratório de Referência Nacional em Vetores das Riquetsioses—Secretaria de Vigilância em Saúde/ Ministério da Saúde, Instituto Oswaldo Cruz/ Fundação Oswaldo Cruz, Rio de Janeiro, Brazil
Secretaria de Vigilância em Saúde—Ministério da Saúde, Brasilia, Brazil

Reginaldo P. Brazil
Laboratório de Doenças Parasitária, Instituto Oswaldo Cruz/Fundação Oswaldo Cruz, Rio de Janeiro, Brazil

Hélder Tiago da Silva Lopes
Laboratory of Landscape, Heritage and Territory, Department of Geography, University of Minho, Braga, Portugal
Group of Climatology, Water Research Institute, Department of Geography, University of Barcelona, Spain

Paula Cristina Almeida Cadima Remoaldo and Vitor Ribeiro
Laboratory of Landscape, Heritage and Territory, Department of Geography, University of Minho, Braga, Portugal

Index

www.ingramcontent.com/pod-product-compliance
Lightning Source LLC
LaVergne TN
LVHW060555210926
848485LV00005B/1315